FOUNDATIONS OF COLLEGE MATH

JAMES A. PAGE
Mountain View College

GLENCOE
McGraw-Hill

New York, New York   Columbus, Ohio   Woodland Hills, California   Peoria, Illinois

This book is dedicated to the memory of my father, Allen Page, and to my son, Clinton.

© Richard D. Irwin, Inc., 1993. Copyright transferred in 1997 to Glencoe/McGraw-Hill.

All rights reserved. No part of this publication may be reproduced, stored in a retrieval system, or transmitted, in any form or by any means, electronic, mechanical, photocopying, recording, or otherwise, without the prior written permission of the publisher.

Send all inquiries to:
Glencoe/McGraw-Hill
936 Eastwind Drive
Westerville, OH 43081

ISBN: 0-256-14567-9
 0-256-14945-3 (Instructor's Edition)

Printed in the United States of America.
 4 5 6 7 8 9 10   071/055   03 02 01 00 99 98

PREFACE

Mathematics is a common bond among all people in every country in the world. A thorough understanding of math concepts and principles is important not only on the job but also for daily living. *Foundations of College Math* is designed to help students develop or reinforce math skills that are essential in a modern technological world.

WHO CAN USE THIS BOOK?

This book can be used by students studying either independently or in a traditional classroom setting. It can be used by students who need only one math course or as a foundation for more advanced math courses.

BENEFITS OF USING THIS BOOK

Students register for math classes for a variety of reasons. A strong foundation in math is necessary for success in many other fields: accounting, nursing, bookkeeping, education, music, business, manufacturing, and sales to name a few. Good math skills are also needed for daily living: balancing a checkbook, budgeting, calculating a discount on a purchase, or computing change due. One objective of this text is to provide examples of how math may be used both on the job and at home. A second objective is to provide a math textbook that can be read and understood by the average student. The goal is to ease the "math anxiety" that many people, not just students, experience. As with any worthwhile project, learning mathematics requires work. Working diligently through this text will help students alleviate their "math anxiety" while at the same time preparing them to make the transition from basic computational math to higher mathematics.

SKILLS THAT WILL BE TAUGHT

Upon completion of this course, students will have developed or reinforced the following skills.

- Reading and writing numbers (whole numbers, decimal numbers, common fractions)
- Applying the four basic math operations—addition, subtraction, multiplication, and division—to whole numbers, decimal numbers, common fractions, and negative numbers.
- Rounding whole numbers and decimal numbers.
- Solving problems requiring more than one step by using the Order of Operations Agreement.
- Solving application problems by using a logical, orderly approach to analyze the problem and seek a realistic solution.
- Writing ratios and proportions and using them to solve problems.

- Solving problems using percent, including how to analyze a percent problem to identify rate, base, or amount.
- Determining which form of a number is best suited for solving a particular problem: common fraction, decimal fraction, or percent.
- Working with variables (a letter used to represent a number in an equation), and solving for the unknown in different types of equations.
- Using the distributive property to simplify equations.
- Recognizing the features of two measurement systems currently in use—the U.S. Customary System (USCS) and the metric system—converting units from one measurement system to the other, and solving problems using either system.
- Identifying and labeling angles and lines, and identifying plane and solid geometric figures.
- Using formulas to determine the perimeter and area of each geometric figure and the volume of each geometric solid.

FEATURES OF THIS BOOK

The pedagogy of the book has been planned to make teaching easy for the instructor and learning easy for the student.

Chapter Outline Each chapter is introduced with a brief outline of the major topics that are included in that chapter.

Skills Preview Located at the beginning of each chapter, this pretest can be used to determine how much work needs to be done in the chapter.

Key Terms Key terms appear in bold type where they are defined in the chapter.

Boxed Features Rules, procedures, and strategies are boxed for easy reference and review.

Examples Each example has a twofold purpose: to show students how to work a problem and to help students develop reasoning skills. Examples are explained step-by-step so students can see how a problem should be solved from beginning to end in a logical manner.

Comprehension Checkpoints These problems are located at the end of each section. They should be worked immediately after discussing a section to determine if students understand the material in that section. If students complete these problems successfully, they are ready for the Practice Sets.

Algebra Connection This boxed feature is included in each chapter to show how algebra is a continuation of basic math. It is algebra that allows anyone to solve a problem. All problems are algebraic in the sense that there is an unknown quantity; if there is no unknown, there is no problem. The Algebra Connection in the last chapter is designed to encourage students to reflect on just how far they have progressed in their studies. This feature demonstrates that algebra is a bridge between basic computational math and higher-level applied mathematics.

Practice Sets Practice Sets A and B at the end of each section provide students an opportunity to drill and apply the skills they learned in that section.

Summary of Key Concepts At the conclusion of each chapter is a list of the key terms and the key rules discussed in that chapter which should help students when reviewing the chapter material.

Practice Tests

Practice Tests A and B, which follow the Summary of Key Concepts, can be used to prepare for chapter quizzes or as chapter quizzes. Students in independent study should find these especially helpful for measuring their progress.

Skills Review

The final set of exercises in each chapter is a Skills Review, which provides a cumulative selection of problems from previous chapters. Working these problems allows students to reinforce skills developed previously.

Solutions

Solutions to all Skills Preview, Comprehension Checkpoint, Practice Set A, Practice Test A, and Skills Review problems are located at the end of each chapter. In addition, all problems in each Practice Set A have been worked completely. This feature allows students to check both their answers and their reasoning skills.

SUPPLEMENTS

This text is supported by a practical supplements package.

Instructor's Edition

This edition of the student text includes solutions to all problems, printed in the second color.

Instructor's Manual with Printed Test Bank

This carefully prepared Instructor's Manual offers curriculum guidance, including a course syllabus, chapter objectives, and detailed lesson plans. Supplementary exercises, chapter tests, a final exam, and answer keys for each are also included.

Computerized Test Bank

The Computerized Test Bank allows instructors to create their own tests from hundreds of prepared problems.

Tutorial Software

This interactive instructional software offers students additional instruction and practice. Students may use it as a "makeup" session when they have missed a class, as additional instruction and practice on a problem area, or as a pretest review session.

ACKNOWLEDGMENTS

The fine comments, observations, questions, and suggestions of the following reviewers were instrumental in the development of the manuscript.

Michele B. Bach	Kansas City (Kansas) Community College
Nell K. Bishop	Phillips Junior College
Elizabeth A. Gambuto	Johnson & Wales University
Frank Goulard	Portland Community College
Robert Isenhower	Spartanburg Technical College
Marcia S. Klingbeil	Davis College
Linda Kuchenbecker	National College
Stephanie L. Lakey	Executive Secretarial School
Cindy Lenhart	Blue Mountain Community College
Gary W. Martin	DeVry Institute of Technology, Decatur
Thomas M. O'Keefe	Bucks County Community College
Ed Petrunak	New Castle School of Trades
Evelyn L. Plummer	DeVry Institute of Technology, Phoenix
Margaret Schwabel	Bryant & Stratton Corp.

Finally, I would like to express my sincere appreciation to those at Richard D. Irwin, Inc., who have helped in the development and production of this book. Carol Long, Executive Editor, and Jean Roberts, Senior Developmental Editor, Irwin Career Education Division, offered guidance, counseling, and most important, patience and understanding, while this project was in progress. Jean Lou Hess, project editor; Heidi Baughman, designer; Heather Burbridge, art coordinator; and Ann Cassady, production manager, provided a beautiful, functional design and ensured smooth sailing throughout the production process.

James A. Page

CONTENTS

CHAPTER 1   **WHOLE NUMBERS**   1

     1.1   Reading and Writing Whole Numbers   3
     1.2   Rounding Whole Numbers   11
     1.3   Adding Whole Numbers   17
     1.4   Subtracting Whole Numbers   25
     1.5   Multiplying Whole Numbers   32
     1.6   Dividing Whole Numbers   41
     1.7   Word Problems   51
     1.8   Prime and Composite Numbers   57
     1.9   Order of Operations   65
          Summary of Key Concepts   73
          Practice Tests   77
          Chapter 1 Solutions   81

CHAPTER 2   **FRACTIONS AND MIXED NUMBERS**   91

     2.1   Introduction to Fractions   93
     2.2   Improper Fractions and Mixed Numbers   104
     2.3   Equivalent Fractions   111
     2.4   Adding Fractions and Mixed Numbers   121
     2.5   Subtracting Fractions and Mixed Numbers   131
     2.6   Multiplying Fractions and Mixed Numbers   139
     2.7   Dividing Fractions and Mixed Numbers   147
     2.8   Order of Operations   155
          Summary of Key Concepts   161
          Practice Tests   165
          Skills Review, Chapters 1–2   169
          Chapter 2 Solutions   171

CHAPTER 3   **DECIMALS**   183

     3.1   Decimal Numbers and Place Value   185
     3.2   Rounding Decimals   193
     3.3   Adding and Subtracting Decimal Numbers   199
     3.4   Multiplying Decimal Numbers   207
     3.5   Dividing Decimal Numbers   215
     3.6   Scientific Notation   225
     3.7   Order of Operations   233

	Summary of Key Concepts	236
	Practice Tests	241
	Skills Review, Chapters 1–3	245
	Chapter 3 Solutions	247

CHAPTER 4 RATIOS AND PROPORTIONS — 255

4.1	Ratios	257
4.2	Proportions	263
4.3	Solving Proportions	269
	Summary of Key Concepts	277
	Practice Tests	279
	Skills Review, Chapters 1–4	283
	Chapter 4 Solutions	285

CHAPTER 5 PERCENTS — 289

5.1	Introduction to Percents	290
5.2	Calculating the Amount	299
5.3	Calculating the Rate	305
5.4	Calculating the Base	311
5.5	Percent Increase and Decrease	319
5.6	Percent Applications	327
	Summary of Key Concepts	333
	Practice Tests	335
	Skills Review, Chapters 1–5	339
	Chapter 5 Solutions	341

CHAPTER 6 CONVERTING FRACTIONS — 347

6.1	Equivalent Common Fractions, Decimal Fractions, and Percents	349
6.2	Terminating and Repeating Decimals	357
6.3	Using Conversions to Solve Problems	363
	Summary of Key Concepts	371
	Practice Tests	373
	Skills Reviews, Chapters 1–6	377
	Chapter 6 Solutions	379

CHAPTER 7 SIGNED NUMBERS — 383

7.1	Introduction to Signed Numbers	385
7.2	Adding Signed Numbers	395
7.3	Subtracting Signed Numbers	403

	7.4	Multiplying Signed Numbers	409
	7.5	Dividing Signed Numbers	417
	7.6	Order of Operations	423
		Summary of Key Concepts	429
		Practice Tests	431
		Skills Review, Chapters 1–7	435
		Chapter 7 Solutions	439

CHAPTER 8 — INTRODUCTION TO ALGEBRA — 449

8.1	Terminology	451
8.2	Solving Equations by Addition	459
8.3	Solving Equations by Division	465
8.4	Solving Equations by Multiplication	473
8.5	The Distributive Property	481
8.6	Solving Multistep Equations	489
	Summary of Key Concepts	497
	Practice Tests	499
	Skills Review, Chapters 1–8	503
	Chapter 8 Solutions	507

CHAPTER 9 — MEASUREMENT — 519

9.1	Length Measurement	520
9.2	Weight Measurement	531
9.3	Liquid Measurement	539
9.4	Time and Temperature	547
9.5	Working with Denominate Numbers	555
	Summary of Key Concepts	565
	Practice Tests	569
	Skills Review, Chapters 1–9	573
	Chapter 9 Solutions	575

CHAPTER 10 — INTRODUCTION TO GEOMETRY — 583

10.1	Terminology	587
10.2	Polygons	595
10.3	Circles	607
10.4	Area	615
10.5	Volume	627
	Summary of Key Concepts	635
	Practice Tests	639
	Skills Review, Chapters 1–10	647
	Chapter 10 Solutions	651

APPENDIX A	**ADDITION AND MULTIPLICATION TABLES**	**A-1**
APPENDIX B	**TABLE OF PRIME FACTORS OF NUMBERS 1 THROUGH 100**	**A-3**
APPENDIX C	**TABLE OF SQUARE ROOTS**	**A-5**
APPENDIX D	**USING A CALCULATOR**	**A-7**
INDEX		**I-1**

1 WHOLE NUMBERS

OUTLINE

1.1 Reading and Writing Whole Numbers

1.2 Rounding Whole Numbers

1.3 Adding Whole Numbers

1.4 Subtracting Whole Numbers

1.5 Multiplying Whole Numbers

1.6 Dividing Whole Numbers

1.7 Word Problems

1.8 Prime and Composite Numbers

1.9 Order of Operations

Numbers are an important part of our daily lives. What is the cost of a new television set? How many credit hours are you taking this semester? What is the price of a new dress? You may answer all of these questions by using numbers. Chapter 1 is a discussion of whole numbers and how to solve problems using them.

SKILLS PREVIEW

1. Write 64,790,371 in words.
   _____

2. Write six hundred twelve billion, four hundred eighteen million, five hundred thirty-three thousand, eight hundred fifty-four in numerical form. _____

3. Consider the number 27,465. What digit is in the:
   a. Hundreds place?   *b.* Thousands place?   *c.* Tens place?
   a. _____   *b.* _____   *c.* _____

4. Write 423,792 in expanded notation.
   _____

5. Round 642,973,451 to the indicated place.
   a. nearest thousand _____
   b. nearest ten million _____
   c. nearest hundred _____

Add:

6. 　42
　　 78
　+ 39
   ─────

7. 　4,295
　　　387
　 16,473
　+ 2,401
   ─────

Subtract:

8. 　9,743
　− 8,412
   ─────

9. 　4,056
　− 3,487
   ─────

Multiply:

10. $5 \times 9 \times 2 \times 3 =$

11. 　49
　 × 83
   ─────

12. 　806
　 × 274
   ─────

Divide:

13. 8)336   14. 32)2,752

15. 23)7,015

16. Identify the prime numbers in the following list:

    17   31   42   53   65 _____

17. Write the prime factorization of 96. _____

18. Calculate: $5 + 7 \times 4 \div 2 - 3$ _____

1.1 READING AND WRITING WHOLE NUMBERS

Sets of Numbers

It is often convenient to refer to certain groups of numbers by name. A **set** is a particular group of numbers. One such set of numbers is {1, 2, 3, 4, ... }. This set is called the **natural numbers** (or **counting numbers**). Notice that the set is enclosed in braces and there are three dots at the end. The dots are called an *ellipsis*. They are read "and so on" and indicate that the list continues indefinitely.

When 0 is added to the set of natural numbers, as in {0, 1, 2, 3, ... }, the set is referred to as the set of **whole numbers.**

Place Value

Our number system contains 10 digits: 0, 1, 2, 3, 4, 5, 6, 7, 8, 9. The value of each digit depends on the place it occupies in the number. The following chart gives the place names.

← Billions →			← Millions →			← Thousands →			← Units →		
HUNDRED BILLIONS	TEN BILLIONS	BILLIONS	HUNDRED MILLIONS	TEN MILLIONS	MILLIONS	HUNDRED THOUSANDS	TEN THOUSANDS	THOUSANDS	HUNDREDS	TENS	UNITS

Look at the number 47,385. If you place this number in the boxes at the bottom of the chart, you will see the place value of each digit.

Billions			Millions			Thousands			Units		
HUNDRED BILLIONS	TEN BILLIONS	BILLIONS	HUNDRED MILLIONS	TEN MILLIONS	MILLIONS	HUNDRED THOUSANDS	TEN THOUSANDS	THOUSANDS	HUNDREDS	TENS	UNITS
							4	7	3	8	5

The number 47,385 is read "forty-seven thousand, three hundred eighty-five." Notice that you do not use the word *and* when reading whole numbers. There are 4 ten thousands, 7 thousands, 3 hundreds, 8 tens, and 5 units. This number could be considered the same as

$$40,000 + 7,000 + 300 + 80 + 5$$

Writing a number like this is called **expanded notation** and clearly shows the value of each place in the number. Seeing a number written in this form may help you to understand what a number represents.

EXAMPLE 1 Consider the number 426,091 and answer the following questions:

a. What digit is in the ten thousands place?
b. What digit is in the tens place?
c. What digit is in the hundred thousands place?
d. What digit is in the hundreds place?

SOLUTIONS

a. 2   *b.* 9   *c.* 4   *d.* 0

EXAMPLE 2 Write 586,923 in expanded notation.

SOLUTION

$500,000 + 80,000 + 6,000 + 900 + 20 + 3$

Any number may be read by dividing it into groups of three digits and applying the names from the place value chart. How do you read the number 37,438,279,863?

Billions	Millions	Thousands	Units
37	438	279	863

The number is read "thirty-seven billion, four hundred thirty-eight million, two hundred seventy-nine thousand, eight hundred sixty-three." This is known as the *written* form of the number.

Notice that commas are used to separate digits into groups of three and that all numbers between 21 and 99 (except multiples of 10) are written with hyphens. Remember that the word *and* is never used when writing whole numbers or when reading them.

EXAMPLE 3   Write 84,062,389 in words.

SOLUTION

eighty-four million, sixty-two thousand, three hundred eighty-nine

You should also be able to convert a number from the written form to the numerical form (written in digits).

EXAMPLE 4   Write fifty-six billion, four hundred fifteen million, six hundred twenty-eight thousand, nine hundred sixty-seven in digits.

SOLUTION

56,415,628,967

Being able to read and write whole numbers is the first essential element in studying mathematics. Try the Comprehension Checkpoint problems to assure yourself that you understand these basics. If you have trouble with these problems, review the appropriate material before going on to the Practice Sets.

COMPREHENSION CHECKPOINT

1. Consider the number 95,681,472.
   a. What digit is in the thousands place? _____
   b. What digit is in the millions place? _____
   c. What digit is in the tens place? _____

2. Write 24,795 in expanded notation.

3. Express 593,674,021 in written form.

4. Express four hundred seventy-two thousand, three hundred eighty-nine in numerical form.

1.1A PRACTICE SET

Refer to the number 8,467,302,195 to answer questions 1–5.

1. What digit is in the ten thousands place? _____

2. What digit is in the hundred millions place? _____

3. What digit is in the hundreds place? _____

4. What digit is in the hundred thousands place? _____

5. What digit is in the billions place? _____

In problems 6–10, write each number in expanded notation.

6. 395 _____

7. 4,263 _____

8. 869,475 _____

9. 26,891 _____

10. 7,683,467 _____

Express each number in problems 11–15 in written form.

11. 437 _____

12. 906 _____

13. 8,642 _____

14. 97,405 _____

15. 268,793 _____

In problems 16–20, convert each written number to numerical form.

16. five hundred forty-two _____

17. twenty-three thousand, eight hundred twelve _____

18. nine million, four hundred thirty-three thousand, two hundred fifty-six _____

19. seven hundred six thousand, three hundred sixty-eight _____

20. sixteen billion, six hundred eighteen million, four hundred eighty-nine _____

1.1B PRACTICE SET

Refer to the number 4,607,598,213 to answer questions 1–5.

1. What digit is in the tens place? _____

2. What digit is in the hundred thousands place? _____

3. What digit is in the ten millions place? _____

4. What digit is in the thousands place? _____

5. What digit is in the units place? _____

In problems 6–10, write each number in expanded notation.

6. 817 _____

7. 9,324 _____

8. 84,615 _____

9. 376,489 _____

10. 4,382,078 _____

Express each number in problems 11–15 in written form.

11. 708 _____

12. 1,329 _____

13. 27,685 _____

14. 793,824 _____

15. 24,967,451 _____
_____

In problems 16–20, convert each written number to numerical form.

16. three hundred nineteen _____

17. seven thousand, six hundred eight _____

18. forty-two thousand, nine hundred eighty-two _____

19. nine hundred twenty-seven thousand, four hundred seventy _____

20. thirty-five million, five hundred forty-three thousand, two hundred eleven

_____

1.2 ROUNDING WHOLE NUMBERS

There are many times when an approximate number is preferable to an exact number. **Rounding** is the process of determining the approximate number to use.

You may hear that a major university has 58,000 students when in reality there are 57,862 students. The actual number of students has been rounded to the nearest thousand. The distance from the earth to the moon is often stated as 240,000 miles. The actual distance is between 221,463 at the closest point and 252,710 at the greatest point. The approximate number is convenient to work with and gives a good representation of the distance.

A number may be rounded to any place. The **rounding place** is the place that is to be rounded to, just as the number of students in the preceding paragraph was rounded to the nearest thousand. The **determining digit** is the digit to the right of the rounding place. It determines what action to take.

TO ROUND A NUMBER

1. Determine which digit is in the rounding place and which digit is the determining digit.
2. The rounding place digit is
   a. Unchanged if the determining digit is 4 or less.
   b. Increased by 1 if the determining digit is 5 or larger.
3. Change all digits to the right of the rounding place to 0s.

EXAMPLE 1   Round 673 to the nearest hundred.

SOLUTION

The 6 is in the rounding place. The 7 is the determining digit. It is larger than 5, so add 1 to the 6 and change the other digits to 0s.

$$700 = 673 \text{ rounded to the nearest hundred}$$

EXAMPLE 2   Round 43,289 to the nearest thousand.

SOLUTION

The 3 is in the rounding place, and the determining digit is 2, which is less than 4, so the 3 is unchanged and all digits to its right are changed to 0s.

$$43,000 = 43,289 \text{ rounded to the nearest thousand}$$

EXAMPLE 3   Round 5,974 to the nearest hundred.

SOLUTION

The 9 is in the rounding place; the determining digit is 7, which is larger than 5, so add 1 to the rounding place. Since 9 is in the rounding place and $9 + 1 = 10$, the rounding place becomes 0 and 1 is carried and added to the 5. Change the digits to the right of the rounding place to 0s.

$$6,000 = 5,974 \text{ rounded to the nearest hundred}$$

Work the Comprehension Checkpoint problems to measure your understanding of rounding. If you have difficulty with these problems, review the material in this section before going on to the Practice Sets.

COMPREHENSION CHECKPOINT

1. Round 42,573 to the nearest thousand. _____

2. Round 576,813 to the nearest hundred. _____

3. Round 384 to the nearest ten. _____

4. Round 298,407 to the nearest ten thousand. _____

5. Round 44,763,095 to the nearest ten million. _____

1.2A PRACTICE SET

Round each number to the indicated place.

	Nearest Ten	Nearest Hundred	Nearest Thousand
1. 892	_____	_____	_____
2. 509	_____	_____	_____
3. 368	_____	_____	_____
4. 2,484	_____	_____	_____
5. 4,675	_____	_____	_____
6. 8,953	_____	_____	_____
7. 3,269	_____	_____	_____
8. 56,743	_____	_____	_____
9. 84,917	_____	_____	_____
10. 76,495	_____	_____	_____

11. Round 24,576,893 to the nearest hundred thousand. _____

12. Round 37,904,382 to the nearest ten thousand. _____

13. Round 84,765,309 to the nearest million. _____

14. Round 63,870,959 to the nearest ten million. _____

15. Round 52,957,806 to the nearest hundred thousand. _____

Applications

16. The utility bill for a domed stadium is $17,640 per month. Round this amount to the nearest thousand dollars.

_____

17. The distance from Dallas to Chicago is 931 miles. Round this distance to the nearest hundred miles.

_____

18. The population of New York State in 1980 was 17,558,072. Round this figure to the nearest hundred thousand.

_____

19. A diesel tractor-trailer has a fuel tank that holds 225 gallons of fuel. Round this amount to the nearest ten gallons.

_____

20. Attendance at a rock concert was 127,649. Round this figure to the nearest ten thousand.

_____

1.2B PRACTICE SET

Round each number to the indicated place.

	Nearest Ten	Nearest Hundred	Nearest Thousand
1. 768	_____	_____	_____
2. 274	_____	_____	_____
3. 925	_____	_____	_____
4. 4,885	_____	_____	_____
5. 9,546	_____	_____	_____
6. 3,712	_____	_____	_____
7. 1,895	_____	_____	_____
8. 48,374	_____	_____	_____
9. 59,572	_____	_____	_____
10. 87,685	_____	_____	_____

11. Round 65,703,274 to the nearest hundred thousand. _____

12. Round 25,642,953 to the nearest ten million. _____

13. Round 53,396,492 to the nearest ten thousand. _____

14. Round 86,265,718 to the nearest thousand. _____

15. Round 19,564,073 to the nearest million. _____

Applications

16. Attendance at a football game was 64,590. Round this figure to the nearest thousand.

_____

17. The estimated population of Los Angeles in 1988 was 3,352,710. Round this figure to the nearest hundred thousand.

_____

18. The cost of a new refrigerator is $649. Round this to the nearest hundred dollars.

_____

19. The New Cornelia Tailings Dam in Arizona has a volume of 274,015 cubic yards. Round this figure to the nearest ten thousand.

_____

20. The distance from Seattle to Miami is 3,389 miles. Round this to the nearest hundred miles.

_____

1.3 ADDING WHOLE NUMBERS

Addition is the process of grouping or combining terms that are alike. **Addends** are the numbers being added. The answer in an addition problem is called the **sum** or **total**. The symbol + is used to indicate addition.

Before you are able to add numbers with more than one digit, you must be able to add single-digit numbers. For your convenience and review, a Table of Addition Facts is included in Appendix A. You should memorize these basic facts so that you can use them without hesitation. This will lead to greater success with addition.

To add numbers, you must align them properly. **Aligning** numbers means placing them in columns with the units over the units, tens over tens, hundreds over hundreds, and so on. Once you have the digits aligned, add each column, beginning with the units and working to the left, and place the sum at the bottom of the column. Failure to align the digits correctly could easily result in getting an incorrect sum.

EXAMPLE 1 Add: 25
              + 43

SOLUTION

```
   25
 + 43
 ────
   68
```
Add each column. Place the sum of each column under that column.

The sum of 25 and 43 is 68.

The number of digits in each addend does not affect the addition process. You must still align the digits in columns and add each column beginning on the right (units column).

EXAMPLE 2 Add: 132
              4,524
            + 2,341

SOLUTION

```
    132
  4,524
+ 2,341
 ──────
  6,997
```
Add each column. Place the sum of each column under the correct column.

The sum of $132 + 4{,}524 + 2{,}341 = 6{,}997$.

In these first two examples, the sum of each column was a single-digit number. The sum could also be a two-digit number. The second digit in that case must be **carried** to the next column and added in with the digits already in that column.

EXAMPLE 3   Add: 479
                    + 896

SOLUTION

```
    1  →   Carried digit.
  479
+ 896
─────
    5       Add the units column. 9 + 6 = 15. Write the 5 and carry the 1 to
            the next column.
   11  →   Carried digits.
  479
+ 896
─────
   75       Add the tens column. 7 + 9 + 1 (the 1 carried) = 17. Write the 7
            and carry the 1 to the next column.
   11  →   Carried digits.
  479
+ 896
─────
1,375       Add the hundreds column. 4 + 8 + 1 (the 1 carried) = 13. Since
            there are no more columns to be added, write the 13 down.
```

The sum of 479 and 896 is 1,375.

Addition is defined as the process of grouping like terms. In Example 3, when the 9 and the 6 in the units column were added, the result was 15. This is 1 ten and 5 units. The 1 must be carried to the next column because it represents 10 and can be added only to other tens. Similarly, the 7 and the 9 in the tens column represent 70 (7 tens) and 90 (9 tens). Thus, $1 + 7 + 9 = 17$, but this is 17 tens, or 170. The 1 represents 100 and must be carried to the hundreds column. This process is continued until all columns have been added.

The number of addends in a problem does not affect the addition process. Align the digits in columns and add each column beginning on the right with the units and working your way to the left.

EXAMPLE 4   Add: 4,263
                         319
                       52,735
                      + 3,064

SOLUTION

```
  11 12 →   Carried digits.
   4,263
     319
  52,735
+  3,064
────────
  60,381    Begin with the units column and add each column. Carry as
            needed.
```

The sum of $4,263 + 319 + 52,735 + 3,064$ is 60,381.

In each of the examples in this section, the addends were already aligned in columns. Sometimes the addends will be arranged horizontally. This does not change the addition process. Align the numbers in vertical columns and add.

Adding Whole Numbers

> **a + b = x**
>
> ## ALGEBRA CONNECTION
>
> One of the things that causes concern among many students is the use of symbols, usually letters, in algebra. These letters stand in for a number. Once that number has been determined, it can replace the letter in the problem. Look at Example 5 below.
>
> $$24 + 75 + 356 = \underline{?}$$
>
> Many basic math textbooks will place a question mark (?) or a blank space (_____) to indicate that the answer goes in that place. The same problem could be written
>
> $$24 + 75 + 356 = n$$
>
> where the n takes the place of the question mark. Both are symbols indicating that you are looking for a number to put in that place.
>
> Remember that there is an unknown quantity in every math problem. Without an unknown, there is no math problem to be solved. In algebra, the unknown is represented by a letter rather than a question mark or a blank space. This is in preparation for higher mathematics, where the use of symbols is more convenient than the question mark or the blank.

EXAMPLE 5   Add: $24 + 75 + 356 = \underline{?}$

SOLUTION

```
  1 1   →   Carried digits.
   24
   75
+ 356
 ----
  455        Begin with the units column and add each column. Carry as
             needed.
```

Work the Comprehension Checkpoint problems to test your addition skills. If you have difficulty with these problems, review the appropriate material before going on to the Practice Sets.

COMPREHENSION CHECKPOINT

Find the sum in each problem.

1.   24
   + 17

2.   138
      206
   + 495

3. 1,863
    2,471
      357
   +   92

4. $275 + 1{,}064 + 43 + 4{,}908 =$ _____

1.3A PRACTICE SET

Find the sum in each problem.

1. 82
    + 15

2. 64
    + 32

3. 28
    + 47

4. 95
    + 74

5. 58
    + 49

6. 375
    + 488

7. 164
    + 329

8. 509
    + 793

9. 4,310
    + 8,695

10. 7,256
    784
    + 4,939

11. 18,406
    2,461
    + 24,597

12. 42,875
    364,980
    + 68,417

13. 4,719
    28,604
    319
    + 1,408

14. 95,006
    2,784
    13,309
    + 27

15. 4,342
    809
    427,602
    + 39,855

16. 48 + 95 + 37 = _____

17. 872 + 4,316 + 25 + 970 = _____

18. 24,612 + 917 + 1,423 = _____

19. 560 + 49 + 89,600 + 356,183 = _____

20. 243,517 + 9,346 + 295 + 24,617 = _____

Applications

21. Enrollment in last semester's chemistry classes was 13, 21, 18, and 24 students. What was the total enrollment in the four classes?

    _____

22. Michael drove his new car 105 miles the first day he had it. On the following days he drove 84 miles, 92 miles, 65 miles, and 137 miles. How many miles total did he drive his car the first five days?

    _____

1.3B PRACTICE SET

Find the sum in each problem.

1. 38
    + 61
    ___

2. 86
    + 25
    ___

3. 79
    + 46
    ___

4. 39
    + 54
    ___

5. 63
    + 91
    ___

6. 453
    + 819
    ___

7. 845
    + 160
    ___

8. 443
    + 561
    ___

9. 8,764
    + 3,493
    ___

10. 3,812
    672
    + 5,298
    ___

11. 94,305
    37,991
    + 5,827
    ___

12. 68,164
    4,892
    + 94,871
    ___

13. 4,835
    86,710
    9,675
    +   384
    ___

14. 57,413
    2,864
    342,918
    + 26,445
    ___

15. 30,690
    45,843
    268
    + 817,695
    ___

16. 473 + 68 + 3,719 = _____

17. 6,945 + 844 + 5,612 = _____

18. 47,963 + 3,425 + 84,609 = _____

19. 246,017 + 4,982 + 38,640 = _____

20. 193,695 + 48,519 + 7,634 + 893 = _____

Applications

21. Required reading in an American novel class included *1984,* 245 pages; *The Adventures of Tom Sawyer,* 218 pages; *Two Years before the Mast,* 371 pages; and *All the King's Men,* 438 pages. What was the total number of pages of required reading?

   _____

22. Maria drove from San Francisco to Miami. She went to Phoenix first, a distance of 800 miles. Then she traveled 1,158 miles to Houston, 365 miles to New Orleans, and finally 892 miles to Miami. How many miles total did Maria drive?

   _____

1.4 SUBTRACTING WHOLE NUMBERS

Subtraction is the opposite of addition. It is the process of finding the difference between two numbers. You find this difference by taking one number away from another. Just as you had to know the addition facts to work addition problems, you must also use these same facts to work a subtraction problem.

Consider the addition fact

$$3 + 5 = 8$$

This is a basic addition fact. Suppose you wanted to subtract 5 from 8. The problem would be written

$$8 - 5 = ?$$

The question is, What do you have to add to 5 to get 8? The answer is found in the basic addition fact illustrated previously: 3.

If you wanted to subtract 3 from 8, the answer would be 5, because $3 + 5 = 8$.

In a subtraction problem, the **subtrahend** is the bottom number. It is subtracted from the **minuend,** or the top number. The answer in a subtraction problem is called the **difference.** The sign or symbol that indicates subtraction is −.

$$\begin{array}{r} 9 \rightarrow \textit{minuend} \\ -\,6 \rightarrow \textit{subtrahend} \\ \hline 3 \rightarrow \textit{difference} \end{array}$$

Setting up a subtraction problem is similar to setting up an addition problem. The digits must be properly aligned. Begin with the right column and subtract units from units, tens from tens, hundreds from hundreds, and so on.

EXAMPLE 1   Subtract: $\begin{array}{r} 86 \\ -\,25 \\ \hline \end{array}$

SOLUTION

$\begin{array}{r} 86 \\ -\,25 \\ \hline 61 \end{array}$ *Begin with the units column and subtract each column. Place the difference from each subtraction under the correct column.*

The difference between 86 and 25 is 61.

Sometimes a digit in the subtrahend will be larger than the digit above it in the minuend. This requires *borrowing.* Recall from Section 1.1 that each place is 10 times larger than the place to its right. **Borrowing** is shifting 1 unit from any place to the place on its right. Borrowing 1 from the tens place and shifting it to the right adds 10 units to the units column. If you borrow 1 unit from the hundreds place and shift it to the tens place, you are adding 10 tens to that place. Anytime you borrow, you must remember to reduce the place borrowed from by 1. Study the explanation for Example 2 thoroughly.

EXAMPLE 2   Subtract: $\begin{array}{r} 62 \\ -\,47 \\ \hline \end{array}$

SOLUTION

$$\begin{array}{r} \overset{5\ 12}{\cancel{6}\cancel{2}} \\ -\ 47 \\ \hline 5 \end{array}$$

Begin with the units column. 7 will not subtract from 2, so you must borrow from the tens column. Borrow 1 ten and change it to 10 units. Add this ten to the 2 units already there to get 12 units. Now subtract 7 from 12. The difference is 5. Place this 5 in the units place in the answer.

$$\begin{array}{r} \overset{5\ 12}{\cancel{6}\cancel{2}} \\ -\ 47 \\ \hline 15 \end{array}$$

Now subtract the tens column. Subtract 4 from the 5 that is still in the tens column. The difference is 1. Place this in the tens place in the answer.

The difference between 62 and 47 is 15.

The borrowing process may be used as often as needed to complete a problem.

EXAMPLE 3 Subtract: $\begin{array}{r} 5{,}063 \\ -\ 2{,}495 \end{array}$

SOLUTION

$$\begin{array}{r} \overset{9\ 15}{\overset{4\ 10\ 5\ 13}{5{,}063}} \\ -\ 2{,}495 \\ \hline 2{,}568 \end{array}$$

Begin with the units column and subtract. Borrow as needed until each digit in the subtrahend has been subtracted from the corresponding digit in the minuend. Remember that when you borrow, the 1 unit borrowed is changed to 10 units of the column to the right before being added to the number of units in that column.

Subtracting 2,495 from 5,063 gives 2,568.

Any subtraction problem can be checked for accuracy. This is accomplished by addition. Adding the difference to the subtrahend should give you the minuend.

EXAMPLE 4 Subtract: $\begin{array}{r} 5{,}063 \\ -\ 2{,}495 \end{array}$

SOLUTION   Check:

$$\begin{array}{r} 5{,}063 \\ -\ 2{,}495 \\ \hline 2{,}568 \end{array} \qquad \begin{array}{r} 2{,}568 \\ +\ 2{,}495 \\ \hline 5{,}063 \end{array} \begin{array}{l} \rightarrow difference \\ \rightarrow subtrahend \\ \rightarrow sum\ (minuend) \end{array}$$

Subtracting Whole Numbers

Work the Comprehension Checkpoint problems to test your understanding of subtraction. If you have trouble with these problems, review the material in this section before going on to the Practice Sets.

COMPREHENSION CHECKPOINT

Find the difference in each problem.

1.   57
    − 34

2.   862
    − 451

3.   6,493
    − 2,765

4.   65,982
    − 7,094

1.4A PRACTICE SET

Find the difference in each problem.

1. 48
   − 31

2. 94
   − 63

3. 85
   − 69

4. 917
   − 738

5. 450
   − 87

6. 300
   − 184

7. 3,673
   − 2,197

8. 6,451
   −   964

9. 5,820
   − 4,909

10. 49,005
    − 28,764

11. 76,405
    − 58,793

12. 54,571
    −   983

13. 276,491
    − 95,876

14. 825,937
    − 168,043

15. 397,644
    −   955

16. Subtract 825 from 4,163. _____

17. Subtract 4,987 from 9,402. _____

18. Subtract 14,607 from 95,000. _____

19. Subtract 67,415 from 229,608. _____

20. Subtract 448,716 from 560,913. _____

Applications

21. The Sears Tower in Chicago is 1,454 feet tall. The Empire State Building in New York City is 1,250 feet tall. How much taller is the Sears Tower than the Empire State Building?

    _____

22. The distance from Buffalo to Boston is 457 miles, while the distance from Buffalo to Cleveland is 192 miles. How much farther is it to Boston from Buffalo than it is to Cleveland from Buffalo?

    _____

1.4B PRACTICE SET

Find the difference in each problem.

1. 75
    − 23

2. 57
    − 39

3. 91
    − 26

4. 394
    − 217

5. 801
    − 79

6. 574
    − 389

7. 1,810
    − 995

8. 6,451
    − 964

9. 7,682
    − 7,099

10. 86,443
    − 9,873

11. 11,604
    − 9,412

12. 45,008
    − 24,134

13. 842,375
    − 389,604

14. 590,473
    − 28,691

15. 900,864
    − 100,989

16. Subtract 563 from 981. _____

17. Subtract 4,495 from 6,178. _____

18. Subtract 5,764 from 77,212. _____

19. Subtract 42,698 from 90,025. _____

20. Subtract 271,892 from 404,160. _____

Applications

21. The regular price of a man's suit is $165, but the suit is on sale for $139. What is the amount of the price reduction?

    _____

22. The over-the-road distance from Kansas City, Missouri, to Denver, Colorado, is 616 miles. The air distance between the same two cities is 558 miles. How many more miles would you travel by road than by air?

    _____

1.5 MULTIPLYING WHOLE NUMBERS

Multiplication is often called *repeated addition.* It is a quicker method for adding the same number several times. Look at the following diagram:

```
* * * * *
* * * * *
* * * * *
* * * * *
```

There are 4 rows with 5 asterisks in each row. What is the total number of asterisks in the diagram? This problem could be solved by addition:

$$5 + 5 + 5 + 5 = 20$$

Notice that the addition was repeated until all the asterisks had been counted. The problem could also have been solved by multiplication:

$$4 \text{ (rows)} \times 5 \text{ (asterisks per row)} = 20$$

It may appear to you that there was not a great deal of time or effort saved by using multiplication rather than addition. Try another example. Suppose that a group of 18 people plan to attend a concert and tickets are $39 each. What will be the total cost of the 18 tickets? Using the addition method, you would have $39 + $39 + $39 + · · · until you had $39 a total of 18 times. Using multiplication, you have

$$\$39 \times 18 = \$702$$

It is easy to see from this problem that multiplication is faster than addition.

Numbers being multiplied are called **factors.** The answer in a multiplication problem is called the **product.**

$$\begin{array}{r} 9 \to factor \\ \times\ 8 \to factor \\ \hline 72 \to product \end{array}$$

Many texts refer to the top number in a multiplication problem as the *multiplicand* and the bottom number as the *multiplier.* You can see that $9 \times 8 = 72$ and $8 \times 9 = 72$, so that it does not matter which number is on top and which is on bottom. The product of the same two numbers will always be the same. Calling both of them factors allows for this changing of positions without being concerned about what to call them.

There are several methods of indicating multiplication:

with a times sign ($\times$): 5×4
with parentheses around each factor: $(5)(4)$
with parentheses around one factor: $(4)5$ or $4(5)$
with a dot between factors: $5 \cdot 4$

Success in multiplying any numbers requires that you know the **multiplication facts.** These include all single-digit multiplication equations from $0 \times 0 = 0$ to $9 \times 9 = 81$. A complete Table of Multiplication Facts is located in Appendix A.

Every multiplication problem involves the basic multiplication facts. Example 1 demonstrates this.

EXAMPLE 1 Multiply: $\begin{array}{r} 321 \\ \times\ \ \ 3 \\ \hline \end{array}$

Multiplying Whole Numbers

SOLUTION

$$\begin{array}{r} 321 \\ \times3 \\ \hline 963 \end{array}$$

Each digit in the top factor must be multiplied by each digit in the bottom factor. $3 \times 1 = 3$; put the 3 in the units place. $3 \times 2 = 6$; put the 6 in the tens place. $3 \times 3 = 9$; put the 9 in the hundreds place.

The result of multiplying 321 and 3 is 963.

Because multiplication is repeated addition, you will often have to carry numbers.

EXAMPLE 2 Multiply: $\begin{array}{r}458 \\ \times6\end{array}$

SOLUTION

$$\begin{array}{r} 4 \rightarrow \\ 458 \\ \times6 \\ \hline 8 \end{array}$$ Carried digits

Multiply the units digit by 6. $6 \times 8 = 48$. Put the 8 in the units place, and carry the 4 to the next column.

$$\begin{array}{r} 3\,4 \rightarrow \\ 458 \\ \times6 \\ \hline 48 \end{array}$$ Carried digits.

Multiply the next digit, 5, by 6. $6 \times 5 = 30$. $30 + 4$ (carried from the previous column) $= 34$. Put the 4 in the tens place and carry the 3.

$$\begin{array}{r} 3\,4 \rightarrow \\ 458 \\ \times6 \\ \hline 2{,}748 \end{array}$$ Carried digits.

Multiply the next digit, 4, by 6. $6 \times 4 = 24$. $24 + 3$ (carried from the previous column) $= 27$. Put the 27 in the answer.

In each of the first two examples, you can see the necessity of knowing the multiplication facts. Every multiplication problem is a collection of several of these facts. The next examples involve multiplication problems where each factor has more than one digit.

EXAMPLE 3 Multiply: $\begin{array}{r}56 \\ \times\,87\end{array}$

SOLUTION

$$\begin{array}{r} 56 \\ \times\,87 \\ \hline 392 \end{array}$$

Multiply 56 by 7 using the steps explained in Examples 1 and 2. The result of multiplying 7×56 is 392.

```
    56
  × 87     Next multiply 56 by 8. The result is 448. Remember that the 8 is in
   392     the tens column. When you multiply 8 × 6, you are actually
   448     multiplying 80 × 6, getting 480. Put the 8 in the tens column and
           carry the 4. It is not necessary to write down the 0 in the units
           column.

    56
  × 87
   392
  4 48
  4,872    Add the two products to get the final product.
```

The product of 56 and 87 is 4,872.

Example 3 demonstrates that when you multiply a multidigit number by a multidigit number, you have a series of single-digit multiplication problems. The result of multiplying 7 by 56 is 392. The result of multiplying 8 by 56 is 448. Each of these products is called a **partial product.** The partial products are added to get the final product in the problem. One important item to note is how the partial products are aligned. Because they are going to be added, the partial products must be aligned in columns just as addition problems were. *The digit on the right in each partial product is placed under the digit in the bottom factor that was used to get it.* Thus, the 2 in the 392 is placed in the column under the 7, while the 8 in the 448 is placed in the column under the 8. Correct alignment is crucial to finding the correct product.

Study Example 4. Notice how the partial products are aligned for addition.

EXAMPLE 4   Multiply: 374
                           × 635

SOLUTION

```
      374     Multiply each digit in the top factor by each digit in the bottom
    × 635     factor.
     1 870    first partial product (5 × 374)
    11 22     second partial product (3 × 374)
   224 4      third partial product (6 × 374)
   237,490    product of 374 and 635
```

If one or both factors end in 0 (or 0s), the 0s may be ignored while you multiply the other numbers. After multiplying the other numbers, attach the total number of 0s at the end of the factors.

EXAMPLE 5   Multiply: 40
                           × 6

SOLUTION

```
    40
  ×  6
   240     Multiply: 4 × 6 = 24. Since there is one zero at the end of the top
           factor, add one zero to the right of the 24. Thus, 6 × 40 = 240.
```

EXAMPLE 6 Multiply: 400
              × 60

SOLUTION

400
× 60
────
24,000 *Multiply: 6 × 4 = 24. In this example, there are two zeros in the top factor and one zero in the bottom factor. Add three zeros to the right of the product. 400 × 60 = 24,000*

To check a multiplication problem, reverse the order of the factors and multiply again. You should get the same product.

EXAMPLE 7 Multiply: 46
              × 25

SOLUTION:

46   Check: 25
× 25   × 46
────   ────
230   150
92   1 00
────   ────
1,150   1,150

The products are the same, so the solution is correct.

Example 7 illustrates a point made earlier in this section. The order of the factors does not make a difference. The product of two numbers will be the same regardless of the order of the factors. Also, Examples 3, 4, 5, 6, and 7 show that multiplying multidigit numbers is actually a series of single-digit problems. *It is essential that you know the multiplication facts to multiply correctly.*

Exponents

Sometimes the same factor will appear in a problem more than once. For example, you may have a problem that requires you to multiply $3 \times 3 \times 3 \times 3$. This problem can be simplified by the use of exponents. An **exponent** is a number written above and to the right of a **base** number that tells how many times the base is a factor. A number written with an exponent is written in **exponential notation.** Thus, $3 \times 3 \times 3 \times 3$ could be written as 3^4. In this expression, 3 is the base and 4 is the exponent. This is read "three to the fourth power."

EXAMPLE 8 Write $4 \times 4 \times 4 \times 4 \times 4$ in exponential notation.

SOLUTION

4^5. The 4 is a factor five times.

You could write a number that is in exponential notation as factors.

EXAMPLE 9 Write 7^8 as factors.

SOLUTION

$7 \times 7 \times 7 \times 7 \times 7 \times 7 \times 7 \times 7$. The 7 is a factor eight times.

You can compute the value of a number written in exponential notation by multiplying.

EXAMPLE 10 Find the value of each of the following:

a. 2^3 *b.* 4^2 *c.* 5^5

SOLUTIONS

a. $2 \times 2 \times 2 = 8$ 2 is a factor three times.
b. $4 \times 4 = 16$ 4 is a factor two times.
c. $5 \times 5 \times 5 \times 5 \times 5 = 3{,}125$ 5 is a factor five times.

A base with an exponent of 2 is called a *square*. Thus, 4^2 is read "four squared." A base with an exponent of 3 is called a *cube*. Thus, 5^3 is read "five cubed."

Test your multiplication skills by working the Comprehension Checkpoint problems. If you have difficulty with these problems, review the material in this section before going on to the Practice Sets.

COMPREHENSION CHECKPOINT

Find the product in each problem.

1. 93 × 8
2. 64 × 38
3. 241 × 374

4. 500 × 60
5. 2,407 × 942

6. Write $8 \times 8 \times 8 \times 8 \times 8$ in exponential notation. _____

7. Write 9^5 as factors. _____

8. What is the value of 6^3? _____

1.5A PRACTICE SET

Find the product in each problem.

1. 47 × 8
2. 685 × 3
3. 508 × 9

4. 83 × 24
5. 56 × 19
6. 93 × 72

7. 34 × 52
8. 67 × 90
9. 18 × 35

10. 206 × 48
11. 864 × 39
12. 483 × 52

13. 275 × 807
14. 812 × 346
15. 560 × 425

16. 2,815 × 300
17. 4,693 × 581
18. 3,520 × 908

19. (5)(7)(3)(2) = _____

20. (8)(9)(4)(6) = _____

21. (24)(25)(10) = _____

22. (35)(74) = _____

23. (60)(20) = _____

24. 412 · 500 = _____

25. 8,000 · 400 = _____

In problems 26–28, write each multiplication problem in exponential notation.

26. $4 \times 4 \times 4 \times 4 \times 4$ _____

27. $10 \times 10 \times 10 \times 10 \times 10 \times 10 \times 10$ _____

28. $(6)(6)(6)(6)(6)(6)(6)(6)(6)$ _____

In problems 29–31, write each exponential expression as factors.

29. 2^8 _____

30. 9^6 _____

31. 12^2 _____

In problems 32–34, find the value of each exponential expression.

32. 2^6 _____

33. 10^5 _____

34. 15^3 _____

Applications

35. Leon travels 35 miles to work each morning and 35 miles back again at the end of the day. If Leon goes to work 4 days a week, how many miles does he travel each week?

_____

36. A marathon race is approximately 26 miles. If Denise participated in 14 marathons one year, how many miles total did she run in those races?

_____

1.5B PRACTICE SET

Find the product in each problem.

1. 74
   × 6

2. 38
   × 4

3. 724
   × 8

4. 96
   × 34

5. 83
   × 92

6. 41
   × 80

7. 78
   × 54

8. 29
   × 62

9. 47
   × 12

10. 350
    × 68

11. 517
    × 86

12. 800
    × 250

13. 472
    × 887

14. 935
    × 212

15. 713
    × 654

16. 4,719
    × 500

17. 8,243
    × 1,604

18. 3,768
    × 4,310

19. (6)(8)(2)(5) = _____

20. (9)(4)(3)(10) = _____

21. (45)(90) = _____

22. (50)(40) = _____

23. (25)(16)(100) = _____

24. (600)(48) = _____

25. (9,000)(5,000) = _____

In problems 26–28, write each multiplication problem in exponential notation.

26. $5 \times 5 \times 5$ _____

27. $12 \times 12 \times 12 \times 12 \times 12$ _____

28. $(7)(7)(7)(7)(7)(7)(7)$ _____

In problems 29–31, write each exponential expression as factors.

29. 18^2 _____

30. 25^9 _____

31. 50^4 _____

In problems 32–34, find the value of each exponential expression.

32. 3^6 _____

33. 5^8 _____

34. 12^3 _____

Applications

35. A video store rented movies for $3 each. During one week there were a total of 2,569 rentals. How much money did the store get from rentals that week?

_____

36. A restaurant offered a prom night special that included an entire meal (including tip) for $35. If 22 students ordered the special, how much money did the restaurant make on the special?

_____

1.6 DIVIDING WHOLE NUMBERS

Division may be defined as the opposite of multiplication. It is the process of determining how many times one number is contained in another number. Just as you had to know your addition facts to work a subtraction problem successfully, you must know your multiplication facts to work a division problem successfully.

Consider the multiplication fact

$$3 \times 5 = 15$$

The factors here are 3 and 5; the product is 15. If 15 is divided by 5, the answer is 3. If 15 is divided by 3, the answer is 5. To divide, you must know what factors multiply together to give a product.

You can label the parts of a division problem:

$$\text{divisor} \leftarrow 5\overline{)15} \begin{array}{l} \rightarrow \text{quotient} \\ \rightarrow \text{dividend} \end{array}$$

The **divisor** is the number used to divide another number. The **dividend** is the number being divided. The **quotient** is the answer in a division problem. The **remainder,** which will be described later in Section 1.6, is the part of a dividend that cannot be divided by the divisor.

As with multiplication, there is more than one way to indicate division.

$$\text{divisor}\overline{)\text{dividend}}^{\text{quotient}}$$

$$\frac{\text{dividend}}{\text{divisor}} = \text{quotient}$$

$$\text{dividend} \div \text{divisor} = \text{quotient}$$

It is easier to understand division if you know exactly what is happening in a division problem.

Consider the problem $6 \div 3 = ?$ In this problem, 6 units are being divided into 3 groups. How many will be in each group?

```
|←——— 6 objects ———→|
  *  *  |  *  *  |  *  *
     2       2       2
```

You can see that there are 2 objects in each of the 3 groups. Thus, $6 \div 3 = 2$.

Dividing by a Single-Digit Number

Problems such as $12 \div 4 = 3$, $24 \div 6 = 4$, and $81 \div 9 = 9$ are merely the inverse of the basic multiplication facts. The only way to divide successfully is to know the multiplication facts. What the equations shown have in common is a dividend that is an exact multiple of the divisor. These are the simplest division problems, because they require only a reversal of the multiplication facts. Other problems will require a longer process. Study the examples carefully to make sure you understand the steps required to solve a division problem.

EXAMPLE 1   Divide: $29 \div 6$

SOLUTION

$$6\overline{)29}^{4}$$

How many groups of 6 are in 29? From multiplication facts, you know that $4 \times 6 = 24$ and $5 \times 6 = 30$. Thus there are 4 groups of 6 in 29. The 4 goes above the 9 since the 6 is being divided into a two-digit number.

$$\begin{array}{r} 4 \\ 6{\overline{\smash{\big)}\,29}} \\ 24 \end{array}$$ *Multiply 4 by 6. Place the product, 24, under the dividend, 29.*

$$\begin{array}{r} 4 \\ 6{\overline{\smash{\big)}\,29}} \\ \underline{24} \\ 5 \end{array}$$ *Subtract 24 from 29. The result, 5, is the remainder.*

The quotient is written as 4 R 5, meaning that 6 divides 29 four times with 5 remaining.

To check a division problem, you will use multiplication, because division and multiplication are inverse operations. Multiply the quotient by the divisor. If there is a remainder, add it to the product of the multiplication. The answer should be the dividend. To check the solution in Example 1:

$$\begin{array}{c} \text{quotient} \times \text{divisor} + \text{remainder} = \text{dividend} \\ 4 \times 6 + 5 \stackrel{?}{=} 29 \\ 24 + 5 \stackrel{?}{=} 29 \\ 29 = 29 \end{array}$$

You can determine whether the digit placed in the quotient is correct. Consider Example 1 again. What would happen if you placed a 5 in the quotient instead of the 4?

$$\begin{array}{r} 5 \\ 6{\overline{\smash{\big)}\,29}} \\ 30 \end{array}$$ *$5 \times 6 = 30$; 30 is larger than the number it is to be subtracted from, 29, so the 5 is too large. Try a smaller digit in the quotient.*

If you had placed a 3 in the quotient instead of the 4, you would have

$$\begin{array}{r} 3 \\ 6{\overline{\smash{\big)}\,29}} \\ \underline{18} \\ 11 \end{array}$$ *The remainder after subtracting is larger than the divisor. This means the digit in the quotient is too small; try a larger digit in the quotient.*

EXAMPLE 2   Divide: $125 \div 8$

SOLUTION

$$\begin{array}{r} 15 \\ 8{\overline{\smash{\big)}\,125}} \\ \underline{8} \\ 4 \end{array}$$ *How many 8s are in 12? There is one.*
$1 \times 8 = 8$
$12 - 8 = 4$
So far this is identical to Example 1.

$$\begin{array}{r} 15 \\ 8{\overline{\smash{\big)}\,125}} \\ \underline{8} \\ 45 \\ \underline{40} \\ 5 \end{array}$$ *Bring down the next digit in the dividend and repeat the same steps.*
How many 8s are in 45? There are 5.
$5 \times 8 = 40$
$45 - 40 = 5 = $ remainder

Dividing Whole Numbers

Check: $15 \times 8 + 5 \stackrel{?}{=} 125$
$120 + 5 \stackrel{?}{=} 125$
$125 = 125$

The solution is 15 R 5.

Dividing by a Multidigit Divisor

The steps required when the divisor has more than one digit are the same as with single-digit divisors. Study Example 3 and notice its similarities to the first two examples.

EXAMPLE 3 Divide: $561 \div 26$

SOLUTION

$\;\;2$
$26\overline{)561}\quad$ *How many times will 26 go into 56?*
$52\quad\;\;\,$ *$2 \times 26 = 52$; 52 is less than 56; go on.*
$\overline{}$
$4\quad\;\;\;\,$ *$56 - 52 = 4$; 4 is less than 26; go on.*

21
$26\overline{)561}$
52
$\overline{}$
$41\quad\;\,$ *Bring down the 1. How many 26s are in 41? 1*
$26\quad\;\,$ *$1 \times 26 = 26$*
$\overline{}$
$15\quad\;\,$ *$41 - 26 = 15$; 15 is the remainder, since all digits in the dividend have been divided.*

The solution is 21 R 15.

Check: $21 \times 26 + 15 \stackrel{?}{=} 561$
$546 + 15 \stackrel{?}{=} 561$
$561 = 561$

The following example shows that division can be a trial-and-error process.

EXAMPLE 4 Divide: $1{,}754 \div 38$

SOLUTION

$38\overline{)1{,}754}\quad$ *38 will not divide into 17, so you must divide it into 175. How many times will 38 go into 175? For a good estimation, divide 17 by 3. How many 3s are in 17? 5; try 5 in the quotient.*

$\phantom{38)1{,}}5$
$38\overline{)1{,}754}\quad$ *Because you are dividing 175 by 38, place the 5 in the quotient over the 5 in the dividend. $5 \times 38 = 190$, which is too large; try 4 in the quotient.*
$1\,90$

$\phantom{38)1{,}}4$
$38\overline{)1{,}754}$
$1\,52\quad$ *$4 \times 38 = 152$*
$\phantom{38)1{,}}\overline{}$
$\phantom{38)1{,}}23\quad$ *$175 - 152 = 23$; 23 is smaller than 38, so continue.*

$\phantom{38)1{,}}47$
$38\overline{)1{,}754}$
$1\,52$
$\phantom{38)1{,}}\overline{}$
$\phantom{38)1{,}}234\quad$ *Bring down the 4; $234 \div 38 = ?$ How many 3s are in 23? 7*
$\phantom{38)1{,}}266\quad$ *$7 \times 38 = 266$, which is too large, try 6.*

```
      46
38)1,754
    1 52
    ‾‾‾‾
      234
      228    6 × 38 = 228
      ‾‾‾
        6    234 − 228 = 6 = remainder
```

The solution is 46 R 6. The check is left for you to do.

Sometimes the divisor will not divide into the dividend. Study Example 5.

EXAMPLE 5   Divide: 8,154 ÷ 27

SOLUTION

```
      3
27)8,154     81 ÷ 27 = 3
   8 1       3 × 27 = 81
   ‾‾‾
     0       81 − 81 = 0


     30
27)8,154
   8 1
   ‾‾‾
     05      Bring down the 5. 27 will not divide into 5, so place a 0 in the
             quotient. This 0 is important to the answer and must be used as a
             placeholder.


     302
27)8,154
   8 1
   ‾‾‾
     054     Bring down the 4. 54 ÷ 27 = 2.
      54     2 × 27 = 54.
     ‾‾‾
       0     There is no remainder.
```

When a problem does not have a remainder, as in Example 5, it is said to *come out even*. This means that the divisor goes into the dividend a whole number of times.

One final comment about division. You learned in Section 1.1 that 0 is a whole number. If 0 is divided by any other number, the result is always 0.

$$0 \div 5 = 0$$

$$0 \div 12 = 0$$

However, if 0 is the divisor, the problem cannot be done.

$$5 \div 0 = \text{cannot be done}$$

$$12 \div 0 = \text{cannot be done}$$

One way to remember this is to think about $1. If you divide $1 by 4, you have 4 quarters. If you divide $1 by 10, you have 10 dimes. If you divide $1 by 20, you have 20 nickels. What do you have if you divide $1 by 0? It cannot be done!

Dividing Whole Numbers

Work the Comprehension Checkpoint problems to test your understanding of division. If you have difficulty with these problems, review the material in this section before going on to the Practice Sets.

COMPREHENSION CHECKPOINT

Find the quotient in each problem.

1. 28 ÷ 4 = _____
2. 360 ÷ 24 = _____
3. 504 ÷ 42 = _____
4. 1,347 ÷ 56 = _____
5. 3,675 ÷ 35 = _____
6. 0 ÷ 16 = _____
7. 15 ÷ 0 = _____

1.6A PRACTICE SET

Find the quotient in each problem. Identify any remainders.

1. 7)84
2. 9)270
3. 6)132

4. 8)448
5. 4)256
6. 3)192

7. 18)259
8. 23)127
9. 38)461

10. 26)793
11. 56)675
12. 84)2,189

13. 42)257
14. 75)2,632
15. 48)5,040

16. 25)5,152
17. 63)1,325
18. 34)17,309

19. $64 \div 8 =$ _____

20. $72 \div 9 =$ _____

21. $54 \div 6 =$ _____

22. $32 \div 4 =$ _____

23. $42 \div 6 =$ _____

24. $24 \div 3 =$ _____

25. $50 \div 2 =$ _____

Applications

26. Tuition for 14 credit hours at a community college is $210. What is the cost per credit hour?

_____

27. Darlene scored 832 points in 26 basketball games. Assuming that Darlene scored the same number of points in each game, how many points did she score in each game?

_____

1.6B PRACTICE SET

Find the quotient in each problem. Identify any remainders.

1. 8)96
2. 4)492
3. 3)615

4. 9)171
5. 6)402
6. 7)151

7. 25)775
8. 31)372
9. 43)1,032

10. 28)1,433
11. 64)5,379
12. 14)1,680

13. 80)2,889
14. 24)6,295
15. 57)1,147

16. 70)3,441
17. 82)1,479
18. 97)10,180

19. 24 ÷ 6 = _____

20. 40 ÷ 8 = _____

21. 56 ÷ 7 = _____

22. 21 ÷ 3 = _____

23. 63 ÷ 9 = _____

24. 48 ÷ 8 = _____

25. 39 ÷ 3 = _____

Applications

26. An apartment rents for $4,500 for one year. What is the amount of the monthly rent payment?

     _____

27. Westview Community College offers 14 class sections of freshman English. If the 322 students enrolled are divided evenly among the 14 sections, how many are in each class?

     _____

1.7 WORD PROBLEMS

Many students cringe at the thought of having to solve a word or application problem. The ability to solve word problems is a learned skill. People who are very good at what they do are good because they are persistent and they learn from their mistakes. You can learn how to solve word problems if you are willing to apply yourself to the challenge.

You should develop a strategy for solving word problems. Consider the following strategy.

> **STRATEGY FOR SOLVING WORD PROBLEMS**
> 1. Read the problem completely before you attempt to solve it.
> 2. Write down the facts that you know.
> 3. Write down what it is you are trying to find.
> 4. Draw a sketch or diagram to help you visualize the problem.
> 5. Determine how the given information can be used to find the desired answer.
> 6. Solve the problem.
> 7. Check your answer to make sure it answers the question asked and is reasonable.

Study the examples to see how these steps are helpful.

EXAMPLE 1   Jose bought 3 shirts for $19 each. What was the total cost of the shirts?

SOLUTION

Step 1   Read the problem carefully.
Step 2   Facts: He bought 3 shirts. Each shirt cost $19.
Step 3   What are you trying to find? The total cost of the 3 shirts.
Step 4   (May not help on this problem.)
Step 5   You are looking for the total cost. Addition gives you the total. Multiplication is a faster form of addition.
Step 6   3 (number of shirts) × $19 (cost per shirt) = $57. The total cost of the shirts is $57.
Step 7   The answer is a reasonable solution to the problem.

EXAMPLE 2   Jennifer budgeted $900 for her vacation. She planned to be gone for 12 days. How much did she budget for each day?

SOLUTION

Step 1   Read the problem carefully.
Step 2   Facts: total budget, $900; total days, 12.
Step 3   What are you trying to find? The amount allocated for each day.
Step 4   (May not help in this problem.)
Step 5   You know the total budget and the number of days it must last. You want to divide the total amount so that you will have an equal amount each day. Use division.
Step 6 $900 ÷ 12 = $75. Jennifer has allowed $75 per day.
Step 7   The answer is a reasonable solution to the problem.

Some problems may require more than one step to arrive at the desired answer.

EXAMPLE 3 Marie attends a technical school. She goes to school 5 hours each day for 4 days per week. The total number of hours in the program is 1,000. How many weeks will it take her to complete the program?

SOLUTION

Step 1   Read the problem carefully.
Step 2   Facts: She is in school 5 hours each day. She goes to school 4 days each week. She must attend 1,000 hours to complete the program.
Step 3   What are you trying to find? The number of weeks she must attend to complete the program.
Step 4   (May not help with this problem.)
Step 5   If you knew how many hours she attended each week, you could divide the total hours required, 1,000, by the number of hours each week to find the number of weeks.
Step 6   5 (hours per day) × 4 (days per week) = 20 hours attended each week. 1,000 (total hours) ÷ 20 (hours per week) = 50 weeks attended.
Step 7   The answer appears to be a reasonable solution to the problem.

You can learn to solve a word problem if you practice. It is a skill that most people can use daily. You will eventually get to the point where it will not be necessary to write down each step. With enough practice you will be able to do some of the solution process mentally.

It is important to note that as you learn more math, you may be able to solve the same problem in different ways. There is not necessarily a best way to solve a problem. The ultimate objective is to solve it using a logical approach that you understand.

Try the Comprehension Checkpoint problems to test your ability to solve a word problem. If you have difficulty with these problems, review the material in this section before attempting the Practice Sets.

COMPREHENSION CHECKPOINT

1. A case of canned green beans contains 24 cans. How many cans would be in 12 cases?

2. Kim's expenses for one month included $350 for rent, $125 for utilities, $47 for the telephone, $168 for groceries, and $31 for miscellaneous expense. She had budgeted $750 for these expenses. How much did she have left after paying all of her bills?

3. Tickets for a concert sold for $15, $20, and $25. A total of 22,500 tickets were sold. If 12,000 tickets sold for $15 and 6,000 tickets sold for $20, what was the total revenue from ticket sales?

1.7A PRACTICE SET

Solve each application problem.

1. Katie spent $259 for a dress, $95 for shoes, and $137 for other accessories for her senior prom. What was her total expenditure for the prom?

2. At the beginning of the school year the freshman class at State U. included 3,450 students. Other class figures were: sophomore, 2,912 students; junior, 3,120 students; senior, 2,389 students; graduate, 1,463 students. What was the total enrollment?

3. Census figures for 1980 show California with a population of 23,667,902, New York with a population of 17,558,072, and Texas with a population of 14,229,191. What was the total population in these three states?

4. In 1988 Oakland, California, had an estimated population of 356,860. Across the bay, San Francisco had an estimated population of 731,600. How many more people lived in San Francisco than in Oakland?

5. A new car cost $14,610 at one dealership. The same car at a competing dealership was priced at $13,895. How much could you save if you purchased the car from the second dealer?

6. A utility customer's electric meter reading one month was 28,082 units. The next month the reading was 29,609. How many units of electricity did the customer use for the month?

7. Mark's new car gets 38 miles per gallon. If the tank holds 14 gallons of gasoline, how far can Mark drive on a full tank of gas?

8. The salesman told Lucy that she could finance a new car for 54 months at a monthly payment of $265. What is the total amount that Lucy will have to pay?

9. A textbook publisher printed 8,500 copies of a book containing 416 pages. What was the total number of pages printed?

10. A group of 58 people took a bus tour of a historic park. They paid a total of $696. What was the cost for each person?

11. A laser printer produced a 280-page document in 35 minutes. How many pages were printed each minute?

12. Each of the 28 teams in the National Football League has the same number of players. If there are a total of 1,372 players in the league, how many are on each team?

13. The organizers of a political fund-raiser set up 25 tables with 8 chairs at each table. The charge to attend was $75 per person. If each chair was occupied, how much revenue did the fund-raiser generate?

14. Weight Loss Anonymous made a special offer to new members. When 6 people signed up for the offer, their combined weight was 1,170 pounds. After 3 months, the 6 people weighed a total of 1,068 pounds. Assuming that each person lost the same amount of weight, how much did each person lose?

1.7B PRACTICE SET

Solve each application problem.

1. A company leased office space in three different buildings. It had 4,560 square feet in one building, 6,715 square feet in a second building, and 4,802 square feet in the third building. How many total square feet did the company lease?

2. Attendance figures for the five games of a softball tournament were 870, 695, 1,037, 912, and 1,309. What was the total attendance for the five games?

3. A company employs 14,746 people in the United States, 2,506 people in Canada, 10,035 people in Europe, and 9,112 people in Asia. How many employees total does the company have?

4. Carol paid $825 for tuition this semester. Amanda paid $680. How much more did Carol pay than Amanda?

5. In 1787, Delaware was the first former colony to become a state in the United States. The last territory to become a state was Hawaii in 1959. How many years apart did these states enter the Union?

6. The fence in a baseball park is 385 feet from home plate. A batter, standing at home plate, hit a ball 512 feet. How far past the fence did the ball land?

7. Lindsey can take shorthand at 120 words per minute. How many words could she write in shorthand in 24 minutes?

8. A luxury suite at a hotel rents for $189 per night. Last year the suite was rented 217 nights. How much money did the hotel collect from renting that suite?

9. An airline route from Honolulu to Chicago is 4,250 miles. If a pilot flew round-trip from Honolulu to Chicago once each week for one year, how many miles total would the pilot fly?

_____

10. Bryan used 15 gallons of gasoline to go 480 miles. How many miles did he go on each gallon of gas?

_____

11. If a dozen roses cost $48, what is the cost of each rose?

_____

12. How many hours would it take to drive 1,870 miles if you drove 55 miles each hour?

_____

13. Caesario, a pilot, traveled 575 miles per hour for a trip that was 2,300 miles long. He then flew back over the same route but could average only 460 miles per hour because of strong head winds. What was his total flying time?

_____

14. George Washington became president in 1789. George Bush left the presidency in 1993. A total of 40 men have served as president of the United States. Of those 40, 12 served 8 years (two terms) and 17 served 4 years (one term). What was the total number of years served by presidents who did not serve either one or two full terms?

_____

1.8 PRIME AND COMPOSITE NUMBERS

Every natural number may be classified as either prime or composite. A **prime number** is a natural number greater than 1 that has no natural-number factors except itself and 1. By this definition, 2 is a prime number, because it has natural-number factors of only 1 and 2. Similarly, 3 is prime, since its only natural-number factors are 1 and 3. A partial list of prime numbers includes

$$2, 3, 5, 7, 11, 13, 17, 19, 23, \ldots$$

Recall from Section 1.1 that the three dots (called *ellipses*) indicate that the list goes on indefinitely.

EXAMPLE 1 Identify the prime numbers in the following list:

25, 31, 33, 43

SOLUTION

31 and 43 are prime. 25 and 33 both have factors other than themselves and 1.

A **composite number** is a number that has other natural-number factors besides 1 and itself.

EXAMPLE 2 Identify the composite numbers from the following list:

24, 41, 51, 59

SOLUTION

Both 24 and 51 can be divided by 3. They are composite numbers. The other numbers, 41 and 59, are both prime.

It should be noted that 1 is neither prime nor composite.

Any composite number may be shown to be the product of a series of prime factors. This is known as the **prime factorization** of a number. You know from the basic multiplication facts that $2 \times 3 = 6$. Both 2 and 3 are prime numbers. The prime factorization of 6 is written as $6 = 2 \times 3$.

EXAMPLE 3 Determine the prime factorization of 12.

SOLUTION

$2 \overline{)12}$ gives 6. *Divide 12 by the lowest prime factor that will divide it evenly.*

$2 \overline{)6}$ gives 3. *Divide the quotient of the first division by the lowest prime factor that will divide it evenly.*
When the quotient is itself a prime number, the problem is complete.

The prime factorization of a number is all the factors that were used as divisors and the final prime quotient.

The prime factorization of $12 = 2 \times 2 \times 3$.

Any time the dividend is an even number (2, 4, 6, 8, . . .), the prime factor used as a divisor will be 2. If the dividend is not divisible by 2, go to the next lowest prime number and see if it divides the dividend evenly. Continue this process until the quotient is a prime number. Notice in Example 4 how the continued division is shown.

EXAMPLE 4   Determine the prime factorization of 30.

SOLUTION

```
                         5   prime number quotient
second divisor    3)15       first quotient
first divisor      2)30      original composite number
```

The prime factorization of 30 = 2 × 3 × 5.

Remember that each factor in a prime factorization must be a prime number. To check your work, multiply the prime factors. The product should be the original composite number. Checking Example 4 gives

$$2 \times 3 \times 5 = 30$$
$$6 \times 5 = 30$$
$$30 = 30$$

One more example will illustrate that even larger composite numbers may be shown to be the product of prime factors.

EXAMPLE 5   Determine the prime factorization of 594.

SOLUTION

```
                          11    prime-number quotient
fourth divisor    3)  33        third quotient
third divisor     3)  99        second quotient
second divisor    3)297         first quotient
first divisor     2)594         original composite number
```

The prime factorization of 594 = 2 × 3 × 3 × 3 × 11.

Consider one example with an odd composite number.

EXAMPLE 6   Determine the prime factorization of 105.

SOLUTION

```
         7      Since 105 cannot be divided by 2, try the next lowest prime, 3. 105
    5) 35       ÷ 3 = 35. 35 cannot be divided by 3, so try 5, the next lowest
    3)105       prime. 35 ÷ 5 = 7. 7 is prime, so the division process is complete.
```

The prime factorization of 105 = 3 × 5 × 7.

Prime and Composite Numbers

Many numbers share common factors. For example, 3 is a factor of 6, 9, 12, 18, 30, and many other numbers. However, it is important to note that every composite number has a unique prime factorization.

Try the Comprehension Checkpoint problems to measure your understanding of prime and composite numbers. If you have difficulty with these problems, review the material in this section before going on to the Practice Sets.

COMPREHENSION CHECKPOINT

1. Circle the prime numbers in the following list:

   18, 37, 53, 63, 71

2. What is the prime factorization of 360? _____

3. What is the prime factorization of 207? _____

1.8A PRACTICE SET

There are two prime numbers in each group in problems 1–5. Circle those prime numbers.

1. 27, 39, 41, 57, 83

2. 42, 97, 87, 101, 95

3. 71, 51, 61, 81, 91

4. 63, 73, 89, 99, 54

5. 109, 111, 114, 117, 103

In problems 6–20, determine the prime factorization of each number.

6. 16 = _____

7. 35 = _____

8. 48 = _____

9. 56 = _____

10. 63 = _____

11. 84 = _____

12. 90 = _____

13. 80 = _____

14. 42 = _____

15. 225 = _____

16. 336 = _____

17. 120 = _____

18. 210 = _____

19. 1,188 = _____

20. 2,000 = _____

1.8B PRACTICE SET

There are two prime numbers in each group in problems 1–5. Circle those prime numbers.

1. 21, 31, 41, 51, 81

2. 27, 57, 97, 77, 47

3. 33, 43, 55, 53, 88

4. 83, 57, 49, 63, 29

5. 59, 87, 63, 67, 91

In problems 6–20, determine the prime factorization of each number.

6. 8 = _____

7. 24 = _____

8. 33 = _____

9. 36 = _____

10. 44 = _____

11. 54 = _____

12. 60 = _____

13. 75 = _____

14. 108 = _____

15. 160 = _____

16. 250 = _____

17. 750 = _____

18. 654 = _____

19. 2,520 = _____

20. 3,000 = _____

1.9 ORDER OF OPERATIONS

Many problems will require more than one operation before they can be solved completely. Consider the problem

$$12 - 4 \div 2$$

Two operations are indicated: subtraction and division. If you subtract first, you get

$$12 - 4 \div 2$$
$$8 \div 2$$
$$4$$

If you divide first, you get

$$12 - 4 \div 2$$
$$12 - 2$$
$$10$$

It is important to have a set of rules to govern this situation. The rules governing the order of operations are listed in the following box.

RULES FOR THE ORDER OF OPERATIONS

1. Do all multiplication and division in order first, working from left to right.
2. After completing all multiplication and division, do addition and subtraction in order, from left to right.
3. If grouping symbols such as ( ) or [ ] are included, operations within these symbols should be performed first, using the same steps. After simplifying inside the grouping symbols, begin again using the first two steps for the operations that remain.

EXAMPLE 1   Simplify $6 \times 3 + 7$.

SOLUTION

$6 \times 3 + 7$ *multiply first*

$18 + 7$ *add after multiplying*

25

EXAMPLE 2 Simplify $5 + 6 \div 3 \times 4 - 1$.

SOLUTION

$5 + \underbrace{6 \div 3}\ \times 4 - 1$ *divide*

$5 + \underbrace{2\ \times 4}\ - 1$ *multiply*

$\underbrace{5 +\qquad 8}\ - 1$ *add*

$\underbrace{13\ - 1}$ *subtract*

12

EXAMPLE 3 Simplify $(3 + 7 \times 6) \div 5$.

SOLUTION

$(3 + \underbrace{7 \times 6}) \div 5$ *multiply*

$(\underbrace{3 +\ 42}) \div 5$ *add inside ( ) before dividing*

$\underbrace{45 \div 5}$ *divide*

9

EXAMPLE 4 Simplify $(20 + 10 \div 5)(40 \div 4 - 6)$.

SOLUTION

$(20 + \underbrace{10 \div 5})(\underbrace{40 \div 4} - 6)$ *Work within each set of ( ).*

$(\underbrace{20 + 2})\quad (\underbrace{10 - 6})$

$(22)(4)$ *Remember that two ( ) ( ) written side by side indicate multiplication.*

88

Try the Comprehension Checkpoint problems to check your understanding of the order of operations. If you have difficulty with these problems, review this section before going on to the Practice Sets.

COMPREHENSION CHECKPOINT

Evaluate each expression.

1. $5 \times 9 + 3 = $ _____

2. $6 \div 2 + 5 \times 4 = $ _____

3. $(8 \div 2 - 3) + 5 = $ _____

4. $(25 - 3 \times 5) \div (10 \div 2) = $ _____

1.9A PRACTICE SET

Evaluate each expression.

1. $2 + 3 \times 5 =$ _____

2. $4 \div 2 \times 6 =$ _____

3. $16 + 8 - 5 =$ _____

4. $10 \div 5 \times 6 - 3 =$ _____

5. $8 - 4 \times 2 =$ _____

6. $(6)(8) \div 12 =$ _____

7. $(10)(5) - (4)(6) =$ _____

8. $15 \div 3 - 4 + 20 \div 10 =$ _____

9. $(24 + 18 \div 9) \div 13 =$ _____

10. $28 \div 14 + (4 \times 3) =$ _____

11. $(12 \times 3 + 6 \div 2) + 8 \times 3 =$ _____

12. $9 \times (18 \div 2) + 6 =$ _____

13. $24 \div 3 \times 5 + 25 =$ _____

14. $3 + 8 \times (9 - 3) - 6 =$ _____

15. $(20)(10) - (25)(8) =$ _____

16. $42 \div 6 - 39 \div 13 =$ _____

17. $[(40)(8) \div 20] + 18 \times 5 =$ _____

18. $8 + (5)(7) - 12 =$ _____

19. $(25 + 12 \times 5)(2)(3) =$ _____

20. $[(28)(3) \div 14] + 168 \div 24 =$ _____

Applications

21. On a shopping trip, Marie bought two dresses for $89 each and a pair of shoes for $45. What was the total amount of her purchases?

   _____

22. One oil well produces 20 barrels of oil each day. A second well produces 27 barrels of oil each day. What would the total production be for a week in which the first well operated 6 days and the second well operated only 4 days?

   _____

23. Bill left his house with $100. He spent $15 on gasoline and $28 on groceries and paid his $42 phone bill. How much money did he have when he returned home?

   _____

1.9B PRACTICE SET

Evaluate each expression.

1. $7 \times 6 + 8 =$ _____

2. $10 \div 2 \times 5 =$ _____

3. $14 + 7 - 3 \times 6 =$ _____

4. $12 + 6 \div 3 \times 2 =$ _____

5. $15 - 3 \times 5 =$ _____

6. $24 + 6 \times 8 \div 12 =$ _____

7. $(7)(8) \div 2 =$ _____

8. $(10)(12) - (8)(6) =$ _____

9. $18 \div 9 + 4 \times 5 =$ _____

10. $(5 + 7) \times 3 =$ _____

11. $28 \div (7 \times 2) =$ _____

12. $(8)(9) \div (3)(6) =$ _____

13. $(10 + 6 \times 3) \div 4 =$ _____

14. $(8 + 12)(15 - 10) =$ _____

15. $9 \times (14 + 11) - 50 =$ _____

16. $48 + 20 \div 5 - 3 \times 14 =$ _____

17. $(55 + 10) \div 13 - 4 =$ _____

18. $15 \times (5 + 6) \div 55 =$ _____

19. $(24)(3) \div (18)(4) =$ _____

20. $(36 + 40 \div 2) \div (8 \div 2) =$ _____

Applications

21. Luis bought three notebooks at $4 each, two pens at $3 each, and a lab manual for $9. What was the total cost of his purchases?

_____

22. Mrs. Sanchez drives 74 miles round-trip to work and back 5 days each week. At the beginning of one week her odometer reading was 24,690. What was her odometer reading at the end of the week?

_____

23. During a workout, Sean bench-pressed 225 pounds 15 times. In another exercise he lifted 310 pounds 18 times. What was the total amount of weight he lifted during the workout?

_____

SUMMARY OF KEY CONCEPTS

KEY TERMS

set (1.1): A set is a particular group of numbers.

natural numbers (counting numbers) (1.1): The set of numbers that includes {1, 2, 3, 4, . . . }.

whole numbers (1.1): The set of numbers that includes {0, 1, 2, 3, . . . }.

expanded notation (1.1): A method of writing a number that clearly shows the value of each place in the number.

rounding (1.2): Using an approximate number rather than an exact number.

rounding place (1.2): The place being rounded to (nearest ten, nearest thousand, etc.).

determining digit (1.2): The digit to the right of the rounding place.

addition (1.3): The process of grouping or combining terms that are alike.

addends (1.3): The numbers being added in an addition problem.

addition facts (1.3): All single-digit addition equations from $0 + 0 = 0$ to $9 + 9 = 18$.

sum (or total) (1.3): The answer in an addition problem.

aligning numbers (1.3): Placing one number over another with the units over the units, tens over tens, hundreds over hundreds, and so on, to prepare for addition or subtraction.

carrying (1.3): In addition and multiplication, the process of placing one digit over the next column, to be added to the other digits in that column.

subtraction (1.4): The process of finding the difference between two numbers. It is the opposite of addition.

subtrahend (1.4): The bottom number in a subtraction problem.

minuend (1.4): The top number in a subtraction problem.

difference (1.4): The answer in a subtraction problem.

borrowing (1.4): Shifting 1 unit from one column and adding it to the column to the right. Before the unit borrowed can be used, it must be renamed as 10 units of the same value as the column to its right.

multiplication (1.5): The process of repeated addition. A quicker method of adding numbers several times.

factors (1.5): The numbers being multiplied together in a multiplication problem.

product (1.5): The answer in a multiplication problem.

multiplication facts (1.5): All single-digit multiplication equations from $0 \times 0 = 0$ to $9 \times 9 = 81$.

partial product (1.5): The product of a single factor and the top factor in a multidigit multiplication problem.

exponent (1.5): A number written above and to the right of a base number. It tells how many times the base is a factor.

base (1.5): A base is a number being used as a factor.

exponential notation (1.5): Writing a number with an exponent.

division (1.6): The opposite of multiplication—the process of determining how many times one number is contained in another number.

divisor (1.6): The number being used to divide another number.

dividend (1.6): The number being divided.

quotient (1.6): The answer in a division problem.

remainder (1.6): The part of a dividend that cannot be divided by the divisor.

prime number (1.8): A natural number greater than 1 with no natural-number factors except itself and 1.

composite number (1.8): A number that has natural-number factors other than itself and 1.

prime factorization (1.8): Showing a composite number as a series of factors in which each factor is a prime number.

KEY RULES

Rounding (1.2):

Find the rounding place and the determining digit. If the determining digit is 4 or less, change it and all digits to its right to zero. If the determining digit is 5 or more, add 1 to the rounding place and change all digits to the right of the rounding place to zero.

Addition (1.3):

Align the addends in columns with units over units, tens over tens, and so on, and then add each column.

Subtraction (1.4):

Align the subtrahend and minuend in columns and subtract units from units, tens from tens, and so on.

Multiplication (1.5):

Each digit in one factor must be multiplied by each digit in another factor. Carefully align each partial product. Add the partial products to get the final product.

Division (1.6):

Divide the divisor into the dividend using the least number of digits that the divisor will go into. Continue the process until the entire dividend has been divided.

Strategy for solving word problems (1.7):

1. Read the problem completely before you attempt to solve it.
2. Write down the facts that you know.
3. Write down what it is that you are trying to find.
4. Draw a sketch or diagram, if possible, to help you visualize the problem.

5. Determine how the given information can be used to find the desired answer.
6. Solve the problem.
7. Check your answer to make sure it answers the question asked and is reasonable.

Prime factorization (1.8):

Divide a composite number by the smallest prime number that will divide it evenly. Divide the quotient by the smallest prime number that will divide it evenly. Continue this process until the quotient is itself a prime number. The prime factorization is all the prime-number divisors and the prime-number quotient.

Order of operations (1.9):

When a problem requires more than one operation, multiply or divide from left to right first and then add or subtract from left to right. Perform operations inside grouping symbols first and then work with the numbers outside the grouping symbols.

CHAPTER 1

PRACTICE TEST 1A

1. Write 74,680,173 in words.
   _____

2. Write twenty-seven thousand, six hundred eighty-nine in numerical form.
   _____

3. Consider the number 604,892. What digit is in the:
   a. hundreds place? *b.* thousands place? *c.* hundred thousands place?

   a. _____   *b.* _____   *c.* _____

4. Write 8,674 in expanded notation. _____

5. Round 895,706 to the indicated place.
   a. nearest hundred _____
   b. nearest thousand _____
   c. nearest ten _____

Add:

6.   27
       43
    + 64
    _____

7.   17,064
        295
      1,603
    + 12,187
    _____

Subtract:

8.   4,768
    − 2,454
    _____

9. 10,409
    − 8,270
    _____

Multiply:

10. 6 × 3 × 7 × 8 = _____

11.   75
    × 28
    ____

12.   350
    × 408
    _____

13. Write $10 \times 10 \times 10 \times 10$ in exponential notation. _____

14. Write 24^6 as factors. _____

15. What is the value of 20^4? _____

Divide:

16. $9\overline{)648}$ $24\overline{)1{,}344}$

18. $45\overline{)9{,}315}$

19. Circle the prime numbers in the following list:
    18   23   47   61   77

20. Write the prime factorization of 72. _____

21. Simplify $9 \times 4 \div 12 - 2$. _____

22. A small cargo plane can carry 2,800 pounds. If the pilot weighs 160 pounds, how many 30-pound boxes of freight can the plane carry?

    _____

CHAPTER 1

PRACTICE TEST 1B

1. Write 54,976 in words. _____

2. Write eight million, four hundred seventeen thousand, sixty-three in numerical form. _____

3. Consider the number 789,065. What digit is in the:

   a. hundreds place? *b.* thousands place? *c.* hundred thousands place?

   a. _____   *b.* _____   *c.* _____

4. Write 92,743 in expanded notation. _____

5. Round 689,543 to the indicated place.

   a. nearest hundred _____

   b. nearest thousand _____

   c. nearest ten _____

Add:

6. $\quad$ 91
    $\quad$ 68
    $+\ 17$

7. $\quad$ 24,120
    $\qquad\ $ 42
    $\quad\ $ 3,893
    $+\ 54,075$

Subtract:

8. $\quad$ 8,245
    $-\ 6,041$

9. $\quad$ 47,871
    $-\ 36,980$

Multiply:

10. $4 \times 5 \times 8 \times 10 =$ _____

11. $\quad$ 83
    $\times\ 19$

12. $\quad$ 912
    $\times\ 604$

13. Write 8 × 8 × 8 × 8 × 8 in exponential notation. _____

14. Write 16^7 as factors. _____

15. What is the value of 7^3? _____

Divide:

16. 6)492   63)1,512

18. 37)4,033

19. Circle the prime numbers in the following list:
    49   89   61   101   56

20. Write the prime factorization of 150. _____

21. Simplify 18 + 6 − 4 × 3. _____

22. Lorenzo bought a tire for $55. He also paid $3 to have it balanced and $4 tax. He gave the salesperson four $20 bills. How much change did he receive?

_____

CHAPTER 1 SOLUTIONS

Skills Preview

1. Sixty-four million, seven hundred ninety thousand, three hundred seventy-one
2. 612,418,533,854
3. *a.* 4 *b.* 7 *c.* 6
4. 400,000 + 20,000 + 3,000 + 700 + 90 + 2
5. *a.* 642,973,000
   b. 640,000,000
   c. 642,973,500
6. 159
7. 23,556
8. 1,331
9. 569
10. 270
11. 4,067
12. 220,844
13. 42
14. 86
15. 305
16. 17, 31, 53
17. 2 × 2 × 2 × 2 × 2 × 3
18. 16

Section 1.1 Comprehension Checkpoint

1. *a.* 1
   b. 5
   c. 7
2. 20,000 + 4,000 + 700 + 90 + 5
3. five hundred ninety-three million, six hundred seventy-four thousand, twenty-one
4. 472,389

1.1A Practice Set

Use the following place value chart to determine what digit is in each place in problems 1–5.

Billions			Millions			Thousands			Units		
HUNDRED BILLIONS	TEN BILLIONS	BILLIONS	HUNDRED MILLIONS	TEN MILLIONS	MILLIONS	HUNDRED THOUSANDS	TEN THOUSANDS	THOUSANDS	HUNDREDS	TENS	UNITS
		8	4	6	7	3	0	2	1	9	5

1. 0
2. 4
3. 1
4. 3
5. 8
6. 300 + 90 + 5
7. 4,000 + 200 + 60 + 3
8. 800,000 + 60,000 + 9,000 + 400 + 70 + 5
9. 20,000 + 6,000 + 800 + 90 + 1
10. 7,000,000 + 600,000 + 80,000 + 3,000 + 400 + 60 + 7
11. four hundred thirty-seven
12. nine hundred six
13. eight thousand, six hundred forty-two
14. ninety-seven thousand, four hundred five
15. two hundred sixty-eight thousand, seven hundred ninety-three
16. 542
17. 23,812
18. 9,433,256
19. 706,368
20. 16,618,000,489

Section 1.2   Comprehension Checkpoint

1. 43,000   2. 576,800   3. 380   4. 300,000   5. 40,000,000

1.2A   Practice Set

	Nearest Ten	Nearest Hundred	Nearest Thousand
1.	890	900	1,000
2.	510	500	1,000
3.	370	400	0
4.	2,480	2,500	2,000
5.	4,680	4,700	5,000
6.	8,950	9,000	9,000
7.	3,270	3,300	3,000
8.	56,740	56,700	57,000

Solutions

Nearest Ten	Nearest Hundred	Nearest Thousand
9. 84,920	84,900	85,000
10. 76,500	76,500	76,000

11. 24,600,000
12. 37,900,000
13. 85,000,000
14. 60,000,000
15. 53,000,000
16. $18,000
17. 900
18. 17,600,000
19. 230
20. 130,000

Section 1.3 Comprehension Checkpoint

1. 41 **2.** 839 **3.** 4,783 **4.** 6,290

1.3A Practice Set

1. $\begin{array}{r} 82 \\ +\ 15 \\ \hline 97 \end{array}$
2. $\begin{array}{r} 64 \\ +\ 32 \\ \hline 96 \end{array}$
3. $\begin{array}{r} 28 \\ +\ 47 \\ \hline 75 \end{array}$
4. $\begin{array}{r} 95 \\ +\ 74 \\ \hline 169 \end{array}$
5. $\begin{array}{r} 58 \\ +\ 49 \\ \hline 107 \end{array}$

6. $\begin{array}{r} 375 \\ +\ 488 \\ \hline 863 \end{array}$
7. $\begin{array}{r} 164 \\ +\ 329 \\ \hline 493 \end{array}$
8. $\begin{array}{r} 509 \\ +\ 793 \\ \hline 1{,}302 \end{array}$
9. $\begin{array}{r} 4{,}310 \\ +\ 8{,}695 \\ \hline 13{,}005 \end{array}$
10. $\begin{array}{r} 7{,}256 \\ 784 \\ +\ 4{,}939 \\ \hline 12{,}979 \end{array}$

11. $\begin{array}{r} 18{,}406 \\ 2{,}461 \\ +\ 24{,}597 \\ \hline 45{,}464 \end{array}$
12. $\begin{array}{r} 42{,}875 \\ 364{,}980 \\ +\ 68{,}417 \\ \hline 476{,}272 \end{array}$
13. $\begin{array}{r} 4{,}719 \\ 28{,}604 \\ 319 \\ +\ 1{,}408 \\ \hline 35{,}050 \end{array}$
14. $\begin{array}{r} 95{,}006 \\ 2{,}784 \\ 13{,}309 \\ +\ 27 \\ \hline 111{,}126 \end{array}$
15. $\begin{array}{r} 4{,}342 \\ 809 \\ 427{,}602 \\ +\ 39{,}855 \\ \hline 472{,}608 \end{array}$

16. $\begin{array}{r} 48 \\ 95 \\ +\ 37 \\ \hline 180 \end{array}$
17. $\begin{array}{r} 872 \\ 4{,}316 \\ 25 \\ +\ 970 \\ \hline 6{,}183 \end{array}$
18. $\begin{array}{r} 24{,}612 \\ 917 \\ +\ 1{,}423 \\ \hline 26{,}952 \end{array}$
19. $\begin{array}{r} 560 \\ 49 \\ 89{,}600 \\ +\ 356{,}183 \\ \hline 446{,}392 \end{array}$
20. $\begin{array}{r} 243{,}517 \\ 9{,}346 \\ 295 \\ +\ 24{,}617 \\ \hline 277{,}775 \end{array}$

21. $\begin{array}{r} 13 \\ 21 \\ 18 \\ +\ 24 \\ \hline 76 \end{array}$
22. $\begin{array}{r} 105 \\ 84 \\ 92 \\ 65 \\ +\ 137 \\ \hline 483 \end{array}$

Section 1.4   Comprehension Checkpoint

1. 23   **2.** 411   **3.** 3,728   **4.** 58,888

1.4A   Practice Set

1. 48
   − 31
   ―――
    17

2. 94
   − 63
   ―――
    31

3. 85
   − 69
   ―――
    16

4. 917
   − 738
   ―――
    179

5. 450
   − 87
   ―――
    363

6. 300
   − 184
   ―――
    116

7. 3,673
   − 2,197
   ―――
    1,476

8. 6,451
   −   964
   ―――
    5,487

9. 5,820
   − 4,909
   ―――
      911

10. 49,005
   − 28,764
   ―――
    20,241

11. 76,405
   − 58,793
   ―――
    17,612

12. 54,571
   −   983
   ―――
    53,588

13. 276,491
   − 95,876
   ―――
    180,615

14. 825,937
   − 168,043
   ―――
    657,894

15. 397,644
   −   955
   ―――
    396,689

16. 4,163
   −   825
   ―――
    3,338

17. 9,402
   − 4,987
   ―――
    4,415

18. 95,000
   − 14,607
   ―――
    80,393

19. 229,608
   − 67,415
   ―――
    162,193

20. 560,913
   − 448,716
   ―――
    112,197

21. 1,454
   − 1,250
   ―――
      204

22.   457
   − 192
   ―――
     265

Section 1.5   Comprehension Checkpoint

1. 744   **2.** 2,432   **3.** 90,134   **4.** 30,000

5. 2,267,394   **6.** 8^5 **7.** $9 \times 9 \times 9 \times 9 \times 9$ **8.** 216

1.5A   Practice Set

1.   47
   × 8
   ―――
    376

2. 685
   ×   3
   ―――
    2,055

3. 508
   ×   9
   ―――
    4,572

4.   83
   × 24
   ―――
    332
   1 66
   ―――
   1,992

5.   56
   × 19
   ―――
    504
    56
   ―――
   1,064

6.   93
   × 72
   ―――
    186
   6 51
   ―――
   6,696

7.   34
   × 52
   ―――
    68
   1 70
   ―――
   1,768

8.   67
   × 90
   ―――
   6,030

9.   18
   × 35
   ―――
    90
    54
   ―――
    630

10. 206
   × 48
   ―――
   1 648
   8 24
   ―――
   9,888

11. 864
   × 39
   ―――
   7 776
   25 92
   ―――
   33,696

12. 483
   × 52
   ―――
    966
   24 15
   ―――
   25,116

Solutions

13. $\begin{array}{r} 275 \\ \times\ 807 \\ \hline 1\ 925 \\ 220\ 00\ \ \\ \hline 221{,}925 \end{array}$
14. $\begin{array}{r} 812 \\ \times\ 346 \\ \hline 4\ 872 \\ 32\ 48\ \\ 243\ 6\ \ \\ \hline 280{,}952 \end{array}$
15. $\begin{array}{r} 560 \\ \times\ 425 \\ \hline 2\ 800 \\ 11\ 20\ \\ 224\ 0\ \ \\ \hline 238{,}000 \end{array}$
16. $\begin{array}{r} 2815 \\ \times\ \ 300 \\ \hline 844{,}500 \end{array}$

17. $\begin{array}{r} 4{,}693 \\ \times\ \ \ 581 \\ \hline 4\ 693 \\ 375\ 44\ \\ 2\ 346\ 5\ \ \\ \hline 2{,}726{,}633 \end{array}$
18. $\begin{array}{r} 3{,}520 \\ \times\ \ \ 908 \\ \hline 28\ 160 \\ 3\ 168\ 00\ \ \\ \hline 3{,}196{,}160 \end{array}$
19. $(5)(7)(3)(2) =$
    $5 \times 7 = 35$
    $35 \times 3 = 105$
    $105 \times 2 = 210$

20. $(8)(9)(4)(6) =$
    $8 \times 9 = 72$
    $72 \times 4 = 288$
    $288 \times 6 = 1{,}728$
21. $(24)(25)(10) =$
    $24 \times 25 = 600$
    $600 \times 10 = 6{,}000$
22. $(35)(74) =$ $\begin{array}{r} 35 \\ \times\ 74 \\ \hline 140 \\ 2\ 45\ \\ \hline 2{,}590 \end{array}$

23. $(60)(20) =$ $\begin{array}{r} 60 \\ \times\ 20 \\ \hline 1{,}200 \end{array}$
24. $412 \cdot 500 =$ $\begin{array}{r} 412 \\ \times\ 500 \\ \hline 206{,}000 \end{array}$
25. $8{,}000 \cdot 400 =$ $\begin{array}{r} 8{,}000 \\ \times\ \ \ 400 \\ \hline 3{,}200{,}000 \end{array}$

26. 4^5
27. 10^7
28. 6^9
29. $2 \times 2 \times 2 \times 2 \times 2 \times 2 \times 2 \times 2$
30. $9 \times 9 \times 9 \times 9 \times 9 \times 9$
31. 12×12
32. $2 \times 2 \times 2 \times 2 \times 2 \times 2 = 64$
33. $10 \times 10 \times 10 \times 10 \times 10 = 100{,}000$
34. $15 \times 15 \times 15 = 3{,}375$

35. $35 \times 2 \times 4 =$
    $35 \times 2 = 70$
    $70 \times 4 = 280$
36. $\begin{array}{r} 26 \\ \times\ 14 \\ \hline 104 \\ 26\ \ \\ \hline 364 \end{array}$

Section 1.6   Comprehension Checkpoint

1. 7
2. 15
3. 12
4. 24 R 3
5. 105
6. 0
7. cannot be worked

1.6A   Practice Set

1. $\begin{array}{r} 12 \\ 7{\overline{\smash{\big)}\,84}} \\ \underline{7\ \ } \\ 14 \\ \underline{14} \\ 0 \end{array}$
2. $\begin{array}{r} 30 \\ 9{\overline{\smash{\big)}\,270}} \\ \underline{27\ \ } \\ 00 \end{array}$
3. $\begin{array}{r} 22 \\ 6{\overline{\smash{\big)}\,132}} \\ \underline{12\ \ } \\ 12 \\ \underline{12} \\ 0 \end{array}$
4. $\begin{array}{r} 56 \\ 8{\overline{\smash{\big)}\,448}} \\ \underline{40\ \ } \\ 48 \\ \underline{48} \\ 0 \end{array}$

5. 64
   $4\overline{)256}$
   24
   16
   16
   0

6. 64
   $3\overline{)192}$
   18
   12
   12
   0

7. 14 R 7
   $18\overline{)259}$
   18
   79
   72
   7

8. 5 R 12
   $23\overline{)127}$
   115
   12

9. 12 R 5
   $38\overline{)461}$
   38
   81
   76
   5

10. 30 R 13
    $26\overline{)793}$
    78
    13

11. 12 R 3
    $56\overline{)675}$
    56
    115
    112
    3

12. 26 R 5
    $84\overline{)2{,}189}$
    $1\,68$
    509
    504
    5

13. 6 R 5
    $42\overline{)257}$
    252
    5

14. 35 R 7
    $75\overline{)2{,}632}$
    $2\,25$
    382
    375
    7

15. 105
    $48\overline{)5{,}040}$
    $4\,8$
    240
    240
    0

16. 206 R 2
    $25\overline{)5{,}152}$
    $5\,0$
    152
    150
    2

17. 21 R 2
    $63\overline{)1{,}325}$
    $1\,26$
    65
    63
    2

18. 509 R 3
    $34\overline{)17{,}309}$
    $17\,0$
    309
    306
    3

19. $64 \div 8 = 8$

20. $72 \div 9 = 8$

21. $54 \div 6 = 9$

22. $32 \div 4 = 8$

23. $42 \div 6 = 7$

24. $24 \div 3 = 8$

25. $50 \div 2 = 25$

26. 15
    $14\overline{)210}$ $15 per credit hour
    14
    70
    70
    0

27. 32
    $26\overline{)832}$ 32 points per game
    78
    52
    52
    0

Section 1.7 Comprehension Checkpoint

1. 288
2. $29
3. $412,500

1.7A Practice Set

1. $259
   95
   $+$137
   $\overline{$491}$

2. $3{,}450$
   $2{,}912$
   $3{,}120$
   $2{,}389$
   $+\,1{,}463$
   $\overline{13{,}334}$

3. $23{,}667{,}902$
   $17{,}558{,}072$
   $+\,14{,}229{,}191$
   $\overline{55{,}455{,}165}$

4. $731{,}600$
   $-\,356{,}860$
   $\overline{374{,}740}$

Solutions

5. $14,610
    − 13,895
    $ 715

6. 29,609
    − 28,082
    1,527

7. 38
    × 14
    152
    38
    532

8. $265
    × 54
    1 060
    13 25
    $14,310

9. 8,500
    × 416
    51 000
    85 00
    3 400 0
    3,536,000

10.
        12
    58)696
        58
        116
        116
          0

11.
        8
    35)280
       280
         0

12.
         49
    28)1,372
       1 12
         252
         252
           0

13. $25 \times 8 = 200 =$ total number of chairs
    $200 \times \$75 = \$15,000 =$ revenue generated

14. $1,170 − 1,068 = 102 =$ total weight lost
    $102 \div 6 = 17$ pounds lost per person

Section 1.8   Comprehension Checkpoint

1. 37, 53, 71 are prime
2. $2 \times 2 \times 2 \times 3 \times 3 \times 5$
3. $3 \times 3 \times 23$

1.8A Practice Set

1. Prime numbers: 41, 83
2. Prime numbers: 97, 101
3. Prime numbers: 71, 61
4. Prime numbers: 73, 89
5. Prime numbers: 103, 109

6.
        2
    2)4
    2)8
    2)16
    $16 = 2 \times 2 \times 2 \times 2$

7.
        7
    5)35
    $35 = 5 \times 7$

8.
        3
    2)6
    2)12
    2)24
    2)48
    $48 = 2 \times 2 \times 2 \times 2 \times 3$

9.
        7
    2)14
    2)28
    2)56
    $56 = 2 \times 2 \times 2 \times 7$

10.
        7
    3)21
    3)63
    $63 = 3 \times 3 \times 7$

11.
        7
    3)21
    2)42
    2)84
    $84 = 2 \times 2 \times 3 \times 7$

12.
        5
    3)15
    3)45
    2)90
    $90 = 2 \times 3 \times 3 \times 5$

13.
        5
    2)10
    2)20
    2)40
    2)80
    $80 = 2 \times 2 \times 2 \times 2 \times 5$

14.
$$\begin{array}{r}7\\3\overline{)21}\\2\overline{)42}\end{array}$$ $42 = 2 \times 3 \times 7$

15.
$$\begin{array}{r}5\\5\overline{)25}\\3\overline{)75}\\3\overline{)225}\end{array}$$ $225 = 3 \times 3 \times 5 \times 5$

16.
$$\begin{array}{r}7\\3\overline{)21}\\2\overline{)42}\\2\overline{)84}\\2\overline{)168}\\2\overline{)336}\end{array}$$ $336 = 2 \times 2 \times 2 \times 2 \times 3 \times 7$

17.
$$\begin{array}{r}5\\3\overline{)15}\\2\overline{)30}\\2\overline{)60}\\2\overline{)120}\end{array}$$ $120 = 2 \times 2 \times 2 \times 3 \times 5$

18.
$$\begin{array}{r}7\\5\overline{)35}\\3\overline{)105}\\2\overline{)210}\end{array}$$ $210 = 2 \times 3 \times 5 \times 7$

19.
$$\begin{array}{r}11\\3\overline{)33}\\3\overline{)99}\\3\overline{)297}\\2\overline{)594}\\2\overline{)1{,}188}\end{array}$$ $1{,}188 = 2 \times 2 \times 3 \times 3 \times 3 \times 11$

20.
$$\begin{array}{r}5\\5\overline{)25}\\5\overline{)125}\\2\overline{)250}\\2\overline{)500}\\2\overline{)1{,}000}\\2\overline{)2{,}000}\end{array}$$ $2{,}000 = 2 \times 2 \times 2 \times 2 \times 5 \times 5 \times 5$

Section 1.9   Comprehension Checkpoint

1. 48   2. 23   3. 6   4. 2

1.9A   Practice Set

1. $2 + 3 \times 5 =$
    multiply
   $2 + 15 =$
    add
    17

2. $4 \div 2 \times 6 =$
    divide
   2×6
    multiply
    12

3. $16 + 8 - 5 =$
    add
   $24 - 5$
    subtract
    19

4. $10 \div 5 \times 6 - 3 =$
    divide
   $2 \times 6 - 3$
    multiply
   $12 - 3$
    subtract
    9

Solutions

5. $8 - 4 \times 2 =$
        ⌐⌐ multiply
    $8 - 8$
    ⌐___⌐ subtract
       0

6. $(6)(8) \div 12 =$
    ⌐⌐ multiply
    $48 \div 12$
    ⌐___⌐ divide
       4

7. $(10)(5) - (4)(6) =$
    ⌐⌐   ⌐⌐ multiply
    $50 - 24$
      ⌐____⌐ subtract
        26

8. $15 \div 3 - 4 + 20 \div 10 =$
    ⌐⌐   ⌐___⌐ divide
    $5 - 4 + 2$
    ⌐___⌐ subtract
    $1 + 2$
    ⌐__⌐ add
     3

9. $(24 + 18 \div 9) \div 13 =$
          ⌐___⌐ divide
    $(24 + 2) \div 13$
      ⌐____⌐ add inside ( )
    $26 \div 13$
    ⌐___⌐ divide
     2

10. $28 \div 14 + (4 \times 3) =$
 divide ⌐___⌐   ⌐___⌐ multiply
    $2 + 12$
    ⌐_____⌐ add
     14

11. $(12 \times 3 + 6 \div 2) + 8 \times 3 =$
 multiply ⌐_⌐ ⌐_⌐ divide
    $(36 + 3) + 8 \times 3$
  add ⌐___⌐   ⌐___⌐ multiply
      $39 + 24$
      ⌐____⌐ add
       63

12. $9 \times (18 \div 2) + 6 =$
    ⌐___⌐ divide
    $9 \times 9 + 6$
  ⌐_____⌐ multiply
    $81 + 6$
    ⌐___⌐ add
     87

13. $24 \div 3 \times 5 + 25 =$
  ⌐___⌐ divide
  $8 \times 5 + 25$
    ⌐___⌐ multiply
    $40 + 25$
      ⌐___⌐ add
     65

14. $3 + 8 \times (9 - 3) - 6 =$
        ⌐____⌐ subtract inside ( )
  $3 + 8 \times 6 - 6$
      ⌐____⌐ multiply
  $3 + 48 - 6$
  ⌐____⌐ add
    $51 - 6$
    ⌐___⌐ subtract
     45

15. $(20)(10) - (25)(8) =$
   ⌐_⌐   ⌐_⌐ multiply
   $200 - 200$
     ⌐_____⌐ subtract
      0

16. $42 \div 6 - 39 \div 13 =$
   ⌐_⌐   ⌐___⌐ divide
   $7 - 3$
    ⌐___⌐ subtract
     4

17. $[(40)(8) \div 20] + 18 \times 5$
  └──┘ └──┘ *multiply*
  $[320 \div 20] + 90$
  └──── *divide*
  $16 + 90$
  └────── *add*
  106

18. $8 + (5)(7) - 12 =$
  └──┘ *multiply*
  $8 + 35 - 12$
  └──┘ *add*
  $43 - 12$
  └──── *subtract*
  31

19. $(25 + 12 \times 5)(2)(3) =$
  multiply └──┘ └──┘ *multiply*
  $(25 + 60)(6)$
  └──┘ *add*
  $(85)(6)$
  └──── *multiply*
  510

20. $[(28)(3) \div 14] + 168 \div 24 =$
  multiply └──┘   └──── *divide*
  $[84 \div 14] + 7$
  └──── *divide*
  $6 + 7$
  └── *add*
  13

21. $2 \times 89 + 45 =$
  └── *multiply*
  $178 + 45$
  └──── *add*
  223 Her purchases totaled $223

22. $(20)(6) + (27)(4) =$
  └──┘   └──┘ *multiply*
  $120 + 108$
  └───── *add*
  228 The two wells would have produced 228 barrels of oil for the week.

23. $100 - (15 + 28 + 42) =$
  └──────── *add*
  $100 - 85$
  └───── *subtract*
  15 Bill had $15 remaining when he returned home.

Practice Test   1A

1. seventy-four million, six hundred eighty thousand, one hundred seventy-three

2. 27,689

3. *a.* 8   *b.* 4   *c.* 6

4. $8,000 + 600 + 70 + 4$

5. *a.* 895,700   *b.* 896,000   *c.* 895,710

6. 134

7. 31,149

8. 2,314

9. 2,139

10. 1,008

11. 2,100

12. 142,800

13. 10^4

14. $24 \times 24 \times 24 \times 24 \times 24 \times 24$

15. 160,000

16. 72

17. 56

18. 207

19. 23, 47, 61

20. $2 \times 2 \times 2 \times 3 \times 3$

21. 1

22. 88

FRACTIONS AND MIXED NUMBERS

OUTLINE

2.1 Introduction to Fractions

2.2 Improper Fractions and Mixed Numbers

2.3 Equivalent Fractions

2.4 Adding Fractions and Mixed Numbers

2.5 Subtracting Fractions and Mixed Numbers

2.6 Multiplying Fractions and Mixed Numbers

2.7 Dividing Fractions and Mixed Numbers

2.8 Order of Operations

In Chapter 1, you learned about whole numbers. Very often problems that we encounter daily involve parts of a whole. The price of a shirt is reduced by $\frac{1}{4}$. Each part of a pizza is $\frac{1}{8}$ of the whole. These parts are called *fractions,* and there are different types of fractions. In this chapter you will learn about common fractions and how to add, subtract, multiply, and divide them.

SKILLS PREVIEW

1. In the fraction $\frac{5}{11}$, identify:
   - a. the numerator
   - b. the denominator

   a. _____   b. _____

2. Label the following values as either a proper fraction, an improper fraction, or a mixed number:

   a. $\frac{8}{5}$ b. $3\frac{7}{10}$ c. $\frac{4}{9}$

   a. _____   b. _____   c. _____

3. Use a common fraction to represent the shaded portion of the following drawing:

    _____

4. Find the missing term in each fraction:

   a. $\frac{9}{10} = \frac{?}{50}$ b. $\frac{24}{36} = \frac{6}{?}$

   a. _____   b. _____

5. Reduce $\frac{30}{42}$ to lowest terms. _____

6. What is the greatest common factor of 48 and 72? _____

In problems 7–17, perform the indicated operation. Simplify and reduce your answers to lowest terms.

7. $\frac{2}{3} + \frac{5}{6} + \frac{7}{9}$ _____

8. $3\frac{5}{8} + 2\frac{3}{10}$ _____

9. $\frac{7}{8} - \frac{3}{16}$ _____

10. $4\frac{5}{9} - 2\frac{2}{3}$ _____

11. $\frac{3}{4} \times \frac{5}{12}$ _____

12. $5\frac{3}{7} \times 10\frac{1}{2}$ _____

13. $\frac{4}{9} \div \frac{2}{3}$ _____

14. $3\frac{4}{7} \div 2\frac{1}{2}$ _____

15. $0 \div \frac{5}{6}$ _____

16. $\frac{9}{10} \div 0$ _____

17. Simplify $\left(\frac{1}{2} - \frac{3}{8}\right) \div \frac{1}{4}$. _____

2.1 INTRODUCTION TO FRACTIONS

A **fraction** is a part of a whole amount. A **common fraction** is a whole number divided by a natural number. These numbers are separated by a line called the **fraction bar** (or fraction line). The number above the line is the **numerator.** The number below the line is the **denominator.**

$$\begin{array}{l} a \rightarrow \text{numerator} \\ - \rightarrow \text{fraction bar} \\ b \rightarrow \text{denominator} \end{array}$$

A **proper fraction** is a fraction in which the numerator is less than the denominator. Examples of proper fractions are

$$\frac{3}{10} \quad \frac{4}{5} \quad \frac{8}{11} \quad \frac{1}{2}$$

An **improper fraction** is a fraction in which the numerator is equal to or greater than the denominator. Examples of improper fractions include

$$\frac{5}{5} \quad \frac{7}{4} \quad \frac{9}{2} \quad \frac{12}{11}$$

A **mixed number** is a combination of a whole number and a fraction. Examples of mixed numbers are

$$3\frac{7}{8} \quad 4\frac{2}{3} \quad 12\frac{9}{16} \quad 101\frac{14}{25}$$

The numerator and denominator of a fraction give specific information about the whole amount. The denominator is the number of equal parts the whole is divided into. The numerator is the number of parts being considered.

Figure 2.1 shows three squares. The square on the left is whole. The square in the middle has been divided into four equal parts. The square on the right has been divided into four equal parts, and three of them are shaded. To express the shaded portion of the right square as a common fraction, write the number of shaded parts as the numerator and the total number of parts as the denominator.

FIGURE 2.1

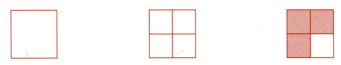

The fraction is

$$\begin{array}{l} \frac{3}{4} \rightarrow \text{numerator, total parts shaded} \\ \phantom{\frac{3}{4}} \rightarrow \text{denominator, total equal parts} \end{array}$$

EXAMPLE 1 Express the shaded part of each drawing as a fraction.

a. b.

SOLUTIONS

a. How many shaded parts? $\frac{7}{10}$ *numerator*
   How many total parts?   *denominator*
b. How many shaded parts? $\frac{11}{12}$ *numerator*
   How many total parts?   *denominator*

Each shaded area in Example 1 is less than the whole amount. The resulting fractions, $\frac{7}{10}$ and $\frac{11}{12}$, are proper fractions.

Figure 2.2 shows how to deal with more than one whole unit. An improper fraction, in which the numerator is equal to or greater than the denominator, must be used to express the shaded area in Figure 2.2. An improper fraction

FIGURE 2.2

indicates that you are dealing with more than one whole unit. To express the shaded area as an improper fraction, observe that each circle is divided into three parts. This will be the denominator of the fraction. A total of seven parts are shaded. This is the numerator. The improper fraction then is

$$\frac{numerator}{denominator} \quad \frac{7}{3} = \frac{\text{total number of shaded parts}}{\text{number of parts in each circle}}$$

The shaded area of Figure 2.2 could also be expressed as a mixed number. Recall that a mixed number is a whole number and a fraction working together to represent a single amount. Two whole circles are shaded, so the whole number will be 2. The third circle is divided into three parts, with one part shaded. The fraction that represents this is $\frac{1}{3}$. The mixed number that represents the shaded portion of Figure 2.2 then is

$$2\frac{1}{3} \;(2 \text{ whole units} + \frac{1}{3} \text{ of another unit})$$

EXAMPLE 2 Express the shaded portion of the drawing as:

a. an improper fraction.
b. a mixed number.

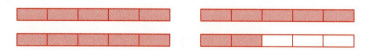

SOLUTIONS

a. Each rectangle is divided into 5 parts. A total of 17 parts are shaded. The improper fraction is $\frac{17}{5}$.

b. Three whole rectangles are shaded. The fourth rectangle is divided into five parts, and two are shaded. The mixed number is $3\frac{2}{5}$.

You should be able to look at a fraction and determine how many parts the whole amount has been divided into (denominator). You should also be able to tell how many of those parts are being considered at any time (numerator).

Introduction to Fractions

EXAMPLE 3   Consider the fraction $\frac{8}{15}$

a.   How many parts has the object been divided into?
b.   How many of those parts are being considered in this fraction?

SOLUTIONS

a.   15   The denominator tells how many parts the whole has been divided into.
b.   8   The numerator tells how many parts are being considered.

You may on occasion be required to determine how many parts are being considered altogether. Look at Example 4.

EXAMPLE 4   There are 11 male and 14 female students in a business class. What fractional part of the class is female?

SOLUTION

The number being considered is 14, the number of women. The denominator is the total class, 25 (11 + 14). The fractional part of the class that is women is $\frac{14}{25}$.

Try the Comprehension Checkpoint problems to make sure you understand fractions. If you have difficulty with these problems, review the material in this section before going on to the Practice Sets.

COMPREHENSION CHECKPOINT

1. In the fraction $\frac{4}{7}$

    a. How many parts total has the whole been divided into?
    b. How many of those parts are being considered?

    a. _____   *b.* _____

2. Express the shaded portion of the drawing as a proper fraction.

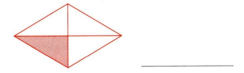

_____

3. Express the shaded portion of the drawing as:

   a. an improper fraction

   b. a mixed number

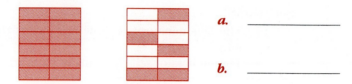

a. _____

b. _____

4. At a company picnic, 229 hamburgers and 148 hot dogs were eaten. What fractional part of the food eaten did the hot dogs represent?

_____

2.1A PRACTICE SET

In problems 1–5, fill in each blank.

	Parts Being Considered	Total Parts
1. $\frac{5}{9}$	_____	_____
2. $\frac{6}{13}$	_____	_____
3. $\frac{5}{8}$	_____	_____
4. $\frac{17}{25}$	_____	_____
5. $\frac{24}{7}$	_____	_____

In problems 6–15, identify each number as either a proper fraction, an improper fraction, or a mixed number.

6. $\frac{1}{2}$ _____   7. $\frac{9}{7}$ _____

8. $2\frac{1}{4}$ _____   9. $\frac{13}{8}$ _____

10. $9\frac{5}{16}$ _____   11. $\frac{12}{25}$ _____

12. $\frac{15}{4}$ _____   13. $7\frac{2}{3}$ _____

14. $\frac{7}{12}$ _____   15. $\frac{5}{5}$ _____

In problems 16–20, express the shaded portion as a proper fraction.

16. _____

CHAPTER 2 Fractions and Mixed Numbers

17.

18.

19.

20.

In problems 21–25, express the total shaded area as:

a. an improper fraction **b.** a mixed number

21. a. _____
    b. _____

22. a. _____
    b. _____

23. a. _____
    b. _____

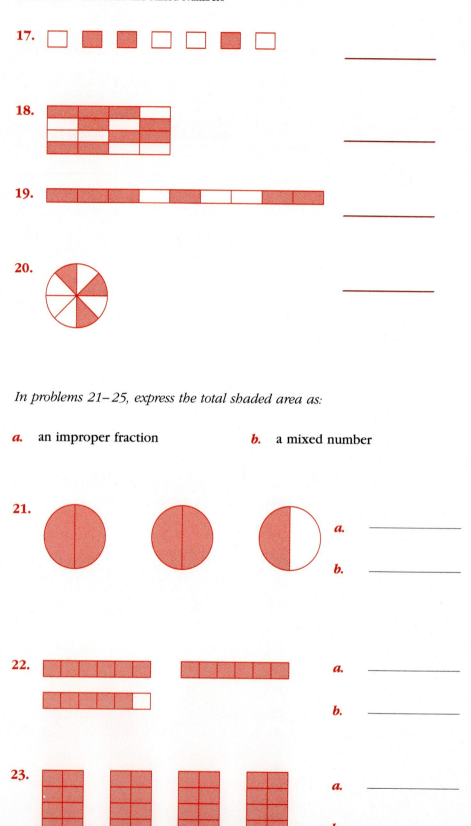

Introduction to Fractions

24.

a. _____

b. _____

25.

a. _____

b. _____

Applications

26. On the golf course, Anne hit her drive 242 yards. The hole is 411 yards long. What fractional part of the length of the hole did she hit her drive?

_____

27. During the lunch hour, a fast-food restaurant normally sells 52 burgers, 48 fries, and 67 large drinks. What fractional part of these sales do the fries represent?

_____

28. A box of computer paper contains 500 sheets of paper. A typed report took 273 pages. What fractional part of the paper did the report use?

_____

2.1B PRACTICE SET

In problems 1–5, fill in each blank.

	Parts Being Considered	Total Parts
1. $\frac{14}{15}$	_____	_____
2. $\frac{28}{19}$	_____	_____
3. $\frac{1}{8}$	_____	_____
4. $\frac{3}{20}$	_____	_____
5. $\frac{35}{6}$	_____	_____

In problems 6–15, identify each number as either a proper fraction, an improper fraction, or a mixed number.

6. $\frac{9}{16}$ _____

7. $3\frac{7}{8}$ _____

8. $8\frac{1}{3}$ _____

9. $\frac{16}{3}$ _____

10. $\frac{25}{18}$ _____

11. $\frac{8}{8}$ _____

12. $\frac{13}{21}$ _____

13. $\frac{11}{20}$ _____

14. $\frac{9}{1}$ _____

15. $4\frac{9}{10}$ _____

In problems 16–20, express the shaded portion as a proper fraction.

16. _____

CHAPTER 2 Fractions and Mixed Numbers

17. _____

18. _____

19. _____

20. _____

In problems 21–25, express the total shaded area as:

a. an improper fraction   **b.** a mixed number

21.   **a.** _____

     b. _____

22.   **a.** _____

     b. _____

23.   **a.** _____

     b. _____

24.   **a.** _____

     b. _____

Introduction to Fractions

25.

a. _____

b. _____

Applications

26. A computer screen displays 24 lines of typed text. Susan typed 13 lines. What fractional part of the screen did she use?

   _____

27. A courier makes 28 stops each day. By noon he has completed 17 stops. Express the number of stops completed by noon as a fractional part of the whole.

   _____

28. An airliner has 25 first-class seats and 290 coach seats. What fractional part of the seats on the plane are first-class seats?

   _____

2.2 IMPROPER FRACTIONS AND MIXED NUMBERS

Converting Improper Fractions to Mixed Numbers

In the previous section, the shaded portions of the circles in Figure 2.2 were represented with both an improper fraction and a mixed number.

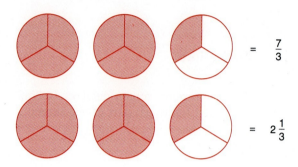

It stands to reason that if the same amount can be represented with two different numbers, then those two numbers must be equal. Thus,

$$\frac{7}{3} = 2\frac{1}{3}$$

Any amount that may be represented by an improper fraction may also be represented by a mixed number. Example 1 shows how to convert an improper fraction to a mixed number with the same value.

EXAMPLE 1 Convert $\frac{9}{5}$ to a mixed number.

SOLUTION

Recall that a fraction indicates division. Converting an improper fraction to a mixed number requires that you perform the indicated division.

$$\begin{array}{r} 1 \\ 5\overline{)9} \\ \underline{5} \\ 4 \end{array}$$ *Instead of writing the remainder as you did in Chapter 1, write it as the numerator of a fraction with the divisor as the denominator.*

Thus, $9 \div 5 = 1\frac{4}{5}$

$$\frac{9}{5} = 1\frac{4}{5}$$

Converting a Mixed Number to an Improper Fraction

Just as you can convert an improper fraction to a mixed number, you can also convert a mixed number to an improper fraction.

EXAMPLE 2 Convert $3\frac{4}{7}$ to an improper fraction.

SOLUTION

$7 \times 3 = 21$ *Multiply the whole number by the denominator of the fraction.*

$21 + 4 = 25$ *Add the numerator of the fraction to the product from the first step.*

Improper Fractions and Mixed Numbers

$\dfrac{25}{7}$ *The numerator of the improper fraction is the sum from step 2. The denominator is the denominator of the original fraction.*

$3\dfrac{4}{7} = \dfrac{25}{7}$

Changing Whole Numbers to Improper Fractions

Any whole number can be written as an improper fraction. Write the whole number as the numerator of a fraction. The denominator is always 1. Any number divided by 1 is the original number.

$$3 = \dfrac{3}{1} \qquad 3 \div 1 = 3$$

EXAMPLE 3   Write each whole number as an improper fraction.

a. 12   *b.* 8   *c.* 115

SOLUTIONS

a. $\dfrac{12}{1}$ *b.* $\dfrac{8}{1}$ *c.* $\dfrac{115}{1}$

The number 1 is something of a special case. It can be converted to an improper fraction by writing the same number as both numerator and denominator.

$$1 = \dfrac{5}{5} \qquad 1 = \dfrac{7}{7} \qquad 1 = \dfrac{10}{10}$$

This will be important for you to remember when subtracting fractions.

EXAMPLE 4   Write 1 as an improper fraction with a denominator of 12.

SOLUTION

$1 = \dfrac{12}{12}$ *The numerator must be the same as the denominator if the fraction = 1.*

EXAMPLE 5   Write 1 as an improper fraction with a numerator of 8.

SOLUTION

$1 = \dfrac{8}{8}$ *The denominator must be the same as the numerator if the fraction = 1.*

Examples 4 and 5 show that any number divided by itself is 1.

Try the Comprehension Checkpoint problems. If you have difficulty with these problems, review the material in this section before going on to the Practice Sets.

COMPREHENSION CHECKPOINT

1. Convert $\frac{23}{7}$ to a mixed number. _____

2. Convert $10\frac{1}{5}$ to an improper fraction. _____

3. Write 16 as an improper fraction. _____

4. Write the number 1 as an improper fraction with a denominator of 15. _____

2.2A PRACTICE SET

In problems 1–10, convert each improper fraction to a mixed number.

1. $\dfrac{12}{5} =$ _____

2. $\dfrac{14}{3} =$ _____

3. $\dfrac{37}{7} =$ _____

4. $\dfrac{13}{2} =$ _____

5. $\dfrac{29}{6} =$ _____

6. $\dfrac{19}{4} =$ _____

7. $\dfrac{43}{8} =$ _____

8. $\dfrac{55}{12} =$ _____

9. $\dfrac{64}{15} =$ _____

10. $\dfrac{88}{9} =$ _____

In problems 11–20, convert each mixed number to an improper fraction.

11. $2\dfrac{4}{5} =$ _____

12. $6\dfrac{7}{8} =$ _____

13. $3\dfrac{2}{3} =$ _____

14. $8\dfrac{9}{10} =$ _____

15. $5\dfrac{7}{12} =$ _____

16. $12\dfrac{6}{7} =$ _____

17. $15\dfrac{1}{8} =$ _____

18. $9\dfrac{11}{15} =$ _____

19. $14\dfrac{1}{7} =$ _____

20. $10\dfrac{3}{4} =$ _____

In problems 21–25, convert the given whole number to an improper fraction.

21. $19 =$ _____

22. $24 =$ _____

23. $42 =$ _____

24. $35 =$ _____

25. 16 = _____

In problems 26 and 27, fill in the missing number.

26. $1 = \dfrac{?}{7}$ _____ **27.** $1 = \dfrac{10}{?} =$ _____

2.2B PRACTICE SET

In problems 1–10, convert each improper fraction to a mixed number.

1. $\dfrac{18}{7} =$ _____

2. $\dfrac{24}{11} =$ _____

3. $\dfrac{15}{4} =$ _____

4. $\dfrac{35}{8} =$ _____

5. $\dfrac{91}{20} =$ _____

6. $\dfrac{65}{12} =$ _____

7. $\dfrac{72}{25} =$ _____

8. $\dfrac{84}{17} =$ _____

9. $\dfrac{101}{3} =$ _____

10. $\dfrac{109}{15} =$ _____

In problems 11–20, convert each mixed number to an improper fraction.

11. $1\dfrac{3}{4} =$ _____

12. $6\dfrac{1}{5} =$ _____

13. $12\dfrac{5}{6} =$ _____

14. $9\dfrac{7}{10} =$ _____

15. $5\dfrac{8}{9} =$ _____

16. $15\dfrac{4}{7} =$ _____

17. $24\dfrac{2}{3} =$ _____

18. $8\dfrac{3}{20} =$ _____

19. $18\dfrac{7}{10} =$ _____

20. $16\dfrac{2}{5} =$ _____

In problems 21–25, convert the given whole number to an improper fraction.

21. $12 =$ _____

22. $6 =$ _____

23. $28 =$ _____

24. $47 =$ _____

25. 84 = _____

In problems 26 and 27, fill in the missing number.

26. $1 = \dfrac{?}{9}$ _____   **27.** $1 = \dfrac{20}{?} =$ _____

2.3 EQUIVALENT FRACTIONS

Equivalent Fractions

Equivalent fractions are fractions that have the same value. Figure 2.3 illustrates equivalent fractions. Rectangle *a* is divided into 2 equal parts with 1

FIGURE 2.3

a. | $\frac{1}{2}$ | $\frac{1}{2}$ |

b. | $\frac{1}{4}$ | $\frac{1}{4}$ | $\frac{1}{4}$ | $\frac{1}{4}$ |

part shaded. The fraction that represents the shaded portion is $\frac{1}{2}$. Rectangle *b* is divided into 4 equal parts with 2 parts shaded. The fraction that represents the shaded portion is $\frac{2}{4}$. Since the same amount of each rectangle is shaded, the two shaded portions must be equal. Thus,

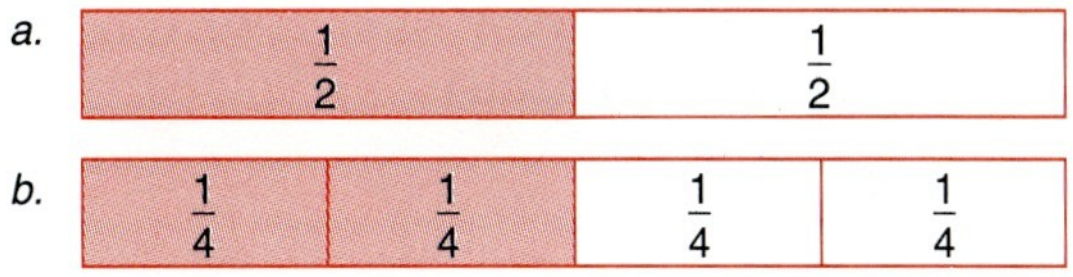

$$\frac{1}{2} = \frac{2}{4}$$

and they are equivalent fractions.

TO DETERMINE WHETHER FRACTIONS ARE EQUIVALENT
1. Cross-multiply (multiply the numerator of one fraction by the denominator of the other fraction).
2. If the two products are the same, the fractions are equivalent.

EXAMPLE 1 Determine whether $\frac{2}{3}$ and $\frac{6}{9}$ are equivalent fractions.

SOLUTION

$\frac{2}{3} \stackrel{?}{=} \frac{6}{9}$ *Multiply the numerator of one fraction by the denominator of the other fraction.*

$(2)(9) = 18$ *The two products are the same, so the fractions are equivalent.*
$(3)(6) = 18$

Changing Fractions to Higher Terms

Multiplying the terms of a fraction by the same nonzero number gives a new fraction in higher terms. *Higher terms* means that the numerator and denominator of the new fraction are greater than the numerator and denominator of the original fraction. Since $1 = \frac{2}{2}, \frac{3}{3}, \frac{4}{4}$, and so on, multiplying the top and bottom terms of a fraction by the same number is the same as multiplying the fraction by 1.

$$\frac{2}{5} \times \frac{3}{3} = \frac{6}{15} \qquad \frac{3}{3} = 1$$

$$\frac{2}{5} \text{ and } \frac{6}{15} \text{ are equivalent fractions}$$

To raise a fraction to higher terms, you must know either the new denominator or the new numerator.

> **TO RAISE FRACTIONS TO HIGHER TERMS**
> 1. Divide the new denominator by the original denominator (or the new numerator by the original numerator).
> 2. Multiply the quotient obtained in step 1 by the original numerator to get the new numerator (or the original denominator to get the new denominator).

EXAMPLE 2 Find the new numerator: $\dfrac{5}{7} = \dfrac{?}{14}$

SOLUTION

$14 \div 7 = 2$ *Divide the new denominator by the original denominator.*

$(5)(2) = 10$ *Multiply the quotient, 2, from step 1 by the original numerator to get the new numerator.*

The new numerator is 10.

$\dfrac{5}{7} = \dfrac{10}{14}$ *These are equivalent fractions.*

EXAMPLE 3 Find the new denominator: $\dfrac{3}{8} = \dfrac{9}{?}$

SOLUTION

$9 \div 3 = 3$ *Divide the new numerator by the original numerator.*

$(8)(3) = 24$ *Multiply the original denominator by the quotient from step 1 to get the new denominator.*

The new denominator is 24.

$\dfrac{3}{8} = \dfrac{9}{24}$ *These are equivalent fractions.*

Changing Fractions to Lower Terms

Changing fractions to lower terms is accomplished by *dividing* the numerator and denominator by the same nonzero number. To be able to change a fraction to lower terms, you must know either the new denominator or the new numerator.

$$\dfrac{9}{12} = \dfrac{?}{4}$$

> **TO CHANGE A FRACTION TO LOWER TERMS**
> 1. Divide the original denominator by the new denominator (or the original numerator by the new numerator).
> 2. Divide the original numerator by the quotient from step 1 to get the new numerator (or the original denominator by the quotient from step 1 to get the new denominator).

Equivalent Fractions

EXAMPLE 4 Find the new numerator: $\frac{10}{12} = \frac{?}{6}$

SOLUTION

$12 \div 6 = 2$ *Divide the original denominator by the new denominator.*

$10 \div 2 = 5$ *Divide the original numerator by the quotient from step 1.*

The new numerator is 5.

$\frac{10}{12} = \frac{5}{6}$ *These are equivalent fractions.*

EXAMPLE 5 Find the new denominator: $\frac{36}{48} = \frac{12}{?}$

SOLUTION

$36 \div 12 = 3$ *Divide the original numerator by the new numerator.*

$48 \div 3 = 16$ *Divide the original denominator by the quotient from step 1.*

The new denominator is 16.

$\frac{36}{48} = \frac{12}{16}$ *These are equivalent fractions.*

Reducing Fractions to Lowest Terms

Reducing a fraction to lowest terms means finding an equivalent fraction that has a numerator and denominator without a common factor between them. This is the same process as changing a fraction to lower terms except that you do not need to know the new numerator or the new denominator to begin the process.

Sometimes reducing a fraction to lowest terms can be done by inspection.

EXAMPLE 6 Reduce $\frac{8}{10}$ to lowest terms.

SOLUTION

$\frac{8}{10}$ *Since both 8 and 10 are even, they can both be divided by 2.*

$\frac{8 \div 2}{10 \div 2} = \frac{4}{5}$ *Because 4 and 5 have no common factor, the original fraction has been reduced to lowest terms.*

Reducing by inspection may require dividing more than once.

EXAMPLE 7 Reduce $\frac{36}{84}$ to lowest terms.

SOLUTION

$\frac{36 \div 2}{84 \div 2} = \frac{18}{42}$ *Both terms are even, so they can be divided by 2.*

$\frac{18 \div 2}{42 \div 2} = \frac{9}{21}$ *Both 18 and 42 can be divided by 2, so lowest terms have not been determined yet.*

$\dfrac{9 \div 3}{21 \div 3} = \dfrac{3}{7}$ *Both 9 and 21 can be divided by 3. Since 3 and 7 do not have a common factor, you now have lowest terms.*

$\dfrac{36}{84}$ reduced to lowest terms is $\dfrac{3}{7}$.

Fractions like the one in Example 7 can be reduced to lowest terms in a single step by dividing by the **greatest common factor (GCF)**. The GCF is the largest number that will divide both the numerator and the denominator evenly.

> **TO FIND THE GREATEST COMMON FACTOR**
> 1. Write the prime factorization of each term.
> 2. The product of the common prime factors is the GCF.

EXAMPLE 8 Determine the greatest common factor for 56 and 98.

SOLUTION

prime factors of 56 = 2 × 2 × 2 × 7
prime factors of 98 = 2 × 7 × 7

The common factors are 2 and 7 (one of each).
(2)(7) = 14 = GCF for 56 and 98

EXAMPLE 9 Reduce $\dfrac{72}{108}$ to lowest terms using the GCF.

SOLUTION

72 = 2 × 2 × 2 × 3 × 3
108 = 2 × 2 × 3 × 3 × 3

The common factors are 2 × 2 × 3 × 3.
The GCF = 2 × 2 × 3 × 3 = 36.
$\dfrac{72 \div 36}{108 \div 36} = \dfrac{2}{3}$

A final method for reducing fractions to lowest terms is canceling prime factors.

> **TO REDUCE TO LOWEST TERMS BY CANCELING PRIME FACTORS**
> 1. Write the prime factorization of each term.
> 2. Cancel like factors.
> 3. Multiply the remaining factors to get the reduced fraction.

ALGEBRA CONNECTION

Learning to work with fractions is essential not only to basic mathematics but also to algebra. The first three sections of this chapter have introduced you to fractions: what they are, what they represent, how to change to higher or lower terms. The next four sections will be concerned with applying the four operations (adding, subtracting, multiplying, and dividing) to fractions. Many students attempt to study algebra without a thorough working knowledge of fractions. If you encounter an equation like

$$\frac{x}{3} = 5$$

you must understand fractions before you can begin to solve it. This particular equation requires multiplication of fractions to find the solution. Some equations may contain more than one fraction:

$$x - \frac{3}{4} = \frac{2}{3}$$

This equation requires addition of fractions to solve it. You must master the material in this chapter before going on to algebra. There will be enough new material for you to learn when you begin algebra that you should not have to review fractions at that time.

EXAMPLE 10 Use the method of canceling prime factors to reduce $\frac{18}{45}$ to lowest terms.

SOLUTION

$\dfrac{18}{45} = \dfrac{2 \times 3 \times 3}{3 \times 3 \times 5}$ *Write the prime factorization of each term.*

$\dfrac{2 \times \cancel{3}^{1} \times \cancel{3}^{1}}{\cancel{3}_{1} \times \cancel{3}_{1} \times 5}$ *Cancel like factors.*

$\dfrac{2 \times 1 \times 1}{1 \times 1 \times 5} = \dfrac{2}{5}$ *Multiply the remaining factors to get the reduced fraction.*

Remember that $\dfrac{3}{3} = 1$. That is why, when you cancel a top and a bottom factor, they are replaced by 1.

Work the Comprehension Checkpoint problems to make sure you have mastered the material in this section. If you have difficulty with these problems, review the material in this section before going on to the Practice Sets.

COMPREHENSION CHECKPOINT

1. Are $\frac{4}{5}$ and $\frac{16}{20}$ equivalent fractions? _____

2. Determine the missing term: $\frac{12}{25} = \frac{?}{75}$ _____

3. Determine the missing term: $\frac{18}{30} = \frac{3}{?}$ _____

4. Reduce $\frac{24}{52}$ to lowest terms. _____

5. What is the GCF for 96 and 144? _____

2.3A PRACTICE SET

In problems 1–8, raise each fraction to the higher terms indicated.

1. $\dfrac{3}{4} = \dfrac{?}{16}$ _____

2. $\dfrac{3}{8} = \dfrac{?}{40}$ _____

3. $\dfrac{9}{11} = \dfrac{?}{66}$ _____

4. $\dfrac{13}{20} = \dfrac{39}{?}$ _____

5. $\dfrac{5}{6} = \dfrac{25}{?}$ _____

6. $\dfrac{4}{9} = \dfrac{?}{36}$ _____

7. $\dfrac{14}{25} = \dfrac{42}{?}$ _____

8. $\dfrac{5}{12} = \dfrac{45}{?}$ _____

In problems 9–16, write each fraction in the lower terms indicated.

9. $\dfrac{40}{16} = \dfrac{20}{?}$ _____

10. $\dfrac{39}{15} = \dfrac{13}{?}$ _____

11. $\dfrac{35}{50} = \dfrac{?}{10}$ _____

12. $\dfrac{16}{48} = \dfrac{?}{3}$ _____

13. $\dfrac{28}{42} = \dfrac{?}{6}$ _____

14. $\dfrac{121}{165} = \dfrac{11}{?}$ _____

15. $\dfrac{57}{95} = \dfrac{?}{5}$ _____

16. $\dfrac{90}{150} = \dfrac{?}{50}$ _____

In problems 17–22, determine the GCF for the given numbers.

17. 8 and 36 _____

18. 24 and 84 _____

19. 45 and 72 _____

20. 80 and 120 _____

21. 76 and 114 _____

22. 135 and 315 _____

In problems 23–32, reduce each fraction to lowest terms.

23. $\dfrac{4}{12} =$ _____

24. $\dfrac{5}{20} =$ _____

25. $\dfrac{15}{75} =$ _____

26. $\dfrac{12}{96} =$ _____

27. $\dfrac{16}{40} =$ _____

28. $\dfrac{72}{90} =$ _____

29. $\dfrac{75}{105} =$ _____

30. $\dfrac{64}{96} =$ _____

31. $\dfrac{36}{63} =$ _____

32. $\dfrac{121}{132} =$ _____

2.3B PRACTICE SET

In problems 1–8, raise each fraction to the higher terms indicated.

1. $\dfrac{2}{5} = \dfrac{?}{15}$ _____

2. $\dfrac{7}{12} = \dfrac{21}{?}$ _____

3. $\dfrac{3}{10} = \dfrac{12}{?}$ _____

4. $\dfrac{11}{15} = \dfrac{?}{75}$ _____

5. $\dfrac{9}{16} = \dfrac{?}{64}$ _____

6. $\dfrac{5}{14} = \dfrac{35}{?}$ _____

7. $\dfrac{17}{20} = \dfrac{85}{?}$ _____

8. $\dfrac{12}{25} = \dfrac{?}{125}$ _____

In problems 9–16, write each fraction in the lower terms indicated.

9. $\dfrac{8}{18} = \dfrac{4}{?}$ _____

10. $\dfrac{16}{24} = \dfrac{?}{6}$ _____

11. $\dfrac{36}{48} = \dfrac{9}{?}$ _____

12. $\dfrac{24}{56} = \dfrac{?}{7}$ _____

13. $\dfrac{45}{12} = \dfrac{15}{?}$ _____

14. $\dfrac{50}{26} = \dfrac{25}{?}$ _____

15. $\dfrac{54}{36} = \dfrac{?}{4}$ _____

16. $\dfrac{96}{80} = \dfrac{?}{5}$ _____

In problems 17–22, determine the GCF for the given numbers.

17. 18 and 72 _____

18. 20 and 35 _____

19. 48 and 120 _____

20. 75 and 200 _____

21. 96 and 128 _____

22. 90 and 126 _____

In problems 23–32, reduce each fraction to lowest terms.

23. $\dfrac{8}{16} =$ _____

24. $\dfrac{16}{24} =$ _____

25. $\dfrac{28}{42} =$ _____

26. $\dfrac{32}{48} =$ _____

27. $\dfrac{48}{60} =$ _____

28. $\dfrac{33}{55} =$ _____

29. $\dfrac{66}{88} =$ _____

30. $\dfrac{132}{144} =$ _____

31. $\dfrac{95}{100} =$ _____

32. $\dfrac{125}{150} =$ _____

2.4 ADDING FRACTIONS AND MIXED NUMBERS

Adding Fractions with Common Denominators

Before you add fractions, they must have **common denominators.** Having common denominators means that all of the fractions to be added have the same denominator. The following fractions have common denominators:

$$\frac{1}{5} \quad \frac{2}{5} \quad \frac{3}{5} \quad \frac{4}{5}$$

TO ADD FRACTIONS WITH COMMON DENOMINATORS
1. Add the numerators and place the sum over the common denominator.
2. Reduce the answer to lowest terms.

EXAMPLE 1 Add: $\frac{2}{9} + \frac{5}{9}$

SOLUTION

$\frac{2}{9} + \frac{5}{9} = \frac{2+5}{9} = \frac{7}{9}$ *This fraction cannot be reduced.*

EXAMPLE 2 Add: $\frac{4}{15} + \frac{2}{15}$

SOLUTION

$\frac{4}{15} + \frac{2}{15} = \frac{4+2}{15} = \frac{6}{15}$ *The numerator and the denominator can each be divided by 3.*

$\frac{6 \div 3}{15 \div 3} = \frac{2}{5}$ *This is the answer in lowest terms.*

Least Common Multiples

Fractions with unlike denominators such as

$$\frac{2}{3} \quad \frac{3}{4} \quad \frac{5}{6}$$

must be converted to equivalent fractions with common denominators before they may be added. It is most convenient to use the **least common multiple (LCM).** The LCM is the smallest natural number that is divisible by each of a group of numbers (remember that the set of natural numbers does not include 0). You should use the LCM of a group of numbers as the common denominator. This will eliminate some reducing when you complete the addition process. The denominators of the three fractions displayed previously are 3, 4, and 6. Each of these numbers will divide 24, 36, 48, 60, and many other numbers. All of these numbers (24, 36, 48, 60) are multiples of 3, multiples of 4, and multiples of 6. The smallest number that 3, 4, and 6 will each divide, however, is 12. Because 12 is the smallest of the common multiples, it is the LCM.

You can use prime factorization to find the LCM of any group of numbers.

> **TO FIND THE LCM**
> 1. Write the prime factorization of each number.
> 2. List each prime factor the *most* times it appears as a factor of any number.
> 3. The product of this list of prime factors is the LCM.

EXAMPLE 3 What is the LCM of 8 and 20?

SOLUTION

$\left.\begin{array}{l} 8 = 2 \cdot 2 \cdot 2 \\ 20 = 2 \cdot 2 \cdot 5 \end{array}\right\}$ *Prime factorization of each number.*

There are 2 different factors: 2 and 5. The 2 appears as a factor of 8 three times and as a factor of 20 two times. It will be listed three times. The 5 is a factor of 20 one time and is not a factor of 8. It will be listed one time.

$2 \cdot 2 \cdot 2 \cdot 5 = 40$

40 is the LCM for 8 and 20. There is no number smaller than 40 that both 8 and 20 will divide evenly.

EXAMPLE 4 What is the LCM of 24, 14, and 84?

SOLUTION

$\left.\begin{array}{l} 24 = 2 \cdot 2 \cdot 2 \cdot 3 \\ 14 = 2 \cdot 7 \\ 84 = 2 \cdot 2 \cdot 3 \cdot 7 \end{array}\right\}$ *Prime factorization of each number.*

$2 \cdot 2 \cdot 2 \cdot 3 \cdot 7 = 168 = $ LCM

Sometimes the LCM will be a large number.

EXAMPLE 5 What is the LCM of 14, 30, and 36?

SOLUTION

$\left.\begin{array}{l} 14 = 2 \cdot 7 \\ 25 = 5 \cdot 5 \\ 30 = 2 \cdot 3 \cdot 5 \end{array}\right\}$ *Prime factorization of each number.*

$2 \cdot 3 \cdot 5 \cdot 5 \cdot 7 = 1{,}260 = $ LCM

Adding Fractions with Different Denominators

Fractions that do not have common denominators must be converted to equivalent fractions with common denominators. It is possible to use any common multiple of the given denominators as the common denominator, but you will find it most convenient to use the LCM of the denominators, or the **least common denominator (LCD)**.

Adding Fractions and Mixed Numbers 123

TO ADD FRACTIONS WITH DIFFERENT DENOMINATORS
1. Find the LCD (this is the LCM) of the given denominators.
2. Convert each original fraction to an equivalent fraction with the LCD as the denominator.
3. Add the fractions and reduce the sum to lowest terms.

EXAMPLE 6 Add: $\dfrac{2}{3} + \dfrac{1}{6}$

SOLUTION

$3 = 1 \cdot 3 \qquad 6 = 2 \cdot 3$

LCM = $2 \cdot 3 = 6$ = LCD. (Sometimes one of the given denominators will be the LCM. You may have noticed this in the example. The method described earlier for finding the LCM may be used to find any LCD.)

$\dfrac{2}{3} = \dfrac{4}{6} \qquad \dfrac{4}{6} + \dfrac{1}{6} = \dfrac{5}{6}$

This answer will not reduce.

EXAMPLE 7 Add: $\dfrac{1}{4} + \dfrac{2}{5}$

SOLUTION

$4 = 2 \cdot 2 \qquad 5 = 1 \cdot 5$

LCM = $2 \cdot 2 \cdot 5 = 20$ = LCD

$\dfrac{1}{4} = \dfrac{5}{20} \qquad \dfrac{2}{5} = \dfrac{8}{20}$

$\dfrac{5}{20} + \dfrac{8}{20} = \dfrac{13}{20}$

EXAMPLE 8 Add: $\dfrac{7}{15} + \dfrac{5}{12} + \dfrac{3}{8}$

SOLUTION

$15 = 3 \cdot 5$

$12 = 2 \cdot 2 \cdot 3$

$8 = 2 \cdot 2 \cdot 2$

LCM = $2 \cdot 2 \cdot 2 \cdot 3 \cdot 5 = 120$ = LCD

$\dfrac{7}{15} = \dfrac{56}{120} \qquad \dfrac{5}{12} = \dfrac{50}{120} \qquad \dfrac{3}{8} = \dfrac{45}{120}$

$\dfrac{56}{120} + \dfrac{50}{120} + \dfrac{45}{120} = \dfrac{151}{120}$

$\dfrac{151}{120} = 1\dfrac{31}{120}$

CHAPTER 2 Fractions and Mixed Numbers

When all of the denominators are prime numbers, the LCM may be found by multiplying the denominators. This is because each prime factorization would be 1 times the denominator. Example 9 demonstrates this.

EXAMPLE 9   Add: $\dfrac{2}{7} + \dfrac{3}{5}$

SOLUTION

$7 = 1 \cdot 7 \qquad 5 = 1 \cdot 5$

$\text{LCM} = 7 \cdot 5 = 35 = \text{LCD}$

$\dfrac{2}{7} = \dfrac{10}{35} \qquad \dfrac{3}{5} = \dfrac{21}{35}$

$\dfrac{10}{35} + \dfrac{21}{35} = \dfrac{31}{35}$

Adding Mixed Numbers

Adding mixed numbers requires two steps: first, add the whole numbers; second, add the fractions.

> **TO ADD MIXED NUMBERS**
> 1. Add the whole numbers.
> 2. If the fractions do not have common denominators, convert them to equivalent fractions with a common denominator.
> 3. Add the fractions and reduce to lowest terms.
> 4. Combine the sum of the fractions and the sum of the whole numbers into a new mixed number.

EXAMPLE 10   Add: $2\dfrac{1}{5} + 3\dfrac{3}{5}$

SOLUTION

$2 + 3 = 5$ *Add the whole numbers.*

$\dfrac{1}{5} + \dfrac{3}{5} = \dfrac{4}{5}$ *Add the fractions.*

$5\dfrac{4}{5}$ *Combine the sum of the whole numbers with the sum of the fractions into a new mixed number that is the sum.*

EXAMPLE 11   Add: $4\dfrac{5}{8} + 6\dfrac{3}{10}$

SOLUTION

$4 + 6 = 10$ *Add the whole numbers.*

$8 = 2 \cdot 2 \cdot 2 \qquad 10 = 2 \cdot 5$

$\text{LCM} = 2 \cdot 2 \cdot 2 \cdot 5 = 40 = \text{LCD}$

$\dfrac{5}{8} = \dfrac{25}{40} \qquad \dfrac{3}{10} = \dfrac{12}{40}$

$\dfrac{25}{40} + \dfrac{12}{40} = \dfrac{37}{40} =$ *Sum of the fractions.*

Adding Fractions and Mixed Numbers

$10\dfrac{37}{40}$ *Combine the sum of the whole numbers with the sum of the fractions to create a new mixed number. This is the final sum.*

Sometimes the sum of the fractions, when simplified, will be a mixed number.

EXAMPLE 12 Add: $3\dfrac{6}{7} + 12\dfrac{2}{9} + 8\dfrac{5}{21}$

SOLUTION

$3 + 12 + 8 = 23$ *Add the whole numbers.*

$7 = 1 \cdot 7 \quad 9 = 3 \cdot 3$
$21 = 3 \cdot 7$
$\text{LCM} = 7 \cdot 3 \cdot 3 = 63 = \text{LCD}$

$\dfrac{6}{7} = \dfrac{54}{63} \quad \dfrac{2}{9} = \dfrac{14}{63} \quad \dfrac{5}{21} = \dfrac{15}{63}$

$\dfrac{54}{63} + \dfrac{14}{63} + \dfrac{15}{63} = \dfrac{83}{63} = 1\dfrac{20}{63} =$ *Sum of the fractions simplified.*

$23 + 1\dfrac{20}{63} = 24\dfrac{20}{63}$

Test your addition skills with the Comprehension Checkpoint problems. If you have difficulty with these problems, review the material in this section before going on to the Practice Sets.

COMPREHENSION CHECKPOINT

Find the LCM for each group of numbers.

1. 8, 12, 16 _____

2. 3, 9, 24 _____

Find the sum in each problem. Reduce all answers to lowest terms.

3. $\dfrac{2}{7} + \dfrac{4}{7} =$ _____

4. $\dfrac{5}{6} + \dfrac{9}{16} + \dfrac{3}{4} =$ _____

5. $5\dfrac{1}{4} + 3\dfrac{2}{7} =$ _____

6. $6\dfrac{7}{10} + 4\dfrac{2}{3} + 8\dfrac{8}{45} =$ _____

2.4A PRACTICE SET

In problems 1–10, find the LCM for the given numbers.

1. 8, 6 _____
2. 12, 18 _____
3. 24, 16 _____
4. 10, 22 _____
5. 30, 45 _____
6. 6, 9, 15 _____
7. 8, 12, 16 _____
8. 10, 15, 20 _____
9. 9, 10, 16 _____
10. 24, 36, 45 _____

In problems 11–25, find the sum. Reduce each answer to lowest terms.

11. $\dfrac{5}{8} + \dfrac{1}{8} =$ _____
12. $\dfrac{9}{16} + \dfrac{3}{16} =$ _____
13. $\dfrac{4}{9} + \dfrac{5}{9} + \dfrac{2}{9} =$ _____
14. $\dfrac{7}{30} + \dfrac{13}{30} + \dfrac{1}{30} =$ _____
15. $\dfrac{9}{32} + \dfrac{7}{32} + \dfrac{13}{32} + \dfrac{3}{32} =$ _____
16. $\dfrac{3}{5} + \dfrac{5}{8} =$ _____
17. $\dfrac{5}{14} + \dfrac{8}{21} =$ _____
18. $\dfrac{3}{10} + \dfrac{7}{16} =$ _____
19. $\dfrac{5}{16} + \dfrac{11}{12} + \dfrac{1}{6} =$ _____
20. $\dfrac{11}{20} + \dfrac{3}{8} + \dfrac{16}{25} =$ _____
21. $2\dfrac{4}{9} + 5\dfrac{3}{7} =$ _____
22. $8\dfrac{2}{5} + 15\dfrac{9}{14} =$ _____
23. $14\dfrac{7}{12} + 26\dfrac{3}{20} =$ _____
24. $52\dfrac{1}{2} + 116\dfrac{4}{5} + 75\dfrac{17}{25} =$ _____
25. $48\dfrac{6}{7} + \dfrac{13}{14} + 63\dfrac{4}{10} =$ _____

Applications

26. As an executive secretary, Susan spends $\frac{1}{6}$ of the first hour each morning getting mentally prepared for work, $\frac{1}{4}$ of an hour preparing her desk, and $\frac{1}{3}$ of an hour responding to messages left on her answering machine. What fractional part of the first hour does she spend on these activities?

   _____

27. David works in his yard each Saturday morning. He spends $1\frac{1}{2}$ hours mowing the lawn, $\frac{2}{3}$ of an hour edging the yard, and $2\frac{5}{8}$ hours working in his flower garden. How many hours total does he work in his yard each Saturday?

   _____

28. Marie studies the following number of hours each week:

   math $\quad 5\frac{1}{4}$ history $\quad 2\frac{1}{2}$

   English $\quad 3\frac{2}{3}$ biology $\quad 4\frac{7}{8}$

   How many hours total does she study each week?

   _____

29. Don, Martha, and Elaine ordered pizza. Martha ate $\frac{1}{8}$ of the pizza, Elaine ate $\frac{1}{4}$ of the pizza, and Don ate $\frac{1}{2}$ of the pizza. What fractional part of the whole pizza did they eat?

   _____

30. A jogger ran $5\frac{4}{5}$ miles one day, $4\frac{9}{10}$ miles the second day, and $8\frac{2}{3}$ miles the third day. How many miles total did she run during the three days?

   _____

2.4B PRACTICE SET

In problems 1–10, find the LCM for the given numbers.

1. 4, 6 _____
2. 5, 8 _____
3. 12, 14 _____
4. 14, 7 _____
5. 20, 24 _____
6. 4, 5, 12 _____
7. 9, 3, 6 _____
8. 10, 18, 5 _____
9. 15, 16, 12 _____
10. 30, 40, 15 _____

In problems 11–25, find the sum. Reduce each answer to lowest terms.

11. $\frac{3}{5} + \frac{1}{5} =$ _____
12. $\frac{5}{12} + \frac{11}{12} =$ _____

13. $\frac{4}{13} + \frac{9}{13} + \frac{6}{13} =$ _____
14. $\frac{23}{25} + \frac{12}{25} + \frac{16}{25} =$ _____

15. $\frac{9}{28} + \frac{11}{28} + \frac{19}{28} + \frac{17}{28} =$ _____
16. $\frac{4}{5} + \frac{5}{9} =$ _____

17. $\frac{3}{7} + \frac{3}{4} =$ _____
18. $\frac{8}{15} + \frac{3}{25} =$ _____

19. $\frac{2}{7} + \frac{1}{4} + \frac{2}{3} =$ _____
20. $\frac{7}{12} + \frac{4}{9} + \frac{5}{6} =$ _____

21. $3\frac{1}{2} + 8\frac{7}{8} =$ _____
22. $8\frac{3}{4} + 25\frac{5}{9} =$ _____

23. $12\frac{5}{12} + 24\frac{11}{15} =$ _____
24. $31\frac{3}{20} + 18\frac{11}{30} + 48\frac{1}{9} =$ _____

25. $25\frac{3}{14} + 53\frac{13}{16} + 37\frac{1}{4} =$ _____

Applications

26. Joan planned to spend $\frac{1}{4}$ of her vacation money on transportation and $\frac{1}{3}$ for a hotel room. What fractional part of her vacation money would go for these two items?

   _____

27. At a fresh fruit market one day, a farmer sold $28\frac{2}{3}$ pounds of apples and $18\frac{4}{5}$ pounds of oranges. What was the total weight of the produce the farmer sold that day?

   _____

28. Bob has $\$28\frac{3}{4}$ to buy his mother a birthday present. His two sisters, Mary and Louise, have $\$31\frac{11}{20}$ and $\$18\frac{3}{10}$, respectively. If they decide to combine their money and buy one gift, how much money will they have for the gift?

   _____

29. Ray works for an overnight delivery service. Last week he delivered four packages to one address. The packages weighed $\frac{2}{5}$ pound, $\frac{1}{3}$ pound, $5\frac{3}{4}$ pounds, and $12\frac{7}{16}$ pounds. What was the total weight of the four packages?

   _____

30. A road race covers a triangular course. The first leg of the race is $25\frac{3}{10}$ miles long. The other two legs of the race are $17\frac{1}{2}$ miles and $20\frac{3}{4}$ miles. What is the total length of the race?

   _____

2.5 SUBTRACTING FRACTIONS AND MIXED NUMBERS

Subtracting Fractions with Common Denominators

Fractions must have common denominators before they can be subtracted.

> **TO SUBTRACT FRACTIONS WITH COMMON DENOMINATORS**
> 1. Subtract the numerators and place the difference over the common denominator.
> 2. Reduce the answer to lowest terms.

EXAMPLE 1 Subtract: $\frac{3}{5} - \frac{1}{5}$

SOLUTION

$$\frac{3}{5} - \frac{1}{5} = \frac{3-1}{5} = \frac{2}{5}$$

EXAMPLE 2 Subtract: $\frac{5}{6} - \frac{1}{6}$

SOLUTION

$$\frac{5}{6} - \frac{1}{6} = \frac{5-1}{6} = \frac{4}{6} \quad \text{\textit{This fraction can be reduced.}}$$

$$\frac{4 \div 2}{6 \div 2} = \frac{2}{3}$$

Subtracting Fractions with Different Denominators

> **TO SUBTRACT FRACTIONS WITH DIFFERENT DENOMINATORS**
> 1. Find the LCD (this is the LCM) of the given denominators.
> 2. Convert each original fraction to an equivalent fraction with the LCD as the new denominator.
> 3. Subtract the fractions and reduce the difference to lowest terms.

EXAMPLE 3 Subtract: $\frac{3}{4} - \frac{1}{6}$

SOLUTION

$4 = 2 \cdot 2 \qquad 6 = 2 \cdot 3$

$\text{LCM} = 2 \cdot 2 \cdot 3 = 12 = \text{LCD}$

$$\frac{3}{4} = \frac{9}{12} \qquad \frac{1}{6} = \frac{2}{12}$$

$$\frac{9}{12} - \frac{2}{12} = \frac{7}{12} \quad \text{\textit{This fraction cannot be reduced.}}$$

EXAMPLE 4 Subtract: $\dfrac{5}{8} - \dfrac{7}{24}$

SOLUTION

The LCD is 24, since 8 will divide 24.

$$\dfrac{5}{8} = \dfrac{15}{24} \qquad \dfrac{15}{24} - \dfrac{7}{24} = \dfrac{8}{24}$$

$\dfrac{8 \div 8}{24 \div 8} = \dfrac{1}{3}$ *This is the answer in reduced terms.*

EXAMPLE 5 Subtract: $1 - \dfrac{4}{5}$

SOLUTION

$1 = \dfrac{5}{5}$ *Change 1 to a fraction with a denominator of 5 so that the fractions will have common denominators.*

$\dfrac{5}{5} - \dfrac{4}{5} = \dfrac{1}{5}$

Remember from Section 2.2 that the number 1 may be converted to a fraction with any number as the denominator. The numerator must be the same as the denominator.

Subtracting Mixed Numbers

The rules for subtracting mixed numbers are similar to the rules for adding mixed numbers.

> **TO SUBTRACT MIXED NUMBERS**
> 1. If the fractions do not already have common denominators, find the LCD and convert each fraction to an equivalent fraction with the LCD as the denominator.
> 2. Subtract the fractions.
> 3. Subtract the whole numbers.
> 4. Simplify and reduce.

EXAMPLE 6 Subtract: $3\dfrac{5}{8} - 2\dfrac{3}{8}$

SOLUTION

$\dfrac{5}{8} - \dfrac{3}{8} = \dfrac{2}{8}$ *Subtract the fractions. They already have common denominators.*

$3 - 2 = 1$ *Subtract the whole numbers.*

$1\dfrac{2}{8} = 1\dfrac{1}{4}$ *Reduce the fraction to lowest terms.*

Subtracting Fractions and Mixed Numbers

EXAMPLE 7 Subtract: $12\dfrac{5}{6} - 4\dfrac{1}{2}$

SOLUTION

The LCD is 6, since 2 will divide 6.

$\dfrac{1}{2} = \dfrac{3}{6}$ $\dfrac{5}{6} - \dfrac{3}{6} = \dfrac{2}{6}$ *Subtract the fractions.*

$12 - 4 = 8$ *Subtract the whole numbers.*

$8\dfrac{2}{6} = 8\dfrac{1}{3}$ *Reduce the fraction to lowest terms.*

Sometimes subtracting mixed numbers will require borrowing. Examples 8 and 9 will illustrate this technique as it applies to fractions.

EXAMPLE 8 Subtract: $24\dfrac{1}{5} - 15\dfrac{3}{4}$

SOLUTION

$5 = 1 \cdot 5$ $4 = 2 \cdot 2$

LCM $= 5 \cdot 2 \cdot 2 = 20 =$ LCD

$24\dfrac{1}{5} = 24\dfrac{4}{20}$ *Since 15 cannot be subtracted from 4, borrow 1 from 24 and change it to a fraction with the same denominator as the other fractions.*

$-15\dfrac{3}{4} = 15\dfrac{15}{20}$

$2\overset{3}{\cancel{4}}\dfrac{4}{20} + \dfrac{20}{20} = 23\dfrac{24}{20}$ *Add the 1 to the fraction. Now you can subtract.*

$\phantom{2\overset{3}{\cancel{4}}\dfrac{4}{20} + \dfrac{20}{20} = 2} -15\dfrac{15}{20}$

$\phantom{2\overset{3}{\cancel{4}}\dfrac{4}{20} + \dfrac{20}{20} = 2\,\,} 8\dfrac{9}{20}$ *This answer will not reduce.*

EXAMPLE 9 Subtract: $18 - 10\dfrac{7}{12}$

SOLUTION

18 *Since there is not a fraction in the top number, borrow 1 from 18 and convert it to a fraction with a denominator the same as the fraction in the bottom number.*

$-10\dfrac{7}{12}$

$\overset{7}{18}\dfrac{12}{12}$ *Now you are ready to subtract. Follow the rules you have already learned for subtracting mixed numbers.*

$-10\dfrac{7}{12}$

$7\dfrac{5}{12}$ *This answer cannot be reduced.*

Check your understanding of subtracting fractions and mixed numbers with the Comprehension Checkpoint problems. If you have difficulty with these problems, review the material in this section before going on to the Practice Sets.

COMPREHENSION CHECKPOINT

Find the difference in each problem. Reduce each answer to lowest terms.

1. $\dfrac{6}{7} - \dfrac{2}{7} =$ _____

2. $\dfrac{17}{24} - \dfrac{7}{24} =$ _____

3. $\dfrac{5}{9} - \dfrac{2}{5} =$ _____

4. $3\dfrac{7}{15} - 1\dfrac{4}{15} =$ _____

5. $14\dfrac{1}{4} - 5\dfrac{7}{10} =$ _____

6. $1 - \dfrac{13}{24} =$ _____

7. $30 - 16\dfrac{9}{14} =$ _____

2.5A PRACTICE SET

Find the difference in each problem. Reduce each answer to lowest terms.

1. $\frac{7}{9} - \frac{3}{9} =$ _____

2. $\frac{9}{10} - \frac{1}{10} =$ _____

3. $\frac{13}{20} - \frac{11}{20} =$ _____

4. $\frac{7}{8} - \frac{1}{3} =$ _____

5. $\frac{5}{6} - \frac{1}{4} =$ _____

6. $\frac{9}{16} - \frac{3}{20} =$ _____

7. $\frac{7}{10} - \frac{3}{5} =$ _____

8. $\frac{12}{25} - \frac{5}{12} =$ _____

9. $\frac{8}{15} - \frac{3}{8} =$ _____

10. $\frac{17}{24} - \frac{5}{18} =$ _____

11. $3\frac{5}{8} - 1\frac{1}{8} =$ _____

12. $8\frac{7}{10} - 5\frac{3}{10} =$ _____

13. $7\frac{2}{3} - 4\frac{5}{12} =$ _____

14. $15\frac{2}{3} - 8\frac{2}{9} =$ _____

15. $21\frac{9}{16} - 4\frac{7}{12} =$ _____

16. $28 - 9\frac{3}{4} =$ _____

17. $48\frac{6}{7} - 30\frac{2}{5} =$ _____

18. $95\frac{5}{8} - 36\frac{9}{22} =$ _____

19. $68\frac{4}{5} - 50 =$ _____

20. $36 - 24\frac{5}{9} =$ _____

21. $40\frac{1}{8} - 24\frac{1}{8} =$ _____

22. $75\frac{2}{15} - 47\frac{7}{9} =$ _____

Applications

23. Alicia bought $\frac{1}{2}$ pound of candy. She gave Robert $\frac{1}{3}$ of a pound. How much candy did Alicia have left?

_____

24. Marie went shopping with $100. She spent 45\frac{1}{10}$ for a blouse. How much money did she have when she returned home?

_____

25. A carpenter had a board 12$\frac{3}{4}$ feet long. He cut off a piece 3$\frac{11}{16}$ feet long. How long was the original piece after he made the cut?

_____

26. Dave jogs 15$\frac{1}{3}$ miles each day. There is a fountain 10$\frac{5}{12}$ miles from the start that he passes each day. How many miles does he still have to run when he passes the fountain?

_____

2.5B PRACTICE SET

Find the difference in each problem. Reduce each answer to lowest terms.

1. $\dfrac{7}{8} - \dfrac{5}{8} =$ _____

2. $\dfrac{15}{16} - \dfrac{7}{16} =$ _____

3. $\dfrac{25}{32} - \dfrac{17}{32} =$ _____

4. $\dfrac{9}{10} - \dfrac{2}{5} =$ _____

5. $\dfrac{4}{15} - \dfrac{3}{25} =$ _____

6. $\dfrac{11}{18} - \dfrac{4}{9} =$ _____

7. $\dfrac{17}{24} - \dfrac{7}{16} =$ _____

8. $\dfrac{19}{20} - \dfrac{3}{8} =$ _____

9. $\dfrac{16}{21} - \dfrac{4}{7} =$ _____

10. $\dfrac{27}{51} - \dfrac{9}{17} =$ _____

11. $4\dfrac{5}{6} - 2\dfrac{1}{6} =$ _____

12. $8\dfrac{9}{10} - 5\dfrac{7}{16} =$ _____

13. $13\dfrac{11}{12} - 10\dfrac{2}{3} =$ _____

14. $22\dfrac{4}{5} - 6\dfrac{5}{9} =$ _____

15. $49\dfrac{9}{20} - 23\dfrac{11}{16} =$ _____

16. $37\dfrac{15}{32} - 18\dfrac{7}{24} =$ _____

17. $45 - 19\dfrac{18}{25} =$ _____

18. $56\dfrac{2}{5} - 31\dfrac{5}{7} =$ _____

19. $28\dfrac{3}{10} - 12 =$ _____

20. $39\dfrac{1}{10} - 29\dfrac{3}{4} =$ _____

21. $100 - 42\dfrac{5}{12} =$ _____

22. $64\dfrac{7}{8} - 12\dfrac{1}{9} =$ _____

Applications

23. A farmer has a field that covers $42\dfrac{5}{8}$ acres. He plants $20\dfrac{1}{2}$ acres of tomatoes each year. How many acres does he have remaining for other crops?

_____

24. Steve paints houses for a living. It takes him $\frac{9}{10}$ of an hour to paint a wall. He has been working for $\frac{1}{6}$ of an hour. What fractional part of an hour will he have to work to finish the job?

_____

25. Terry travels $6\frac{3}{4}$ miles to school each day. She stops for coffee at a convenience store that is $\frac{11}{16}$ of a mile from her apartment. How far does she still have to travel to school?

_____

26. Peggy is saving for a vacation. She needs $250 and has saved 175\frac{7}{10}$. How much more does she need to save to have the desired amount?

_____

2.6 MULTIPLYING FRACTIONS AND MIXED NUMBERS

Multiplying Fractions

Adding or subtracting fractions requires that the fractions have common denominators. It is not necessary to have common denominators when multiplying fractions.

> **TO MULTIPLY FRACTIONS**
> 1. Multiply the numerators to get the numerator of the product.
> 2. Multiply the denominators to get the denominator of the product.
> 3. Reduce the product to lowest terms.

EXAMPLE 1 Multiply: $\dfrac{5}{6} \cdot \dfrac{1}{4}$

SOLUTION

$$\frac{5}{6} \cdot \frac{1}{4} = \frac{5 \cdot 1}{6 \cdot 4} = \frac{5}{24}$$ *This product will not reduce.*

EXAMPLE 2 Multiply: $\dfrac{5}{8} \cdot \dfrac{1}{2} \cdot \dfrac{7}{3}$

SOLUTION

$$\frac{5}{8} \cdot \frac{1}{2} \cdot \frac{7}{3} = \frac{5 \cdot 1 \cdot 7}{8 \cdot 2 \cdot 3} = \frac{35}{48}$$ *This product will not reduce.*

If a numerator and a denominator have a common factor, those factors cancel. **Canceling** means dividing by the same factor. Canceling all common factors before multiplying will give you smaller factors to multiply. It will also eliminate the need to reduce the answer to lowest terms; the answer will already be in lowest terms.

EXAMPLE 3 Multiply: $\dfrac{3}{8} \cdot \dfrac{5}{9}$

SOLUTION

$$\frac{\overset{1}{\cancel{3}}}{8} \cdot \frac{5}{\underset{3}{\cancel{9}}} = \frac{1 \cdot 5}{8 \cdot 3} = \frac{5}{24}$$

The numerator 3 and the denominator 9 have a common factor of 3. Dividing by the common factor is the canceling process. New terms are left in the place of the original terms. After canceling, multiply as usual. Notice that the answer cannot be reduced.

EXAMPLE 4 Multiply: $\dfrac{5}{8} \cdot \dfrac{9}{16} \cdot \dfrac{8}{15} \cdot \dfrac{2}{3}$

SOLUTION

$$\dfrac{\overset{1}{\cancel{5}}}{\underset{1}{\cancel{8}}} \cdot \dfrac{\overset{\overset{1}{\cancel{3}}}{\cancel{9}}}{\underset{8}{\cancel{16}}} \cdot \dfrac{\overset{1}{\cancel{8}}}{\underset{\underset{1}{\cancel{3}}}{\cancel{15}}} \cdot \dfrac{\overset{1}{\cancel{2}}}{\underset{1}{\cancel{3}}} = \dfrac{1 \cdot 1 \cdot 1 \cdot 1}{1 \cdot 8 \cdot 1 \cdot 1} = \dfrac{1}{8}$$

Canceling:
5 and 15 have a factor of 5.
9 and 3 have a factor of 3.
8 and 8 have a factor of 8.
2 and 16 have a factor of 2.
3 and 3 have a factor of 3. (These 3s were the result of canceling the 15 on the bottom and the 9 on top.)

Example 4 demonstrates several important facts about canceling:

1. Any numerator may be canceled with any denominator.
2. A numerator (or denominator) may be canceled more than once. (The 9 and the 15 were each canceled twice.)
3. When a number is itself the factor, like the 8s and the 3s, a 1 takes its place.

Remember that canceling is a division process. When a number is itself the factor, you are dividing the number by itself. Any number divided by itself gives a quotient of 1.

Multiplying Mixed Numbers

TO MULTIPLY MIXED NUMBERS
1. Convert each mixed number to an improper fraction.
2. Multiply the improper fractions as you do other fractions.

EXAMPLE 5 Multiply: $3\dfrac{3}{8} \cdot 5\dfrac{1}{3}$

SOLUTION

$3\dfrac{3}{8} = \dfrac{8 \cdot 3 + 3}{8} = \dfrac{27}{8}$ *Convert the first factor to an improper fraction.*

$5\dfrac{1}{3} = \dfrac{3 \cdot 5 + 1}{3} = \dfrac{16}{3}$ *Convert the second factor to an improper fraction.*

$\dfrac{\overset{9}{\cancel{27}}}{\underset{1}{\cancel{8}}} \cdot \dfrac{\overset{2}{\cancel{16}}}{\underset{1}{\cancel{3}}} = \dfrac{18}{1} = 18$ *Multiply the improper fractions as you would proper fractions and simplify the answer.*

Multiplying Fractions and Mixed Numbers

EXAMPLE 6 Multiply: $4\frac{4}{7} \cdot 2\frac{1}{10}$

SOLUTION

$$4\frac{4}{7} = \frac{7 \cdot 4 + 4}{7} = \frac{32}{7}$$

$$2\frac{1}{10} = \frac{10 \cdot 2 + 1}{10} = \frac{21}{10}$$

Convert each mixed number to an improper fraction.

$$\frac{\overset{16}{\cancel{32}}}{\underset{1}{\cancel{7}}} \cdot \frac{\overset{3}{\cancel{21}}}{\underset{5}{\cancel{10}}} = \frac{48}{5} = 9\frac{3}{5}$$

When the answer in a multiplication problem is an improper fraction, you should convert it to a mixed number as in Example 6.

EXAMPLE 7 Multiply: $5 \cdot 3\frac{12}{25}$

SOLUTION

$$5 = \frac{5}{1} \qquad 3\frac{12}{25} = \frac{25 \cdot 3 + 12}{25} = \frac{87}{25}$$

The whole number, 5, must be converted to a fraction before it can be multiplied. Now multiply as before.

$$\frac{\overset{1}{\cancel{5}}}{1} \cdot \frac{87}{\underset{5}{\cancel{25}}} = \frac{87}{5} = 17\frac{2}{5}$$

In many word problems in mathematics, the word *of* indicates multiplication.

EXAMPLE 8 How much did Mike pay for $\frac{1}{2}$ of a pound of cashew nuts if the price of the nuts is $\$3\frac{3}{5}$ per pound?

SOLUTION

The word *of* indicates that you need to multiply the amount purchased by the cost.

$$\frac{1}{2} \cdot 3\frac{3}{5} = \frac{1}{\underset{1}{\cancel{2}}} \cdot \frac{\overset{9}{\cancel{18}}}{5} = \frac{9}{5} = \$1\frac{4}{5}$$

Mike paid $\$1\frac{4}{5}$ for the cashews.

Use the Comprehension Checkpoint problems to measure your understanding of multiplying fractions and mixed numbers.

COMPREHENSION CHECKPOINT

Find the product.

1. $\dfrac{4}{7} \cdot \dfrac{3}{5} = $ _____

2. $\dfrac{1}{2} \cdot \dfrac{4}{9} = $ _____

3. $\dfrac{6}{11} \cdot \dfrac{4}{15} \cdot \dfrac{33}{16} = $ _____

4. $2\dfrac{1}{2} \cdot 1\dfrac{7}{15} = $ _____

5. $5\dfrac{7}{10} \cdot 4\dfrac{14}{19} = $ _____

6. $3\dfrac{1}{8} \cdot 4\dfrac{2}{3} \cdot 8\dfrac{2}{5} = $ _____

2.6A PRACTICE SET

Find the product. Use cancellation whenever possible, and remember to reduce all answers to lowest terms.

1. $\dfrac{3}{5} \cdot \dfrac{1}{2} =$ _____

2. $\dfrac{5}{8} \cdot \dfrac{3}{4} =$ _____

3. $\dfrac{7}{16} \cdot \dfrac{4}{21} =$ _____

4. $\dfrac{8}{9} \cdot \dfrac{3}{16} =$ _____

5. $\dfrac{4}{5} \cdot \dfrac{15}{28} =$ _____

6. $\dfrac{9}{20} \cdot \dfrac{2}{3} =$ _____

7. $\dfrac{12}{25} \cdot \dfrac{15}{32} =$ _____

8. $\dfrac{7}{10} \cdot \dfrac{25}{42} =$ _____

9. $\dfrac{8}{13} \cdot \dfrac{7}{8} =$ _____

10. $\dfrac{10}{21} \cdot \dfrac{14}{15} =$ _____

11. $\dfrac{8}{25} \cdot \dfrac{2}{3} \cdot \dfrac{15}{16} =$ _____

12. $\dfrac{3}{10} \cdot \dfrac{15}{24} \cdot \dfrac{8}{9} =$ _____

13. $3\dfrac{3}{4} \cdot 5\dfrac{1}{3} =$ _____

14. $4\dfrac{3}{8} \cdot 1\dfrac{3}{5} =$ _____

15. $7\dfrac{1}{12} \cdot 9\dfrac{3}{5} =$ _____

16. $8\dfrac{2}{9} \cdot 5\dfrac{7}{10} =$ _____

17. $10\dfrac{2}{3} \cdot 1\dfrac{23}{32} =$ _____

18. $12\dfrac{1}{2} \cdot 2\dfrac{8}{15} =$ _____

19. $2\dfrac{3}{4} \cdot 3\dfrac{7}{11} =$ _____

20. $20\dfrac{4}{5} \cdot 27\dfrac{1}{2} =$ _____

21. $2\dfrac{3}{16} \cdot 4\dfrac{4}{7} =$ _____

22. $8\dfrac{1}{6} \cdot 4\dfrac{2}{7} \cdot 0 =$ _____

Applications

23. What is $\dfrac{3}{8}$ of $\dfrac{2}{3}$? _____

24. Lori bought $\frac{3}{5}$ of a yard of material to make a dress. The price of the material was 7\frac{1}{2}$ per yard. How much did she pay for the material?

   _____

25. Baseballs cost 6\frac{3}{10}$ each. What would be the total cost of 12 baseballs?

   _____

26. A carpenter has a board $24\frac{7}{12}$ feet long. He needs only $\frac{3}{4}$ of the board. How long a piece of board does the carpenter need?

   _____

27. A college class is scheduled for $\frac{5}{6}$ of an hour. One student left after $\frac{2}{3}$ of the scheduled time. What fractional part of an hour was the student in class?

   _____

2.6B PRACTICE SET

Find the product. Use cancellation whenever possible, and remember to reduce all answers to lowest terms.

1. $\dfrac{3}{8} \cdot \dfrac{3}{4} =$ _____

2. $\dfrac{5}{6} \cdot \dfrac{1}{2} =$ _____

3. $\dfrac{6}{7} \cdot \dfrac{9}{16} =$ _____

4. $\dfrac{5}{12} \cdot \dfrac{12}{25} =$ _____

5. $\dfrac{4}{9} \cdot \dfrac{21}{36} =$ _____

6. $\dfrac{15}{32} \cdot \dfrac{8}{25} =$ _____

7. $\dfrac{7}{16} \cdot \dfrac{10}{21} =$ _____

8. $\dfrac{10}{33} \cdot \dfrac{11}{15} =$ _____

9. $\dfrac{34}{57} \cdot \dfrac{19}{51} =$ _____

10. $\dfrac{7}{18} \cdot \dfrac{24}{35} =$ _____

11. $\dfrac{3}{7} \cdot 0 \cdot \dfrac{4}{9} =$ _____

12. $\dfrac{4}{5} \cdot \dfrac{39}{40} \cdot \dfrac{10}{13} =$ _____

13. $3\dfrac{3}{5} \cdot 2\dfrac{1}{2} =$ _____

14. $4\dfrac{9}{10} \cdot 5\dfrac{5}{7} =$ _____

15. $8\dfrac{4}{5} \cdot 2\dfrac{3}{11} =$ _____

16. $9\dfrac{3}{4} \cdot 3\dfrac{5}{13} =$ _____

17. $12\dfrac{2}{3} \cdot 7\dfrac{1}{2} =$ _____

18. $14\dfrac{1}{7} \cdot 3\dfrac{9}{11} =$ _____

19. $2\dfrac{3}{32} \cdot 16$ _____

20. $15\dfrac{1}{3} \cdot 18\dfrac{3}{4} =$ _____

21. $0 \cdot 4\dfrac{7}{8} \cdot 16\dfrac{9}{24}$ _____

22. $25\dfrac{5}{6} \cdot 2\dfrac{4}{31} =$ _____

Applications

23. What is $\dfrac{5}{8}$ of $17\dfrac{5}{3}$? _____

24. A racetrack is a $2\frac{1}{2}$-mile oval. One driver completed $12\frac{4}{5}$ laps around the track before blowing his engine. How many miles had he driven?

    _____

25. What is the total price of $4\frac{9}{10}$ pounds of ground beef if the ground beef costs $\$1\frac{1}{4}$ per pound?

    _____

26. A brick is $9\frac{3}{8}$ inches long. If a mason uses $11\frac{1}{4}$ bricks for the length of a wall, how long is the wall? (Ignore the mortar between the bricks.)

    _____

27. A minute is $\frac{1}{60}$ of an hour. A second is $\frac{1}{60}$ of a minute. What fractional part of an hour is a second?

    _____

2.7 DIVIDING FRACTIONS AND MIXED NUMBERS

Dividing Fractions

Like multiplication, division of fractions does not require that the fractions have common denominators. Actually, you accomplish division of fractions by converting the problem to a multiplication problem.

Two terms that you need to know to work division problems successfully are **invert** and **reciprocal.** To invert means to turn upside down. The fraction $\frac{2}{5}$ inverted is $\frac{5}{2}$. A whole number may also be inverted, since every whole number may be written as an improper fraction with a denominator of 1. Thus, 6 inverted is $\frac{1}{6}$.

The reciprocal of a number is the inverse of that number. The reciprocal of $\frac{3}{4}$ is $\frac{4}{3}$. The product of reciprocals is always 1.

TO DIVIDE FRACTIONS
1. Invert the divisor and change the operation sign to multiplication.
2. Multiply the fractions.

EXAMPLE 1 Divide: $\frac{3}{8} \div \frac{2}{5}$

SOLUTION

$$\frac{3}{8} \div \frac{2}{5} = \frac{3}{8} \times \frac{5}{2} = \frac{15}{16} \quad \text{quotient}$$

operation sign changed ⟶ divisor inverted

Note: Do not cancel until you have inverted the divisor.

EXAMPLE 2 Divide: $\frac{7}{12} \div \frac{14}{3}$

SOLUTION

$$\frac{7}{12} \div \frac{14}{3} = \frac{\cancel{7}^{1}}{\cancel{12}_{4}} \times \frac{\cancel{3}^{1}}{\cancel{14}_{2}} = \frac{1}{8} \quad \text{quotient}$$

operation sign changed ⟶ divisor inverted

Note: Factors were canceled *after* inverting.

EXAMPLE 3 Divide: $\frac{15}{16} \div 5$

SOLUTION

$$\frac{15}{16} \div 5 = \frac{\cancel{15}^{3}}{16} \times \frac{1}{\cancel{5}_{1}} = \frac{3}{16} \quad \text{quotient}$$

operation sign changed ⟶ divisor inverted

Dividing Mixed Numbers

> **TO DIVIDE MIXED NUMBERS**
> 1. Convert each mixed number to an improper fraction.
> 2. Invert the divisor and change the operation sign to multiplication.
> 3. Multiply the fractions.

EXAMPLE 4 Divide: $4\frac{2}{3} \div 3\frac{1}{2}$

SOLUTION

$$4\frac{2}{3} = \frac{3 \times 4 + 2}{3} = \frac{14}{3}$$
$$3\frac{1}{2} = \frac{2 \times 3 + 1}{2} = \frac{7}{2}$$

Change the mixed numbers to improper fractions.

$$\frac{14}{3} \div \frac{7}{2} = \frac{\cancel{14}^{2}}{3} \times \frac{2}{\cancel{7}_{1}} = \frac{4}{3} = 1\frac{1}{3}$$

operation sign changed ←⎦ ⎣→ divisor inverted

If only one of the terms in the problem is a mixed number, it must still be converted to an improper fraction. After converting to an improper fraction, follow the rules for dividing fractions.

EXAMPLE 5 Divide: $6\frac{2}{3} \div \frac{1}{9}$

SOLUTION

$$6\frac{2}{3} = \frac{3 \times 6 + 2}{3} = \frac{20}{3} \quad \text{Convert the mixed number to an improper fraction.}$$

$$\frac{20}{3} \div \frac{1}{9} = \frac{20}{\cancel{3}_{1}} \times \frac{\cancel{9}^{3}}{1} = \frac{60}{1} = 60$$

operation sign changed ←⎦ ⎣→ divisor inverted

EXAMPLE 6 Divide: $\frac{5}{8} \div 2\frac{7}{24}$

SOLUTION

$$2\frac{7}{24} = \frac{24 \times 2 + 7}{24} = \frac{55}{24} \quad \text{Convert the mixed number to an improper fraction.}$$

$$\frac{5}{8} \div \frac{55}{24} = \frac{\cancel{5}^{1}}{\cancel{8}_{1}} \times \frac{\cancel{24}^{3}}{\cancel{55}_{11}} = \frac{3}{11}$$

operation sign changed ←⎦ ⎣→ divisor inverted

Dividing Fractions and Mixed Numbers

Students often have difficulty understanding an answer in a division problem. Some say the answer does not look right. A word problem may help you understand the answers.

EXAMPLE 7 Sherry has $2\frac{1}{2}$ and decides to play video games. Each video game costs $\frac{1}{4}$. How many video games will she be able to play?

SOLUTION

The problem here is to divide $2\frac{1}{2}$ into quarters ($\frac{1}{4}$ is a quarter). Set up a division problem and solve it.

$$2\frac{1}{2} \div \frac{1}{4} = \frac{5}{2} \times \frac{4}{1} = \frac{10}{1} = 10$$

converted to improper fraction

divisor inverted
operation sign changed

There are 10 quarters in $2\frac{1}{2}$, so she can play 10 games.

The Comprehension Checkpoint problems allow you an opportunity to measure your understanding of dividing fractions and mixed numbers. If you have difficulty with these problems, review the material in this section before going on to the Practice Sets.

COMPREHENSION CHECKPOINT

In problems 1–3, write the reciprocal of each number.

1. $\frac{8}{13}$ _____   2. 12 _____   3. $\frac{1}{4}$ _____

In problems 4–9, determine the quotient.

4. $\frac{9}{16} \div \frac{3}{20} =$ _____   5. $6 \div \frac{2}{3} =$ _____

6. $\frac{8}{15} \div 4 =$ _____   7. $7\frac{1}{2} \div 4\frac{3}{8} =$ _____

8. $9\frac{3}{4} \div 12 =$ _____   9. $\frac{2}{9} \div 0 =$ _____

2.7A PRACTICE SET

Find each quotient. Reduce each answer to lowest terms.

1. $\dfrac{1}{2} \div \dfrac{1}{4} =$ _____

2. $\dfrac{3}{8} \div \dfrac{3}{2} =$ _____

3. $\dfrac{4}{7} \div \dfrac{1}{2} =$ _____

4. $\dfrac{11}{12} \div \dfrac{5}{12} =$ _____

5. $\dfrac{21}{32} \div \dfrac{3}{8} =$ _____

6. $\dfrac{7}{9} \div \dfrac{14}{15} =$ _____

7. $\dfrac{17}{20} \div \dfrac{3}{32} =$ _____

8. $\dfrac{5}{8} \div \dfrac{1}{6} =$ _____

9. $\dfrac{10}{21} \div \dfrac{15}{24} =$ _____

10. $\dfrac{21}{25} \div 0 =$ _____

11. $2\dfrac{1}{2} \div \dfrac{3}{4} =$ _____

12. $8\dfrac{5}{6} \div \dfrac{3}{10} =$ _____

13. $12\dfrac{2}{3} \div \dfrac{1}{3} =$ _____

14. $5 \div 1\dfrac{1}{4} =$ _____

15. $3\dfrac{1}{5} \div 2\dfrac{2}{5} =$ _____

16. $9 \div \dfrac{2}{9} =$ _____

17. $10\dfrac{1}{2} \div 2\dfrac{1}{8} =$ _____

18. $0 \div 16\dfrac{2}{3} =$ _____

19. $6\dfrac{3}{5} \div 3\dfrac{9}{10} =$ _____

20. $5\dfrac{1}{4} \div 8\dfrac{5}{2} =$ _____

Applications

21. A ruler is marked every $\dfrac{1}{8}$ of an inch. How many marks will be on a ruler that is 12 inches long?

    _____

22. Katie bought small cans of fruit that weigh $7\dfrac{1}{4}$ ounces each. The total weight of the cans was $36\dfrac{1}{4}$ ounces. How many cans of fruit did she buy?

    _____

23. A model railroader is building a miniature fence around his railroad table. Each fence post is $\frac{3}{4}$ of an inch tall. How many fence posts will he be able to cut from a stick $20\frac{1}{2}$ inches long?

　　　　　　　　　　　　　　　　　　　_____

24. A radio station plays music for $28\frac{1}{3}$ minutes each hour. The average record is $1\frac{3}{4}$ minutes long. How many records will the station play in one hour?

　　　　　　　　　　　　　　　　　　　_____

2.7B PRACTICE SET

Find each quotient. Reduce each answer to lowest terms.

1. $\dfrac{2}{3} \div \dfrac{1}{2} = $ _____

2. $\dfrac{5}{9} \div \dfrac{1}{3} = $ _____

3. $\dfrac{9}{16} \div \dfrac{3}{32} = $ _____

4. $\dfrac{12}{25} \div \dfrac{3}{5} = $ _____

5. $\dfrac{9}{20} \div \dfrac{18}{15} = $ _____

6. $\dfrac{7}{8} \div 14 = $ _____

7. $\dfrac{25}{18} \div \dfrac{5}{6} = $ _____

8. $\dfrac{4}{7} \div \dfrac{16}{21} = $ _____

9. $24 \div \dfrac{8}{3} = $ _____

10. $\dfrac{33}{48} \div \dfrac{11}{16} = $ _____

11. $\dfrac{5}{3} \div \dfrac{20}{27} = $ _____

12. $0 \div \dfrac{7}{5} = $ _____

13. $6\dfrac{1}{2} \div 2\dfrac{3}{4} = $ _____

14. $3\dfrac{1}{5} \div 4\dfrac{3}{15} = $ _____

15. $1\dfrac{2}{3} \div \dfrac{5}{18} = $ _____

16. $4\dfrac{5}{7} \div 6\dfrac{1}{3} = $ _____

17. $15 \div \dfrac{10}{3} = $ _____

18. $10\dfrac{5}{8} \div 17 = $ _____

19. $9\dfrac{2}{3} \div 1\dfrac{5}{6} = $ _____

20. $8\dfrac{1}{2} \div 0 = $ _____

Applications

21. A piece of manila paper is 24 inches long. Joanne needs strips of paper that are $2\dfrac{7}{8}$ inches wide for an art project. How many strips of paper could she cut from one piece of manila paper? _____

22. Students in Mrs. Long's history class were given $\frac{1}{3}$ of an hour to make a presentation on their term projects. The class was scheduled for $2\frac{1}{2}$ hours. How many presentations were given in one class period?

_____

23. A container of bulk coffee weighs $48\frac{1}{2}$ pounds. How many bags weighing $2\frac{1}{4}$ pounds each could be filled from the bulk container?

_____

24. Jonathan gained 33 yards during the homecoming football game. He averaged $2\frac{3}{4}$ yards every time he carried the ball. How many times did he carry the ball?

_____

2.8 ORDER OF OPERATIONS

The order of operations does not change for different types of numbers (whole numbers, fractions, etc.). As a reminder, the rules governing the order of operations are repeated here.

RULES FOR THE ORDER OF OPERATIONS

1. Do all multiplication and division in order first, working from left to right.
2. After completing all multiplication and division, do addition and subtraction in order, from left to right.
3. If grouping symbols such as ( ) or [ ] are included, operations within these symbols should be performed first, using the first two steps. After simplifying inside the grouping symbols, begin again using the first two steps for the operations that remain.

EXAMPLE 1 Simplify: $\dfrac{2}{3} \cdot \dfrac{1}{2} + \dfrac{1}{4}$

SOLUTION

$\dfrac{2}{3} \cdot \dfrac{1}{2} + \dfrac{1}{4}$ *Multiply.*

$\dfrac{1}{3} + \dfrac{1}{4}$ *Add after finding a common denominator.*

$\dfrac{4}{12} + \dfrac{3}{12}$ *LCD is 12. Change each fraction to an equivalent fraction with a denominator of 12.*

$\dfrac{7}{12}$

EXAMPLE 2 Simplify: $\dfrac{3}{4} - \dfrac{1}{2} \div \dfrac{7}{8}$

SOLUTION

$\dfrac{3}{4} - \dfrac{1}{2} \div \dfrac{7}{8}$ *Divide first.*

$\dfrac{1}{\cancel{2}} \times \dfrac{\cancel{8}^{4}}{7}$ *Invert the divisor; change the sign and multiply.*

$\dfrac{3}{4} - \dfrac{4}{7}$ *Subtract after finding a common denominator.*

$\dfrac{21}{28} - \dfrac{16}{28}$

$\dfrac{5}{28}$

CHAPTER 2 Fractions and Mixed Numbers

EXAMPLE 3 Simplify: $\left(2\dfrac{1}{2} + \dfrac{3}{16}\right) \div \dfrac{9}{2}$

SOLUTION

$\left(2\dfrac{1}{2} + \dfrac{3}{16}\right) \div \dfrac{9}{2}$

Addition is first because it is inside parentheses. The common denominator for the fractions is 16.

$\left(2\dfrac{8}{16} + \dfrac{3}{16}\right) \div \dfrac{9}{2}$

$2\dfrac{11}{16} \div \dfrac{9}{2}$

Divide; invert the divisor and follow the rules for multiplication.

$\dfrac{\overset{}{43}}{\underset{8}{\cancel{16}}} \times \dfrac{\overset{1}{\cancel{2}}}{9}$

$\dfrac{43}{72}$

Work the Comprehension Checkpoint problems to refresh your memory on using the order of operations. If you have any difficulty with these problems, review the material in this section before trying the Practice Sets.

COMPREHENSION CHECKPOINT

Simplify each problem.

1. $\dfrac{5}{9} - \dfrac{2}{5} \cdot \dfrac{15}{16} =$ _____

2. $3\dfrac{1}{5} \cdot \dfrac{5}{12} \div \dfrac{2}{9} =$ _____

3. $\dfrac{7}{8} \cdot \dfrac{16}{15} - \left(2\dfrac{1}{4} \div 4\dfrac{1}{2}\right) =$ _____

2.8A PRACTICE SET

Simplify each expression. Reduce each answer to lowest terms.

1. $\dfrac{2}{3} + \dfrac{3}{5} \cdot \dfrac{5}{9} =$ _____

2. $\left(\dfrac{5}{8} - \dfrac{1}{2}\right) \div \dfrac{1}{4} =$ _____

3. $\dfrac{5}{16} \cdot \dfrac{4}{15} \div 3 =$ _____

4. $2\dfrac{1}{2} + \dfrac{3}{4} - 1\dfrac{5}{8} =$ _____

5. $3\dfrac{1}{5} \div \dfrac{16}{15} + 4\dfrac{5}{7} =$ _____

6. $\dfrac{2}{3} \cdot \left(\dfrac{5}{7} - \dfrac{1}{9}\right) =$ _____

7. $6\dfrac{1}{4} \div 10 - \dfrac{1}{3} =$ _____

8. $\dfrac{5}{12} \div \left(2\dfrac{1}{3} - 1\dfrac{1}{2}\right) =$ _____

9. $1\dfrac{9}{16} \cdot \left(\dfrac{6}{7} + \dfrac{2}{21}\right) =$ _____

10. $\left(\dfrac{5}{8} \cdot 1\dfrac{4}{5}\right) \cdot \left(\dfrac{5}{8} - \dfrac{1}{4}\right) =$ _____

2.8B PRACTICE SET

Simplify each expression. Reduce each answer to lowest terms.

1. $\dfrac{3}{4} - \dfrac{3}{16} \div \dfrac{1}{2} = $ _____

2. $\left(\dfrac{7}{25} + \dfrac{1}{2}\right) \div \dfrac{3}{5} = $ _____

3. $1\dfrac{5}{8} \cdot \dfrac{3}{13} + \dfrac{3}{4} = $ _____

4. $\left(\dfrac{3}{4} - \dfrac{2}{3}\right) \cdot 3\dfrac{1}{5} = $ _____

5. $\dfrac{9}{16} \div \left(2\dfrac{1}{2} + 3\dfrac{1}{4}\right) = $ _____

6. $\dfrac{3}{4} \cdot \dfrac{5}{2} + \dfrac{3}{10} = $ _____

7. $3\dfrac{1}{2} \div \dfrac{7}{8} - \dfrac{1}{2} = $ _____

8. $7 \div \left(5\dfrac{1}{2} - 3\dfrac{3}{4}\right) = $ _____

9. $\left(\dfrac{16}{3} \div \dfrac{4}{9}\right) \cdot \left(\dfrac{1}{2} \div \dfrac{5}{16}\right) = $ _____

10. $2\dfrac{1}{2} \cdot \left(\dfrac{4}{5} - \dfrac{3}{10}\right) \div \dfrac{1}{4} = $ _____

SUMMARY OF KEY CONCEPTS

Key Terms

fraction (2.1): A part of a whole amount.

common fraction (2.1): A whole number divided by a natural number.

numerator (2.1): The top number in a fraction.

denominator (2.1): The bottom number in a fraction.

fraction bar (2.1): The line between the numerator and the denominator.

proper fraction (2.1): A fraction in which the numerator is less than the denominator.

improper fraction (2.1): A fraction in which the numerator is equal to or greater than the denominator.

mixed number (2.1): A combination of a whole number and a fraction.

equivalent fractions (2.3): Fractions that have the same value.

greatest common factor (GCF) (2.3): The largest number that will divide both the numerator and the denominator of a fraction evenly.

common denominators (2.4): Denominators that are the same.

least common multiple (LCM) (2.4): The smallest natural number that is divisible by each of a group of numbers.

least common denominator (LCD) (2.4): The LCM of a group of denominators.

canceling factors (2.6): Reducing fractions by dividing top and bottom by a common factor.

invert (2.7): To turn upside down.

reciprocal (2.7): The inverse of a number.

KEY RULES

— **To convert an improper fraction to a mixed number (2.2):**

Perform the indicated division. Write the remainder as a fraction.

— **To convert a mixed number to an improper fraction (2.2):**

1. Multiply the denominator of the fraction by the whole number and add the numerator of the fraction to the resulting product.
2. Write the result from step 1 as the numerator of the improper fraction.
3. Write the original denominator of the mixed number as the denominator of the improper fraction.

— **To determine whether fractions are equivalent (2.3):**

1. Cross-multiply (multiply the numerator of one fraction by the denominator of the other fraction).
2. If the two products are the same, the fractions are equivalent.

To raise fractions to higher terms (2.3):

1. Divide the new denominator by the original denominator (or the new numerator by the original numerator).
2. Multiply the quotient obtained in step 1 by the original numerator to get the new numerator (or the original denominator to get the new denominator).

To change fractions to lower terms (2.3):

1. Divide the original denominator by the new denominator (or the original numerator by the new numerator).
2. Divide the original numerator by the quotient from step 1 to get the new numerator (or the original denominator by the quotient from step 1 to get the new denominator).

To find the greatest common factor (GCF) (2.3):

1. Write the prime factorization of each term.
2. The product of the common prime factors is the GCF.

To reduce to lowest terms by canceling prime factors (2.3):

1. Write the prime factorization of each term.
2. Cancel like factors.
3. Multiply the remaining factors to get the reduced fraction.

To add fractions with common denominators (2.4):

1. Add the numerators and place the sum over the common denominator.
2. Reduce the answer to lowest terms.

To find the least common multiple (LCM) (2.4):

1. Write the prime factorization of each number.
2. List each prime factor the most times it appears as a factor of any number.
3. The product of this list of prime factors is the LCM.

To add fractions with different denominators (2.4):

1. Find the least common denominator (LCD), which is the LCM of the denominators.
2. Convert each original fraction to an equivalent fraction with the LCD as the denominator.
3. Add the fractions and reduce the sum to lowest terms.

To add mixed numbers (2.4):

1. Add the whole numbers.
2. If the fractions do not have common denominators, convert them to equivalent fractions with a common denominator.
3. Add the fractions and reduce to lowest terms.
4. Combine the sum of the fractions with the sum of the whole numbers into a new mixed number.

Summary of Key Concepts

- **To subtract fractions with common denominators (2.5):**
    1. Subtract the numerators and place the difference over the common denominator.
    2. Reduce the answer to lowest terms.

- **To subtract fractions with different denominators (2.5):**
    1. Find the LCD (this is the LCM) of the given denominators.
    2. Convert each original fraction to an equivalent fraction with the LCD as the new denominator.
    3. Subtract the fractions and reduce the difference to lowest terms.

- **To subtract mixed numbers (2.5):**
    1. If the fractions do not have common denominators, find the LCD and convert each fraction to an equivalent fraction with the LCD as the denominator.
    2. Subtract the fractions.
    3. Subtract the whole numbers.
    4. Simplify and reduce the answer.

- **To multiply fractions (2.6):**
    1. Multiply the numerators to get the numerator of the product.
    2. Multiply the denominators to get the denominator of the product.
    3. Reduce the product to lowest terms.

- **To multiply mixed numbers (2.6):**
    1. Convert each mixed number to an improper fraction.
    2. Multiply the improper fractions as you do proper fractions.

- **To divide fractions (2.7):**
    1. Invert the divisor and change the operation sign to multiplication.
    2. Multiply the fractions.

- **To divide mixed numbers (2.7):**
    1. Convert each mixed number to an improper fraction.
    2. Invert the divisor and change the operation sign to multiplication.
    3. Multiply the improper fractions.

PRACTICE TEST 2A

1. In the fraction $\frac{9}{10}$, identify:

   a. the numerator b. the denominator

   a. _____ b. _____

2. Label the following values as either a proper fraction, an improper fraction, or a mixed number:

   a. $\frac{11}{15}$ b. $3\frac{4}{5}$ c. $\frac{12}{12}$

   a. _____ b. _____ c. _____

3. Express the shaded portion of the drawing with a proper fraction.

   _____

4. Find the missing term in each fraction.

   a. $\frac{4}{5} = \frac{?}{25}$ b. $\frac{40}{75} = \frac{8}{?}$

   a. _____ b. _____

5. Reduce $\frac{44}{99}$ to lowest terms. _____

6. What is the greatest common factor of 45 and 108? ___9___

In problems 7–17, perform the indicated operations. Simplify and reduce answers to lowest terms.

7. $\frac{5}{8} + \frac{3}{4} + \frac{7}{12} = $ _____

8. $4\frac{2}{3} + 5\frac{7}{18} = $ _____

9. $\frac{15}{16} - \frac{5}{48} = $ _____

10. $2\frac{7}{15} - 1\frac{1}{5} = $ _____

11. $\frac{5}{22} \cdot \frac{11}{15} \cdot \frac{3}{2} = $ _____

12. $3\frac{3}{5} \cdot 4\frac{1}{2} = $ _____

165

13. $\dfrac{19}{24} \div \dfrac{2}{3} = $ _____

14. $2\dfrac{5}{8} \div 1\dfrac{3}{4} = $ _____

15. $0 \div \dfrac{5}{9} = $ _____

16. $\dfrac{7}{12} \div 0 = $ _____

17. $\left(\dfrac{5}{16} + \dfrac{3}{8}\right) \div \dfrac{3}{4} = $ _____

PRACTICE TEST 2B

CHAPTER 2

1. In the fraction $\frac{15}{16}$ identify:

   a. the numerator   b. the denominator

   a. _____   b. _____

2. Label the following values as either a proper fraction, an improper fraction, or a mixed number.

   a. $12\frac{3}{16}$ b. $\frac{15}{32}$ c. $\frac{9}{5}$

   a. _____   b. _____   c. _____

3. Express the shaded portion of the drawing with a proper fraction.

   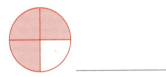

4. Find the missing term in each fraction.

   a. $\frac{20}{24} = \frac{5}{?}$ b. $\frac{3}{7} = \frac{?}{28}$

   a. _____   b. _____

5. Reduce $\frac{35}{63}$ to lowest terms. _____

6. What is the greatest common factor of 60 and 96? _____

In problems 7–17, perform the indicated operations. Simplify and reduce answers to lowest terms.

7. $\frac{3}{5} + \frac{5}{10} + \frac{9}{8} =$ _____

8. $10\frac{2}{7} + 8\frac{1}{9} + 5\frac{8}{21} =$ _____

9. $\dfrac{4}{7} - \dfrac{1}{3} =$ _____

10. $5\dfrac{1}{2} - 3\dfrac{7}{8} =$ _____

11. $\dfrac{5}{8} \cdot \dfrac{4}{25} \cdot \dfrac{18}{7} =$ _____

12. $6\dfrac{3}{4} \cdot 3\dfrac{1}{9} =$ _____

13. $\dfrac{16}{25} \div \dfrac{8}{5} =$ _____

14. $8\dfrac{2}{3} \div 3\dfrac{5}{9} =$ _____

15. $0 \div \dfrac{1}{2} =$ _____

16. $\dfrac{5}{7} \div 0 =$ _____

17. $\dfrac{5}{8} \cdot \left(\dfrac{4}{5} + \dfrac{3}{4}\right) =$ _____

SKILLS REVIEW

1. Write 23,706,483 in words.

   _____

2. Write four hundred seven thousand, six hundred twenty-nine in numerical form. _____

3. Round 37,649 to the nearest thousand. _____

4. Round 295,328 to the nearest ten thousand. _____

In problems 5–8, perform each indicated operation.

5. 42,376
    2,409
    +143,574
    _____

6. 2,459
    − 1,982
    _____

7. 473
    × 95
    _____

8. 26)7,852

9. Identify the prime numbers in the following list:

   9, 41, 67, 15, 17 _____

10. Write the prime factorization of 72. _____

11. In the fraction $\frac{3}{16}$, identify:

    a. the numerator   *b.* the denominator.

    a. _____   *b.* _____

12. Find the missing term in each fraction.

    a. $\frac{5}{8} = \frac{?}{56}$ *b.* $\frac{28}{42} = \frac{4}{?}$

    a. _____   *b.* _____

13. Reduce $\frac{45}{105}$ to lowest terms. _____

In problems 14–19, perform each indicated operation.

14. $\dfrac{3}{8} + \dfrac{5}{16} + \dfrac{4}{5} =$ _____

15. $\dfrac{9}{10} - \dfrac{1}{6} =$ _____

16. $\dfrac{4}{5} \cdot \dfrac{2}{15} \cdot \dfrac{25}{16} =$ _____

17. $3\dfrac{1}{2} \cdot 4\dfrac{2}{3} =$ _____

18. $\dfrac{10}{12} \div \dfrac{1}{2} =$ _____

19. $2\dfrac{1}{4} \div \dfrac{1}{8} =$ _____

20. Simplify: $\left(8 + \dfrac{1}{2} \cdot \dfrac{3}{4}\right) \div 1\dfrac{5}{32} =$ _____

CHAPTER 2   SOLUTIONS

Skills Preview

1. *a.* 5
   b. 11
2. *a.* improper fraction
   b. mixed number
   c. proper fraction
3. $\frac{1}{6}$
4. *a.* 45
   b. 9
5. $\frac{5}{7}$
6. 24
7. $2\frac{5}{18}$
8. $5\frac{37}{40}$
9. $\frac{11}{16}$
10. $1\frac{8}{9}$
11. $\frac{5}{16}$
12. 57
13. $\frac{2}{3}$
14. $1\frac{3}{7}$
15. 0
16. cannot be done
17. $\frac{1}{2}$

Section 2.1   Comprehension Checkpoint

1. *a.* 7
   b. 4
2. $\frac{1}{4}$
3. *a.* $\frac{17}{12}$ *b.* $1\frac{5}{12}$
4. $\frac{148}{377}$

2.1A Practice Set

1. 5, 9
2. 6, 13
3. 5, 8
4. 17, 25
5. 24, 7
6. proper
7. improper
8. mixed number
9. improper
10. mixed number
11. proper
12. improper
13. mixed number
14. proper
15. improper
16. $\frac{2}{3}$
17. $\frac{3}{7}$
18. $\frac{9}{16}$
19. $\frac{6}{9}$
20. $\frac{3}{8}$
21. *a.* $\frac{5}{2}$ *b.* $2\frac{1}{2}$
22. *a.* $\frac{17}{6}$ *b.* $2\frac{5}{6}$
23. *a.* $\frac{39}{10}$ *b.* $3\frac{9}{10}$
24. *a.* $\frac{73}{16}$ *b.* $4\frac{9}{16}$
25. *a.* $\frac{45}{8}$ *b.* $5\frac{5}{8}$
26. $\frac{242}{411}$
27. $\frac{48}{167}$
28. $\frac{273}{500}$

Section 2.2   Comprehension Checkpoint

1. $3\frac{2}{7}$
2. $\frac{51}{5}$
3. $\frac{16}{1}$
4. $\frac{15}{15}$

171

2.2A Practice Set

1. $\frac{12}{5} = 12 \div 5 = 2\frac{2}{5}$
2. $\frac{14}{3} = 14 \div 3 = 4\frac{2}{3}$
3. $\frac{37}{7} = 37 \div 7 = 5\frac{2}{7}$
4. $\frac{13}{2} = 13 \div 2 = 6\frac{1}{2}$
5. $\frac{29}{6} = 29 \div 6 = 4\frac{5}{6}$
6. $\frac{19}{4} = 19 \div 4 = 4\frac{3}{4}$
7. $\frac{43}{8} = 43 \div 8 = 5\frac{3}{8}$
8. $\frac{55}{12} = 55 \div 12 = 4\frac{7}{12}$
9. $\frac{64}{15} = 64 \div 15 = 4\frac{4}{15}$
10. $\frac{88}{9} = 88 \div 9 = 9\frac{7}{9}$
11. $2\frac{4}{5} = \frac{5 \times 2 + 4}{5} = \frac{14}{5}$
12. $6\frac{7}{8} = \frac{8 \times 6 + 7}{8} = \frac{55}{8}$
13. $3\frac{2}{3} = \frac{3 \times 3 + 2}{3} = \frac{11}{3}$
14. $8\frac{9}{10} = \frac{10 \times 8 + 9}{10} = \frac{89}{10}$
15. $5\frac{7}{12} = \frac{12 \times 5 + 7}{12} = \frac{67}{12}$
16. $12\frac{6}{7} = \frac{7 \times 12 + 6}{7} = \frac{90}{7}$
17. $15\frac{1}{8} = \frac{8 \times 15 + 1}{8} = \frac{121}{8}$
18. $9\frac{11}{15} = \frac{15 \times 9 + 11}{15} = \frac{146}{15}$
19. $14\frac{1}{7} = \frac{7 \times 14 + 1}{7} = \frac{99}{7}$
20. $10\frac{3}{4} = \frac{4 \times 10 + 3}{4} = \frac{43}{4}$
21. $19 = \frac{19}{1}$
22. $24 = \frac{24}{1}$
23. $42 = \frac{42}{1}$
24. $35 = \frac{35}{1}$
25. $16 = \frac{16}{1}$
26. $1 = \frac{7}{7}$
27. $1 = \frac{10}{10}$

Section 2.3 Comprehension Checkpoint

1. yes
2. 36
3. 5
4. $\frac{6}{13}$
5. 48

2.3A Practice Set

1. $16 \div 4 = 4$
   $3 \times 4 = 12$
2. $40 \div 8 = 5$
   $3 \times 5 = 15$
3. $66 \div 11 = 6$
   $9 \times 6 = 54$
4. $39 \div 13 = 3$
   $20 \times 3 = 60$
5. $25 \div 5 = 5$
   $6 \times 5 = 30$
6. $36 \div 9 = 4$
   $4 \times 4 = 16$
7. $42 \div 14 = 3$
   $25 \times 3 = 75$
8. $45 \div 5 = 9$
   $12 \times 9 = 108$
9. $40 \div 20 = 2$
   $16 \div 2 = 8$
10. $39 \div 13 = 3$
    $15 \div 3 = 5$
11. $50 \div 10 = 5$
    $35 \div 5 = 7$
12. $48 \div 3 = 16$
    $16 \div 16 = 1$
13. $42 \div 6 = 7$
    $28 \div 7 = 4$
14. $121 \div 11 = 11$
    $165 \div 11 = 15$
15. $95 \div 5 = 19$
    $57 \div 19 = 3$
16. $150 \div 50 = 3$
    $90 \div 3 = 30$

Solutions

17. $8 = 2 \times 2 \times 2$
    $36 = 2 \times 2 \times 3 \times 3$
    GCF $= 2 \times 2 = 4$

18. $24 = 2 \times 2 \times 2 \times 3$
    $84 = 2 \times 2 \times 3 \times 7$
    GCF $= 2 \times 2 \times 3 = 12$

19. $45 = 3 \times 3 \times 5$
    $72 = 2 \times 2 \times 2 \times 3 \times 3$
    GCF $= 3 \times 3 = 9$

20. $80 = 2 \times 2 \times 2 \times 2 \times 5$
    $120 = 2 \times 2 \times 2 \times 3 \times 5$
    GCF $= 2 \times 2 \times 2 \times 5 = 40$

21. $76 = 2 \times 2 \times 19$
    $114 = 2 \times 3 \times 19$
    GCF $= 2 \times 19 = 38$

22. $135 = 3 \times 3 \times 3 \times 5$
    $315 = 3 \times 3 \times 5 \times 7$
    GCF $= 3 \times 3 \times 5 = 45$

23. $\dfrac{4 \div 4}{12 \div 4} = \dfrac{1}{3}$

24. $\dfrac{5 \div 5}{20 \div 5} = \dfrac{1}{4}$

25. $\dfrac{15 \div 15}{75 \div 15} = \dfrac{1}{5}$

26. $\dfrac{12 \div 12}{96 \div 12} = \dfrac{1}{8}$

27. $\dfrac{16 \div 8}{40 \div 8} = \dfrac{2}{5}$

28. $\dfrac{72}{90} = \dfrac{2 \times 2 \times \cancel{2} \times \cancel{3} \times \cancel{3}}{\cancel{2} \times \cancel{3} \times \cancel{3} \times 5} = \dfrac{4}{5}$

29. $\dfrac{75}{105} = \dfrac{\cancel{3} \times \cancel{5} \times 5}{\cancel{3} \times \cancel{5} \times 7} = \dfrac{5}{7}$

30. $\dfrac{64}{96} = \dfrac{\cancel{2} \times \cancel{2} \times \cancel{2} \times \cancel{2} \times \cancel{2} \times 2}{\cancel{2} \times \cancel{2} \times \cancel{2} \times \cancel{2} \times \cancel{2} \times 3} = \dfrac{2}{3}$

31. $\dfrac{36}{63} = \dfrac{2 \times 2 \times \cancel{3} \times \cancel{3}}{\cancel{3} \times \cancel{3} \times 7} = \dfrac{4}{7}$

32. $\dfrac{121}{132} = \dfrac{11 \times \cancel{11}}{2 \times 2 \times 3 \times \cancel{11}} = \dfrac{11}{12}$

Section 2.4   Comprehension Checkpoint

1. LCM = 48
2. LCM = 72
3. $\dfrac{6}{7}$
4. $2\dfrac{7}{48}$
5. $8\dfrac{15}{28}$
6. $19\dfrac{49}{90}$

2.4A Practice Set

1. $8 = 2 \times 2 \times 2$
   $6 = 2 \times 3$
   LCM $= 2 \times 2 \times 2 \times 3 = 24$

2. $12 = 2 \times 2 \times 3$
   $18 = 2 \times 3 \times 3$
   LCM $= 2 \times 2 \times 3 \times 3 = 36$

3. $24 = 2 \times 2 \times 2 \times 3$
   $16 = 2 \times 2 \times 2 \times 2$
   LCM $= 2 \times 2 \times 2 \times 2 \times 3 = 48$

4. $10 = 2 \times 5$
   $22 = 2 \times 11$
   LCM $= 2 \times 5 \times 11 = 110$

5. $30 = 2 \times 3 \times 5$
   $45 = 3 \times 3 \times 5$
   LCM $= 2 \times 3 \times 3 \times 5 = 90$

6. $6 = 2 \times 3$
   $9 = 3 \times 3$
   $15 = 3 \times 5$
   LCM $= 2 \times 3 \times 3 \times 5 = 90$

7. $8 = 2 \times 2 \times 2$
   $12 = 2 \times 2 \times 3$
   $16 = 2 \times 2 \times 2 \times 2$
   LCM $= 2 \times 2 \times 2 \times 2 \times 3 = 48$

8. $10 = 2 \times 5$
   $15 = 3 \times 5$
   $20 = 2 \times 2 \times 5$
   LCM $= 2 \times 2 \times 3 \times 5 = 60$

9. $9 = 3 \times 3$
   $10 = 2 \times 5$
   $16 = 2 \times 2 \times 2 \times 2$
   LCM $= 2 \times 2 \times 2 \times 2 \times 3 \times 3 \times 5 = 720$

10. $24 = 2 \times 2 \times 2 \times 3$
    $36 = 2 \times 2 \times 3 \times 3$
    $45 = 3 \times 3 \times 5$
    LCM $= 2 \times 2 \times 2 \times 3 \times 3 \times 5 = 360$

11. $\dfrac{5}{8} + \dfrac{1}{8} = \dfrac{6}{8} = \dfrac{3}{4}$

12. $\dfrac{9}{16} + \dfrac{3}{16} = \dfrac{12}{16} = \dfrac{3}{4}$

13. $\dfrac{4}{9} + \dfrac{5}{9} + \dfrac{2}{9} = \dfrac{11}{9} = 1\dfrac{2}{9}$

14. $\dfrac{7}{30} + \dfrac{13}{30} + \dfrac{1}{30} = \dfrac{21}{30} = \dfrac{7}{10}$

15. $\dfrac{9}{32} + \dfrac{7}{32} + \dfrac{13}{32} + \dfrac{3}{32} = \dfrac{32}{32} = 1$

16. $\dfrac{3}{5} + \dfrac{5}{8} = \dfrac{24}{40} + \dfrac{25}{40} = \dfrac{49}{40} = 1\dfrac{9}{40}$

17. $\dfrac{5}{14} + \dfrac{8}{21} = \dfrac{15}{42} + \dfrac{16}{42} = \dfrac{31}{42}$

18. $\dfrac{3}{10} + \dfrac{7}{16} = \dfrac{24}{80} + \dfrac{35}{80} = \dfrac{59}{80}$

19. $\dfrac{5}{16} + \dfrac{11}{12} + \dfrac{1}{6} = \dfrac{15}{48} + \dfrac{44}{48} + \dfrac{8}{48} = \dfrac{67}{48} = 1\dfrac{19}{48}$

20. $\dfrac{11}{20} + \dfrac{3}{8} + \dfrac{16}{25} = \dfrac{110}{200} + \dfrac{75}{200} + \dfrac{128}{200} = \dfrac{313}{200} = 1\dfrac{113}{200}$

21. $\begin{aligned} &2\dfrac{4}{9} = 2\dfrac{28}{63} \\ &+ 5\dfrac{3}{7} = 5\dfrac{27}{63} \\ &\hline \\ &7\dfrac{55}{63} \end{aligned}$

22. $\begin{aligned} &8\dfrac{2}{5} = 8\dfrac{28}{70} \\ &+ 15\dfrac{9}{14} = 15\dfrac{45}{70} \\ &\hline \\ &23\dfrac{73}{70} = 24\dfrac{3}{70} \end{aligned}$

23. $\begin{aligned} &14\dfrac{7}{12} = 14\dfrac{35}{60} \\ &+ 26\dfrac{3}{20} = 26\dfrac{9}{60} \\ &\hline \\ &40\dfrac{44}{60} = 40\dfrac{11}{15} \end{aligned}$

24. $\begin{aligned} &52\dfrac{1}{2} = 52\dfrac{25}{50} \\ &116\dfrac{4}{5} = 116\dfrac{40}{50} \\ &+75\dfrac{17}{25} = 75\dfrac{34}{50} \\ &\hline \\ &243\dfrac{99}{50} = 244\dfrac{49}{50} \end{aligned}$

25. $\begin{aligned} &48\dfrac{6}{7} = 48\dfrac{60}{70} \\ &\dfrac{13}{14} = \dfrac{65}{70} \\ &+ 63\dfrac{4}{10} = 63\dfrac{28}{70} \\ &\hline \\ &111\dfrac{153}{70} = 113\dfrac{13}{70} \end{aligned}$

26. $\dfrac{1}{6} + \dfrac{1}{4} + \dfrac{1}{3} = \dfrac{2}{12} + \dfrac{3}{12} + \dfrac{4}{12} = \dfrac{9}{12} = \dfrac{3}{4}$

27. $\begin{aligned} 1\tfrac{1}{2} &= 1\tfrac{12}{24} \\ \tfrac{2}{3} &= \tfrac{16}{24} \\ +\, 2\tfrac{5}{8} &= 2\tfrac{15}{24} \\ \hline &\ 3\tfrac{43}{24} = 4\tfrac{19}{24} \end{aligned}$

28. $\begin{aligned} 5\tfrac{1}{4} &= 5\tfrac{6}{24} \\ 2\tfrac{1}{2} &= 2\tfrac{12}{24} \\ 3\tfrac{2}{3} &= 3\tfrac{16}{24} \\ +\, 4\tfrac{7}{8} &= 4\tfrac{21}{24} \\ \hline &\ 14\tfrac{55}{24} = 16\tfrac{7}{24} \end{aligned}$

29. $\tfrac{1}{8} + \tfrac{1}{2} + \tfrac{1}{4} = \tfrac{1}{8} + \tfrac{4}{8} + \tfrac{2}{8} = \tfrac{7}{8}$

30. $\begin{aligned} 5\tfrac{4}{5} &= 5\tfrac{24}{30} \\ 4\tfrac{9}{10} &= 4\tfrac{27}{30} \\ +\, 8\tfrac{2}{3} &= 8\tfrac{20}{30} \\ \hline &\ 17\tfrac{71}{30} = 19\tfrac{11}{30} \end{aligned}$

Section 2.5 Comprehension Checkpoint

1. $\tfrac{4}{7}$ 2. $\tfrac{5}{12}$ 3. $\tfrac{7}{45}$ 4. $2\tfrac{1}{5}$ 5. $8\tfrac{11}{20}$ 6. $\tfrac{11}{24}$ 7. $13\tfrac{5}{14}$

2.5A Practice Set

1. $\tfrac{7}{9} - \tfrac{3}{9} = \tfrac{4}{9}$

2. $\tfrac{9}{10} - \tfrac{1}{10} = \tfrac{8}{10} = \tfrac{4}{5}$

3. $\tfrac{13}{20} - \tfrac{11}{20} = \tfrac{2}{20} = \tfrac{1}{10}$

4. $\tfrac{7}{8} - \tfrac{1}{3} = \tfrac{21}{24} - \tfrac{8}{24} = \tfrac{13}{24}$

5. $\tfrac{5}{6} - \tfrac{1}{4} = \tfrac{10}{12} - \tfrac{3}{12} = \tfrac{7}{12}$

6. $\tfrac{9}{16} - \tfrac{3}{20} = \tfrac{45}{80} - \tfrac{12}{80} = \tfrac{33}{80}$

7. $\tfrac{7}{10} - \tfrac{3}{5} = \tfrac{7}{10} - \tfrac{6}{10} = \tfrac{1}{10}$

8. $\tfrac{12}{25} - \tfrac{5}{12} = \tfrac{144}{300} - \tfrac{125}{300} = \tfrac{19}{300}$

9. $\tfrac{8}{15} - \tfrac{3}{8} = \tfrac{64}{120} - \tfrac{45}{120} = \tfrac{19}{120}$

10. $\tfrac{17}{24} - \tfrac{5}{18} = \tfrac{51}{72} - \tfrac{20}{72} = \tfrac{31}{72}$

11. $\begin{aligned} &\ 3\tfrac{5}{8} \\ -&\ 1\tfrac{1}{8} \\ \hline &\ 2\tfrac{4}{8} = 2\tfrac{1}{2} \end{aligned}$

12. $\begin{aligned} &\ 8\tfrac{7}{10} \\ -&\ 5\tfrac{3}{10} \\ \hline &\ 3\tfrac{4}{10} = 3\tfrac{2}{5} \end{aligned}$

13. $7\frac{2}{3} = 7\frac{8}{12}$

$-4\frac{5}{12} = 4\frac{5}{12}$

$3\frac{3}{12} = 3\frac{1}{4}$

14. $15\frac{2}{3} = 15\frac{6}{9}$

$-8\frac{2}{9} = 8\frac{2}{9}$

$7\frac{4}{9}$

15. $21\frac{9}{16} = \overset{0}{2\!\!\!/\!1}\frac{27}{48} + \frac{48}{48} = 20\frac{75}{48}$

$-4\frac{7}{12} = 4\frac{28}{48}$

$16\frac{47}{48}$

16. $28 = 28\overset{7}{\!\!\!/\!4}\frac{4}{4}$

$-9\frac{3}{4} = 9\frac{3}{4}$

$18\frac{1}{4}$

17. $48\frac{6}{7} = 48\frac{30}{35}$

$-30\frac{2}{5} = 30\frac{14}{35}$

$18\frac{16}{35}$

18. $95\frac{5}{8} = 95\frac{55}{88}$

$-36\frac{9}{22} = 36\frac{36}{88}$

$59\frac{19}{88}$

19. $68\frac{4}{5}$

-50

$18\frac{4}{5}$

20. $36 = 3\!\!\!/\!6\overset{5}{\frac{9}{9}}$

$-24\frac{5}{9} = 24\frac{5}{9}$

$11\frac{4}{9}$

21. $40\frac{1}{8}$

$-24\frac{1}{8}$

16

22. $75\frac{2}{15} = 7\!\!\!/\!5\overset{4}{\frac{6}{45}} + \frac{45}{45} = 74\frac{51}{45}$

$-47\frac{7}{9} = 47\frac{35}{45}$

$27\frac{16}{45}$

23. $\frac{1}{2} - \frac{1}{3} = \frac{3}{6} - \frac{2}{6} = \frac{1}{6}$

24. $\$100 = 99\frac{10}{10}$

$-45\frac{1}{10} = 45\frac{1}{10}$

$\$54\frac{9}{10}$

25. $12\frac{3}{4} = 12\frac{12}{16}$

$-3\frac{11}{16} = 3\frac{11}{16}$

$9\frac{1}{16}$

26. $15\frac{1}{3} = 1\!\!\!/\!5\overset{4}{\frac{4}{12}} + \frac{12}{12} = 14\frac{16}{12}$

$-10\frac{5}{12} = 10\frac{5}{12}$

$4\frac{11}{12}$ miles

Section 2.6 Comprehension Checkpoint

1. $\dfrac{12}{35}$ 2. $\dfrac{2}{9}$ 3. $\dfrac{3}{10}$ 4. $3\dfrac{2}{3}$ 5. 27 6. $122\dfrac{1}{2}$

2.6A Practice Set

1. $\dfrac{3}{5} \times \dfrac{1}{2} = \dfrac{3}{10}$ 2. $\dfrac{5}{8} \times \dfrac{3}{4} = \dfrac{15}{32}$

3. $\dfrac{\cancel{7}^{1}}{\cancel{16}_{4}} \times \dfrac{\cancel{4}^{1}}{\cancel{21}_{3}} = \dfrac{1}{12}$ 4. $\dfrac{\cancel{8}^{1}}{\cancel{9}_{3}} \times \dfrac{\cancel{3}^{1}}{\cancel{16}_{2}} = \dfrac{1}{6}$

5. $\dfrac{\cancel{4}^{1}}{\cancel{5}_{1}} \times \dfrac{\cancel{15}^{3}}{\cancel{28}_{7}} = \dfrac{3}{7}$ 6. $\dfrac{\cancel{9}^{3}}{\cancel{20}_{10}} \times \dfrac{\cancel{2}^{1}}{\cancel{3}_{1}} = \dfrac{3}{10}$

7. $\dfrac{\cancel{12}^{3}}{\cancel{25}_{5}} \times \dfrac{\cancel{15}^{3}}{\cancel{32}_{8}} = \dfrac{9}{40}$ 8. $\dfrac{\cancel{7}^{1}}{\cancel{10}_{2}} \times \dfrac{\cancel{25}^{5}}{\cancel{42}_{6}} = \dfrac{5}{12}$

9. $\dfrac{\cancel{8}^{1}}{13} \times \dfrac{7}{\cancel{8}_{1}} = \dfrac{7}{13}$ 10. $\dfrac{\cancel{10}^{2}}{\cancel{21}_{3}} \times \dfrac{\cancel{14}^{2}}{\cancel{15}_{3}} = \dfrac{4}{9}$

11. $\dfrac{\cancel{8}^{1}}{\cancel{25}_{5}} \times \dfrac{\cancel{2}^{1}}{\cancel{3}_{1}} \times \dfrac{\cancel{15}^{\cancel{3}\,1}}{\cancel{16}_{\cancel{2}\,1}} = \dfrac{1}{5}$ 12. $\dfrac{\cancel{3}^{1}}{\cancel{10}_{2}} \times \dfrac{\cancel{15}^{\cancel{3}\,1}}{\cancel{24}_{\cancel{8}\,1}} \times \dfrac{\cancel{8}^{1}}{\cancel{9}_{3}} = \dfrac{1}{6}$

13. $3\dfrac{3}{4} \times 5\dfrac{1}{3} = \dfrac{\cancel{15}^{5}}{\cancel{4}_{1}} \times \dfrac{\cancel{16}^{4}}{\cancel{3}_{1}} = \dfrac{20}{1} = 20$ 14. $4\dfrac{3}{8} \times 1\dfrac{3}{5} = \dfrac{\cancel{35}^{7}}{\cancel{8}_{1}} \times \dfrac{\cancel{8}^{1}}{\cancel{5}_{1}} = \dfrac{7}{1} = 7$

15. $7\dfrac{1}{12} \times 9\dfrac{3}{5} = \dfrac{\cancel{85}^{17}}{\cancel{12}_{1}} \times \dfrac{\cancel{48}^{4}}{\cancel{5}_{1}} = \dfrac{68}{1} = 68$ 16. $8\dfrac{2}{9} \times 5\dfrac{7}{10} = \dfrac{\cancel{74}^{37}}{\cancel{9}_{3}} \times \dfrac{\cancel{57}^{19}}{\cancel{10}_{5}} = \dfrac{703}{15} = 46\dfrac{13}{15}$

17. $10\dfrac{2}{3} \times 1\dfrac{23}{32} = \dfrac{32}{3} \times \dfrac{55}{\cancel{32}_{1}} = \dfrac{55}{3} = 18\dfrac{1}{3}$ 18. $12\dfrac{1}{2} \times 2\dfrac{8}{15} = \dfrac{\cancel{25}^{5}}{\cancel{2}_{1}} \times \dfrac{\cancel{38}^{19}}{\cancel{15}_{3}} = \dfrac{95}{3} = 31\dfrac{2}{3}$

19. $2\dfrac{3}{4} \times 3\dfrac{7}{11} = \dfrac{\cancel{11}^{1}}{\cancel{4}_{1}} \times \dfrac{\cancel{40}^{10}}{\cancel{11}_{1}} = \dfrac{10}{1} = 10$ 20. $20\dfrac{4}{5} \times 27\dfrac{1}{2} = \dfrac{\cancel{104}^{52}}{\cancel{5}_{1}} \times \dfrac{\cancel{55}^{11}}{\cancel{2}_{1}} = \dfrac{572}{1} = 572$

21. $2\dfrac{3}{16} \times 4\dfrac{4}{7} = \dfrac{\cancel{35}^{5}}{\cancel{16}_{1}} \times \dfrac{\cancel{32}^{2}}{\cancel{7}_{1}} = \dfrac{10}{1} = 10$ 22. $8\dfrac{1}{6} \times 4\dfrac{2}{7} \times 0 = 0$

23. $\dfrac{\cancel{3}^1}{8} \times \dfrac{\cancel{2}^1}{\cancel{3}_1} = \dfrac{1}{4}$
    _4

24. $\dfrac{3}{5} \times 7\dfrac{1}{2} = \dfrac{3}{\cancel{5}_1} \times \dfrac{\cancel{15}^3}{2} = \dfrac{9}{2} = \$4\dfrac{1}{2}$

25. $6\dfrac{3}{10} \times 12 = \dfrac{63}{\cancel{10}_5} \times \dfrac{\cancel{12}^6}{1} = \dfrac{378}{5} = \$75\dfrac{3}{5}$

26. $\dfrac{3}{4} \times 24\dfrac{7}{12} = \dfrac{\cancel{3}^1}{4} \times \dfrac{295}{\cancel{12}_4} = \dfrac{295}{16} = 18\dfrac{7}{16}$ feet

27. $\dfrac{2}{3} \times \dfrac{5}{\cancel{6}_3}^{1} = \dfrac{5}{9}$

Section 2.7 Comprehension Checkpoint

1. $\dfrac{13}{8}$
2. $\dfrac{1}{12}$
3. 4
4. $3\dfrac{3}{4}$
5. 9
6. $\dfrac{2}{15}$
7. $1\dfrac{5}{7}$
8. $\dfrac{13}{16}$
9. cannot be done

2.7A Practice Set

1. $\dfrac{1}{2} \div \dfrac{1}{4} = \dfrac{1}{\cancel{2}_1} \times \dfrac{\cancel{4}^2}{1} = 2$

2. $\dfrac{3}{8} \div \dfrac{3}{2} = \dfrac{\cancel{3}^1}{\cancel{8}_4} \times \dfrac{\cancel{2}^1}{\cancel{3}_1} = \dfrac{1}{4}$

3. $\dfrac{4}{7} \div \dfrac{1}{2} = \dfrac{4}{7} \times \dfrac{2}{1} = \dfrac{8}{7} = 1\dfrac{1}{7}$

4. $\dfrac{11}{12} \div \dfrac{5}{12} = \dfrac{11}{\cancel{12}_1} \times \dfrac{\cancel{12}^1}{5} = \dfrac{11}{5} = 2\dfrac{1}{5}$

5. $\dfrac{21}{32} \div \dfrac{3}{8} = \dfrac{\cancel{21}^7}{\cancel{32}_4} \times \dfrac{\cancel{8}^1}{\cancel{3}_1} = \dfrac{7}{4} = 1\dfrac{3}{4}$

6. $\dfrac{7}{9} \div \dfrac{14}{15} = \dfrac{\cancel{7}^1}{\cancel{9}_3} \times \dfrac{\cancel{15}^5}{\cancel{14}_2} = \dfrac{5}{6}$

7. $\dfrac{17}{20} \div \dfrac{3}{32} = \dfrac{17}{\cancel{20}_5} \times \dfrac{\cancel{32}^8}{3} = \dfrac{136}{15} = 9\dfrac{1}{15}$

8. $\dfrac{5}{8} \div \dfrac{1}{6} = \dfrac{5}{\cancel{8}_4} \times \dfrac{\cancel{6}^3}{1} = \dfrac{15}{4} = 3\dfrac{3}{4}$

9. $\dfrac{10}{21} \div \dfrac{15}{24} = \dfrac{\cancel{10}^2}{\cancel{21}_7} \times \dfrac{\cancel{24}^8}{\cancel{15}_3} = \dfrac{16}{21}$

10. $\dfrac{21}{25} \div 0 =$ cannot be done (division by 0 cannot be done)

11. $2\dfrac{1}{2} \div \dfrac{3}{4} = \dfrac{5}{\cancel{2}_1} \times \dfrac{\cancel{4}^2}{3} = \dfrac{10}{3} = 3\dfrac{1}{3}$

12. $8\dfrac{5}{6} \div \dfrac{3}{10} = \dfrac{53}{\cancel{6}_3} \times \dfrac{\cancel{10}^5}{3} = \dfrac{265}{9} = 29\dfrac{4}{9}$

Solutions

13. $12\frac{2}{3} \div \frac{1}{3} = \frac{38}{\cancel{3}} \times \frac{\cancel{3}^1}{1} = \frac{38}{1} = 38$

14. $5 \div 1\frac{1}{4} = \frac{5}{1} \div \frac{5}{4} = \frac{\cancel{5}^1}{1} \times \frac{4}{\cancel{5}_1} = \frac{4}{1} = 4$

15. $3\frac{1}{5} \div 2\frac{2}{5} = \frac{16}{5} \div \frac{12}{5} = \frac{\cancel{16}^4}{\cancel{5}_1} \times \frac{\cancel{5}^1}{\cancel{12}_3} = \frac{4}{3} = 1\frac{1}{3}$

16. $9 \div \frac{2}{9} = \frac{9}{1} \times \frac{9}{2} = \frac{81}{2} = 40\frac{1}{2}$

17. $10\frac{1}{2} \div 2\frac{1}{8} = \frac{21}{2} \div \frac{17}{8} = \frac{21}{\cancel{2}_1} \times \frac{\cancel{8}^4}{17} = \frac{84}{17} = 4\frac{16}{17}$

18. $0 \div 16\frac{2}{3} = 0$

19. $6\frac{3}{5} \div 3\frac{9}{10} = \frac{33}{5} \div \frac{39}{10} = \frac{\cancel{33}^{11}}{\cancel{5}_1} \times \frac{\cancel{10}^2}{\cancel{39}_{13}} = \frac{22}{13} = 1\frac{9}{13}$

20. $5\frac{1}{4} \div 8\frac{5}{2} = \frac{21}{4} \div \frac{21}{2} = \frac{\cancel{21}^1}{\cancel{4}_2} \times \frac{\cancel{2}^1}{\cancel{21}_1} = \frac{1}{2}$

21. $12 \div \frac{1}{8} = \frac{12}{1} \times \frac{8}{1} = \frac{96}{1} = 96$

22. $36\frac{1}{4} \div 7\frac{1}{4} = \frac{145}{4} \div \frac{29}{4} = \frac{\cancel{145}^5}{\cancel{4}_1} \times \frac{\cancel{4}^1}{\cancel{29}_1} = \frac{5}{1} = 5$ cans

23. $20\frac{1}{2} \div \frac{3}{4} = \frac{41}{\cancel{2}_1} \times \frac{\cancel{4}^2}{3} = \frac{82}{3} = 27\frac{1}{3}$ He can cut only 27 of the fence posts.

24. $28\frac{1}{3} \div 1\frac{3}{4} = \frac{85}{3} \div \frac{7}{4} = \frac{85}{3} \times \frac{4}{7} = \frac{340}{21} = 16\frac{4}{21}$ The station will play 16 records.

Section 2.8   Comprehension Checkpoint

1. $\frac{13}{72}$ 2. 6 3. $\frac{13}{30}$

2.8A Practice Set

1. $\dfrac{2}{3} + \dfrac{\overset{1}{\cancel{3}}}{\cancel{5}} \times \dfrac{\overset{1}{\cancel{5}}}{\cancel{9}}$
   $\underbrace{}_{\text{multiply}}$
   $\dfrac{2}{3} + \dfrac{1}{3}$
   $\underbrace{}_{\text{add}}$
   1

2. $\left(\dfrac{5}{8} - \dfrac{1}{2}\right) \div \dfrac{1}{4}$
   $\underbrace{}_{\text{subtract}}$
   $\dfrac{5}{8} - \dfrac{4}{8}$
   $\underbrace{}$
   $\dfrac{1}{8} \div \dfrac{1}{4}$
   $\underbrace{}_{\text{divide}}$
   $\dfrac{1}{8} \times \dfrac{\overset{1}{\cancel{4}}}{1}$
   $\underbrace{}_{2}$
   $\dfrac{1}{2}$

3. $\dfrac{\overset{1}{\cancel{5}}}{\cancel{16}} \times \dfrac{\overset{1}{\cancel{4}}}{\cancel{15}} \div 3$
   $\underbrace{}_{\text{multiply}}$
   $\dfrac{1}{12} \div \dfrac{3}{1}$
   $\underbrace{}_{\text{divide}}$
   $\dfrac{1}{12} \times \dfrac{1}{3}$
   $\underbrace{}$
   $\dfrac{1}{36}$

4. $2\dfrac{1}{2} + \dfrac{3}{4} - 1\dfrac{5}{8}$
   $\underbrace{}_{\text{add}}$
   $2\dfrac{2}{4} + \dfrac{3}{4}$
   $\underbrace{}$
   $3\dfrac{1}{4} - 1\dfrac{5}{8}$
   $\underbrace{}_{\text{subtract}}$
   $3\dfrac{2}{8} - 1\dfrac{5}{8}$
   $\underbrace{}$
   $1\dfrac{5}{8}$

5. $3\dfrac{1}{5} \div \dfrac{16}{15} + 4\dfrac{5}{7}$
   $\underbrace{}_{\text{divide}}$
   $\dfrac{\overset{1}{\cancel{16}}}{5} \times \dfrac{\overset{3}{\cancel{15}}}{\cancel{16}}$
   $\underbrace{}_{1}$
   $3 + 4\dfrac{5}{7}$
   $\underbrace{}_{\text{add}}$
   $7\dfrac{5}{7}$

6. $\dfrac{2}{3} \times \left(\dfrac{5}{7} - \dfrac{1}{9}\right)$
   $\underbrace{}_{\text{subtract}}$
   $\dfrac{45}{63} - \dfrac{7}{63}$
   $\underbrace{}$
   $\dfrac{2}{3} \times \dfrac{38}{63}$
   $\underbrace{}_{\text{multiply}}$
   $\dfrac{76}{189}$

Solutions

7. $6\frac{1}{4} \div 10 - \frac{1}{3}$

    └─divide─┘

    $\frac{\overset{5}{\cancel{25}}}{4} \times \frac{1}{\cancel{10}}$
    ${}_{2}$

    └───┘

    $\frac{5}{8} - \frac{1}{3}$

    └─subtract─┘

    $\frac{15}{24} - \frac{8}{24}$

    └────┘

    $\frac{7}{24}$

8. $\frac{5}{12} \div \left(2\frac{1}{3} - 1\frac{1}{2}\right)$

    └─subtract─┘

    $2\frac{2}{6} - 1\frac{3}{6}$

    └────┘

    $\frac{5}{12} \div \frac{5}{6}$

    └─divide─┘

    $\frac{\overset{1}{\cancel{5}}}{\underset{2}{\cancel{12}}} \times \frac{\overset{1}{\cancel{6}}}{\underset{1}{\cancel{5}}}$

    └────┘

    $\frac{1}{2}$

9. $1\frac{9}{16} \times \left(\frac{6}{7} + \frac{2}{21}\right)$

    └─add─┘

    $\frac{18}{21} + \frac{2}{21}$

    └────┘

    $1\frac{9}{16} \times \frac{20}{21}$

    └─multiply─┘

    $\frac{25}{\cancel{16}} \times \frac{\overset{5}{\cancel{20}}}{21}$
    ${}_{4}$

    └────┘

    $\frac{125}{84} = 1\frac{41}{84}$

10. $\left(\frac{5}{8} \times 1\frac{4}{5}\right) \times \left(\frac{5}{8} - \frac{1}{4}\right)$

    multiply └────┘ └──┘ subtract

    $\frac{\overset{1}{\cancel{5}}}{8} \times \frac{9}{\underset{1}{\cancel{5}}}$ $\quad$ $\frac{5}{8} - \frac{2}{8}$

    └────┘ └────┘

    $\frac{9}{8} \quad \times \quad \frac{3}{8}$

    └──────────┘ multiply

    $\frac{27}{64}$

Practice Test 2A

1. a. 9
    b. 10

2. a. proper fraction
    b. mixed number
    c. improper fraction

3. $\frac{9}{16}$

4. a. 20
    b. 15

5. $\frac{4}{9}$

6. 9

7. $1\frac{23}{24}$

8. $10\frac{1}{18}$

9. $\frac{5}{6}$

10. $1\frac{4}{15}$

11. $\frac{1}{4}$

12. $16\frac{1}{5}$

13. $1\frac{3}{16}$

14. $1\frac{1}{2}$

15. 0

16. Cannot be done

17. $\frac{11}{12}$

Skills Review Chapters 1–2

1. twenty-three million, seven hundred six thousand, four hundred eighty-three
2. 407,629
3. 38,000
4. 300,000
5. 188,359
6. 477
7. 44,935
8. 302
9. 41, 67, 17
10. $72 = 2 \times 2 \times 2 \times 3 \times 3$
11. *a.* 3
    b. 16
12. *a.* 35
    b. 6
13. $\frac{3}{7}$
14. $1\frac{39}{80}$
15. $\frac{11}{15}$
16. $\frac{1}{6}$
17. $16\frac{1}{3}$
18. $1\frac{2}{3}$
19. 18
20. $7\frac{9}{37}$

3 DECIMALS

OUTLINE

3.1 Decimal Numbers and Place Value

3.2 Rounding Decimals

3.3 Adding and Subtracting Decimal Numbers

3.4 Multiplying Decimal Numbers

3.5 Dividing Decimal Numbers

3.6 Scientific Notation

3.7 Order of Operations

In Chapter 2 you learned how to express a fractional part of a whole unit with common fractions and mixed numbers. You can also represent part of a whole using decimal fractions. In this chapter, you will learn about decimal numbers and how to work with them.

SKILLS PREVIEW

1. Give the place name for the underlined digit in each number.

   a. 12.4_5_7 _____

   b. 0.893_2_5 _____

   c. 6._1_ _____

2. Write 247.645 in words. _____

3. Write sixteen and twenty-four hundredths in digits. _____

4. Round each decimal number to the indicated place.

   a. 0.8546 to the nearest hundredth _____

   b. 12.351 to the nearest tenth _____

   c. $48.63 to the nearest whole dollar _____

In problems 5–15, perform the indicated operation.

5. 47.62
   304.956
   +   1.7
   _____

6. $0.432 + 81.95 + 241.7 =$ _____

7.   12.095
   $-$ 6.18
   _____

8. $9.6 - 5.423 =$ _____

9.   24.5
   $\times$ 9
   _____

10.   1.37
    $\times$ 0.02
    _____

11. $35.63 \times 100 =$ _____

12. $6\overline{)8.4}$

13. $21.84 \div 3.5 =$ _____

14. $64.9 \div 100 =$ _____

15. Express each number in scientific notation.

    a. 48,000 _____

    b. 0.00083 _____

16. Write each number without exponents.

    a. 8.95×10^7 _____

    b. 5.19×10^{-6} _____

17. Simplify $8.4 \div 2.1 + 3.6 \times 10 =$ _____

3.1 DECIMAL NUMBERS AND PLACE VALUE

Decimal Numbers

Decimal numbers are numbers that include a **decimal point.** A decimal point (.) separates the whole number from the fractional part of a number. A **decimal fraction** is a fraction with a denominator of 10, 100, 1,000, and so forth. Examples of decimal fractions are 0.2, 0.45, and 0.6732. Notice that when writing a decimal fraction, you should place a 0 in the units place.

Remember from Chapter 2 that a mixed number is a combination of a whole number and a common fraction. A **mixed decimal** is a combination of a whole number and a decimal fraction. Examples of mixed decimals are 4.8, 15.98, and 354.058.

When you studied whole numbers in Chapter 1, there was no mention of a decimal point. Every whole number has a decimal point to the right of the units digit. Because it is a whole number, it is not necessary to include the decimal point.

Decimal digits are the digits to the right of the decimal point. A number may have any number of decimal digits. Also, adding zeros to the right of the last decimal digit does not change the value of the number. For example, 0.5 = 0.500, and 0.75 = 0.75000.

It is important to note that there can be only one decimal point in a number. Also notice that commas are not used in writing decimal digits.

Place Value

Each place in a decimal number has a name. Figure 3.1 shows the place names for the first nine decimal places.

FIGURE 3.1

UNITS	DECIMAL POINT	TENTHS	HUNDREDTHS	THOUSANDTHS	TEN THOUSANDTHS	HUNDRED THOUSANDTHS	MILLIONTHS	TEN MILLIONTHS	HUNDRED MILLIONTHS	BILLIONTHS
	.									

Notice that each decimal place name ends with the letters *-ths*. Remember that each whole-number place name ended with an *-s*. More decimal digits could be named, but you should be able to see a pattern. The next column to the right would be ten-billionths, followed by hundred-billionths, and so on.

To determine the place value of a decimal digit, you can place the digits of the number in the boxes at the bottom of the chart. For instance, name the place value of each digit in the number 6.8325. Place the number in the boxes. (You should get to the point where you can visualize the place values and do not depend on the boxes.)

UNITS	DECIMAL POINT	TENTHS	HUNDREDTHS	THOUSANDTHS	TEN THOUSANDTHS	HUNDRED THOUSANDTHS	MILLIONTHS	TEN MILLIONTHS	HUNDRED MILLIONTHS	BILLIONTHS
6	.	8	3	2	5					

The 6 is in the units place.

The 8 is in the tenths place $\left(\frac{8}{10}\right)$.

The 3 is in the hundredths place $\left(\frac{3}{100}\right)$.

The 2 is in the thousandths place $\left(\frac{2}{1,000}\right)$.

The 5 is in the ten-thousandths place $\left(\frac{5}{10,000}\right)$.

You can see from this example that the denominator of a decimal fraction is given by the place name. The tenths column indicates a denominator of 10, the hundredths column indicates a denominator of 100, the thousandths column indicates a denominator of 1,000, and so on. Beginning at the decimal point, as you move to the right, each place has one more 0 in the denominator than the place on its left.

EXAMPLE 1 Name the place value of each underlined digit.

a. 28.43<u>3</u>7   *b.* 1.<u>9</u>0624   *c.* 0.04976<u>2</u>5

SOLUTIONS

a. hundredths (it is two places to the right of the decimal point)
b. tenths (it is the place to the right of the decimal point)
c. millionths (it is six places to the right of the decimal point)

Reading and Writing Decimal Numbers

Suppose you need to read the number

47.63

Begin with the whole number, *forty-seven.* Read the decimal point as *and* and then read the decimal digits. Read these digits as you would a whole number, *sixty-three,* and attach the place name of the right digit—*hundredths,* because it is two places to the right of the decimal point. Thus, 47.63 is read "forty-seven and sixty-three hundredths."

This example also demonstrates how to write a decimal number in words. If you want to write 347.089 in words, write the whole number first, *three hundred forty-seven,* followed by the word *and* for the decimal point. Finally, write the decimal digits, *eighty-nine,* and the place name of the right digit, *thousandths.* Thus, 347.089 written in words is *three hundred forty-seven and eighty-nine thousandths.* (Notice that the word *and* is used only once in reading or writing numbers of any kind.)

Decimal Numbers and Place Value

EXAMPLE 2 Write each decimal number in words.

a. 5.72   *b.* 0.9305   *c.* 29.264925

SOLUTIONS

a. five and seventy-two hundredths
b. nine thousand three hundred five ten-thousandths
c. twenty-nine and two hundred sixty-four thousand nine hundred twenty-five millionths

You should also be able to convert the written form of a number to digits.

EXAMPLE 3 Write each number in digits.

a. fifty-five hundredths
b. thirty-seven and four tenths
c. six and four thousand seven hundred twelve hundred-thousandths

SOLUTIONS

a. 0.55
b. 37.4
c. 6.04712 (the 2 must be in the hundred-thousandths place, so a 0 must be in the tenths place to hold the position)

Determine your understanding of decimal numbers with the Comprehension Checkpoint problems. If you have difficulty with any of these problems, review the material in this section before going on to the Practice Sets.

COMPREHENSION CHECKPOINT

1. Name the place value of each underlined digit.

   a. 56.987<u>4</u>3 _____

   b. 0.2382<u>5</u>6 _____

2. Write each number in words.

   a. 0.872

   _____

   b. 657.91

   _____

3. Write each number in digits.

   a. five and seventy-three hundredths _____

   b. two hundred fourteen ten-thousandths _____

3.1A PRACTICE SET

In exercises 1–8, write the place name of each underlined digit.

1. 0.998<u>6</u>1 _____

2. 12.8<u>7</u>3 _____

3. 93.8256<u>3</u>1 _____

4. 56.<u>0</u>754 _____

5. 783.36487<u>9</u> _____

6. 24.38<u>2</u>79211 _____

7. 0.093136<u>5</u>623 _____

8. 0.952445<u>4</u>02 _____

In exercises 9–16, write each decimal number in words.

9. 0.6 _____

10. 0.003 _____

11. 4.82 _____

12. 12.9532 _____

13. 95.083 _____

14. 967.38221 _____
_____

15. 6.00045 _____

16. 653.583982 _____
_____

In exercises 17–24, write each number in digits.

17. seventeen hundredths _____

18. three thousand seventy-four ten-thousandths _____

19. four and twenty-three hundredths _____

20. fifteen and ninety-five thousandths _____

21. fifty-eight and two hundred nine thousandths _____

22. thirty-six and four thousand seven hundred twenty-five millionths
_____

23. nine hundred twenty-six and five hundred forty-three ten-thousandths
_____

24. seventy-one and six thousand four hundred-thousandths _____

3.1B PRACTICE SET

In exercises 1–8, write the place name of each underlined digit.

1. 0.03<u>4</u> _____

2. 26.<u>9</u>2811 _____

3. 576.091<u>8</u>31 _____

4. 47.628443<u>1</u> _____

5. 10.98328282<u>5</u> _____

6. 914.3<u>7</u>1935 _____

7. 9.8329<u>2</u>904 _____

8. 18.75436<u>3</u>85 _____

In exercises 9–16, write each decimal number in words.

9. 0.68 _____

10. 6.383 _____

11. 8.032 _____

12. 65.6721 _____

13. 38.09834 _____
_____

14. 825.930214 _____
_____

15. 18.4586203 _____
_____

16. 6,842.837212 _____
_____

191

In exercises 17–24, write each number in digits.

17. eight tenths _____

18. seven hundred nineteen thousandths _____

19. nine and forty-one ten-thousandths _____

20. thirty-three and seven tenths _____

21. twenty-four and three thousandths _____

22. four hundred fifty and nineteen thousand two hundred forty-five millionths _____

23. seventeen thousand eighty-four and twelve hundredths _____

24. three thousand, nine hundred twenty-five and seven thousandths _____

3.2 ROUNDING DECIMALS

Rounding decimal numbers is similar to rounding whole numbers.

> **TO ROUND DECIMAL NUMBERS**
> 1. Determine which digit is in the rounding place and which digit is the determining digit.
> 2. The digit in the rounding place is
>   a. Unchanged if the determining digit is 4 or less.
>   b. Increased by 1 if the determining digit is 5 or more.
> 3. Drop all digits to the right of the rounding place.

EXAMPLE 1 Round 3.76 to the nearest tenth.

SOLUTION

3.76

The digit in the rounding place is 7.   *The determining digit is 6. Add 1 to the 7 since the determining digit is more than 5.*

3.76 rounded to the nearest tenth is 3.8.

EXAMPLE 2 Round 1.4836 to the nearest hundredth.

SOLUTION

1.4836

The digit in the rounding place is 8.   *The digit in the determining place is 3, which is less than 5. The rounding place is unchanged, and all digits to the right are dropped.*

1.4836 rounded to the nearest hundredth is 1.48.

One of the most common uses of decimals is in our money system. The decimal point separates the whole dollars from the partial dollars (or cents). Learning to work with dollars and cents is an essential skill in everyday math applications.

EXAMPLE 3 Round $82.49 to the nearest whole dollar.

SOLUTION

$82.49

The digit in the rounding place is 2.   *The determining digit is 4. The rounding place remains unchanged, and the digits to the right of the decimal point are dropped.*

$82.49 rounded to the nearest whole dollar is $82. (Note that the period after the 2 is a punctuation mark ending a sentence and not a decimal point.)

Try the Comprehension Checkpoint problems to assure yourself that you understand how to round decimal numbers. If you have difficulty with these problems, review the material in this section before going on to the Practice Sets.

COMPREHENSION CHECKPOINT

Round each given number to the indicated place.

1. 37.64 to the nearest tenth _____

2. 4.079 to the nearest hundredth _____

3. 0.82765 to the nearest ten-thousandth _____

4. $47.638 to the nearest cent _____

5. $95.25 to the nearest whole dollar _____

3.2A PRACTICE SET

In problems 1–8, round each number to the indicated places.

		Nearest Tenth	Nearest Hundredth	Nearest Thousandth
1.	0.4832	_____	_____	_____
2.	0.0579	_____	_____	_____
3.	0.5308	_____	_____	_____
4.	3.1293	_____	_____	_____
5.	4.9645	_____	_____	_____
6.	2.6791	_____	_____	_____
7.	17.3528	_____	_____	_____
8.	34.7037	_____	_____	_____

9. Round 0.745698231 to the nearest:

   a. ten-thousandth _____

   b. hundred-thousandth _____

   c. millionth _____

10. Round 0.4265874398 to the nearest:

    a. ten-thousandth _____

    b. hundred-thousandth _____

    c. millionth _____

In problems 11–15, round each amount as indicated.

		Nearest Cent	Nearest Ten Cents	Nearest Dollar
11.	$8.495	_____	_____	_____
12.	$12.632	_____	_____	_____
13.	$35.509	_____	_____	_____

	Nearest Cent	Nearest Ten Cents	Nearest Dollar
14. $13.243	_____	_____	_____
15. $50.156	_____	_____	_____

Applications

16. The wholesale price of a loaf of bread is $0.634. If a vendor rounds the price to the nearest whole cent, how much will the vendor charge for one loaf of bread?

_____

17. Sherry's water bill showed consumption of 8,900.45 gallons. The consumption amount is rounded and Sherry is billed only for whole gallons. How many gallons would she be billed for this month?

_____

3.2B PRACTICE SET

In problems 1–8, round each number to the indicated places.

	Nearest Tenth	Nearest Hundredth	Nearest Thousandth
1. 0.6351	_____	_____	_____
2. 0.4595	_____	_____	_____
3. 0.0498	_____	_____	_____
4. 5.9569	_____	_____	_____
5. 9.5872	_____	_____	_____
6. 10.2054	_____	_____	_____
7. 12.3275	_____	_____	_____
8. 16.7491	_____	_____	_____

9. Round 0.8462097315 to the nearest:
   a. ten-thousandth _____
   b. hundred-thousandth _____
   c. millionth _____

10. Round 0.9452168073 to the nearest:
    a. ten-thousandth _____
    b. hundred-thousandth _____
    c. millionth _____

In problems 11–15, round each amount as indicated.

	Nearest Cent	Nearest Ten Cents	Nearest Dollar
11. $18.463	_____	_____	_____
12. $25.957	_____	_____	_____
13. $40.625	_____	_____	_____

CHAPTER 3 Decimals

	Nearest Cent	Nearest Ten Cents	Nearest Dollar
14. $10.095	_____	_____	_____
15. $4.3333	_____	_____	_____

Applications

16. The price of a new coat is $75.60. What is the price rounded to the nearest whole dollar?

    _____

17. The gas tank on Bill's new car holds 18.48 gallons of gas. What is the gasoline capacity rounded to the nearest tenth of a gallon?

    _____

3.3 ADDING AND SUBTRACTING DECIMAL NUMBERS

Adding Decimal Numbers

The procedure for adding decimal numbers is the same as the procedure for adding whole numbers. Tenths can be added only to tenths, hundredths to hundredths, and so on. Like whole numbers, decimal numbers must be aligned properly to make sure they will be added properly. When aligning the digits, you must treat the decimal points like a digit and align them also.

EXAMPLE 1   Add: 12.47 + 3.94

SOLUTION

```
  12.47
+  3.94
```
The digits are aligned with the hundredths over the hundredths, tenths over tenths, decimal point over decimal point, and so on.

```
  12.47
+  3.94
  -----
  16.41
```
Add each column just as you did with whole numbers.

The sum is 16.41.

Not all numbers will have the same number of decimal places.

EXAMPLE 2   Add: 5.76 + 243.9 + 18.072

SOLUTION

```
    5.76
  243.9
+  18.072
```
Notice the alignment. The decimal points are aligned. This helps bring the digits into alignment.

```
    5.76
  243.9
+  18.072
  -------
  267.732
```
Add each column as you would with whole numbers.

The sum is 267.732.

Any numbers can be added: whole numbers, mixed decimals, and decimal fractions.

EXAMPLE 3   Add: 17.643 + 29 + 0.72

SOLUTION

```
  17.643
  29
+  0.72
  ------
  47.363
```
Align the digits into columns and add as you would with whole numbers.

The sum is 47.363.

Subtracting Decimal Numbers

The basic rules that you learned for subtracting whole numbers also apply to decimal numbers. As with addition, the numbers must be aligned properly.

CHAPTER 3   Decimals

EXAMPLE 4   Subtract: 6.85 − 1.73

SOLUTION

$$\begin{array}{r} 6.85 \\ -\ 1.73 \end{array}$$ *Align the numbers, including the decimal points, correctly.*

$$\begin{array}{r} 6.85 \\ -\ 1.73 \\ \hline 5.12 \end{array}$$ *Subtract as you did with whole numbers.*

The difference is 5.12.

If a digit in the subtrahend (the bottom number) is greater than the digit in the minuend (the top number) that it is to be subtracted from, you must borrow just as you learned to do with whole numbers. The next three examples demonstrate how this procedure applies to decimal numbers.

EXAMPLE 5   Subtract: 8.42 − 5.17

SOLUTION

$$\begin{array}{r} 8.42 \\ -\ 5.17 \end{array}$$ *Align the digits properly.*

$$\begin{array}{r} \ \ 3\ 12 \\ 8.\cancel{4}\cancel{2} \\ -\ 5.17 \\ \hline 5 \end{array}$$ *Subtract as you would with whole numbers. The 7 hundredths cannot be subtracted from the 2 hundredths, so borrow 1 tenth from the 4. The 1 borrowed from the 4 is 10 hundredths. Add the 10 hundredths to 2 hundredths and subtract 7 hundredths from the resulting sum.*

$$\begin{array}{r} \ \ 3\ 12 \\ 8.\cancel{4}\cancel{2} \\ -\ 5.17 \\ \hline 3.25 \end{array}$$ *Subtract the rest of the digits as you would in other subtraction problems. No more borrowing is needed.*

The difference is 3.25.

EXAMPLE 6   Subtract: 12.04 − 6.58

SOLUTION

$$\begin{array}{r} 12.04 \\ -\ 6.58 \end{array}$$ *Align the digits properly.*

$$\begin{array}{r} \ \ \ \ 9 \\ 1\ \cancel{1}0\ 14 \\ 1\cancel{2}.\cancel{0}\cancel{4} \\ -\ 6.58 \\ \hline 6 \end{array}$$ *Because 8 cannot be subtracted from 4, you must borrow. You will have to borrow from the 2. The 1 unit becomes 10 tenths. Now you can borrow 1 tenth and convert it to hundredths. This leaves 9 tenths in the tenths column. Add 10 hundredths to the 4 hundredths and subtract 8 from the sum.*

Adding and Subtracting Decimal Numbers

$$\begin{array}{r} \overset{9}{\cancel{1}}\overset{}{\cancel{10}}\,14 \\ 12.04 \\ -\;\;6.58 \\ \hline 5.46 \end{array}$$

Complete the subtraction process as you would in other problems.

The difference is 5.46.

EXAMPLE 7 Subtract: 5.4 − 3.78

SOLUTION

$$\begin{array}{r} 5.4 \\ -\,3.78 \end{array}$$ Align the digits properly.

$$\begin{array}{r} \overset{3}{\cancel{5.4}}\,^{10} \\ -\,3.78 \\ \hline 2 \end{array}$$ There is not a digit for the 8 hundredths to be subtracted from. Borrow 1 tenth and convert it to 10 hundredths. Subtract the 8 hundredths from 10 hundredths.

$$\begin{array}{r} \overset{13}{} \\ \overset{4}{\cancel{\,}}\overset{\cancel{3}}{} \\ \cancel{5.4}\,^{10} \\ -\,3.78 \\ \hline 1.62 \end{array}$$ Continue the subtraction process. 7 tenths cannot be subtracted from the 3 tenths, so borrow 1 unit from the 5 and convert it to 10 tenths. Add the 10 tenths to 3 tenths and subtract 7 tenths from the sum.

The difference is 1.62.

Notice that each example in this section has a decimal point in its answer. The decimal point is part of the answer just as the digits are part of the answer. You should also note that each decimal digit is 10 times as large as the digit to its right. When you borrow 1 tenth and convert it to hundredths, you have 10 hundredths.

The Comprehension Checkpoint problems give you an opportunity to make sure you understand adding and subtracting decimal numbers. If you have difficulty with these problems, review the material in this section before going on to the Practice Sets.

COMPREHENSION CHECKPOINT

Perform the indicated operation in each problem.

1. 2.63 + 5.94 = _____

2. 74.2 + 181.09 + 5.462 + 0.23 = _____

3. 18.48 − 12.23 = _____

4. 0.95 − 0.28 = _____

5. 8.6 − 3.49 = _____

3.3A PRACTICE SET

Find the sum in each of the following problems.

1. $8.95
      3.56
    + 2.38

2.   12.94
    462.731
    + 28.5

3.   76.379
         0.7821
    + 253.53684

4. 18.54 + 0.37 = _____

5. 171.09 + 2.986 + 12.25 = _____

6. $95.29 + $46.87 + $1.72 + $163.68 = _____

7. 0.46 + 0.709 + 0.1812 + 0.985723 = _____

8. $67.00 + $91.50 + $6.93 = _____

Find the difference in each of the following problems.

9. $9.65
    − 3.41

10. 18.47
    − 9.384

11. $354.05
    − 65.85

12. $7.00 − $3.67 = _____

13. 81 − 65.49 = _____

14. 54.9 − 26.09 = _____

15. 389.92485 − 248.5563 = _____

16. 0.41823 − 0.008471 = _____

17. Subtract: 65.27 from 89.83 = _____

18. Subtract: 42.65 from 42.7328 = _____

Applications

19. Heather bought a pound of apples for $0.89, two pounds of bananas for $0.67, and a head of lettuce for $1.25. What was the total cost of her groceries?

   _____

20. If Heather (see problem 19) gave the clerk $5, how much change did she receive?

   _____

21. Roberto wants to put a fence around his yard. Measuring the sides of the yard, he found one side to be 42.8 feet long, a second side to be 64.25 feet long, and a third side to be 50 feet long. How much fencing did Roberto need to go around those three sides of his yard?

   _____

22. Janice received a credit card bill for $75.23. She sent a $37.25 payment. How much did she owe after her payment?

   _____

3.3B PRACTICE SET

Find the sum in each of the following problems.

1. $47.73
       8.93
    + 25.26
    ———

2.   809.64
       45.951
    + 28
    ———

3. 208.73
     10.7483
    + 3.552
    ———

4. 57.26 + 32.98 + 28.8 = _____

5. 11.64 + 0.721 + 3.4213 + 91.832 = _____

6. $61.90 + $98.35 + $89.04 + $7.62 = _____

7. 0.944 + 0.2992 + 0.18462 + 0.5 = _____

8. $86.05 + $191.59 + $38.81 = _____

Find the difference in each of the following problems.

9. $10.76
    −   9.89
    ———

10. 38.3
    − 28.295
    ———

11. $183.00
    − 74.51
    ———

12. $10.00 − $5.38 = _____

13. 90 − 10.56 = _____

14. 472.1 − 267.5 = _____

15. 873.092 − 463.7314 = _____

16. 0.85346 − 0.6513712 = _____

17. Subtract: 4.937 from 6.013 = _____

18. Subtract: 27.852 from 94.82 = _____

Applications

19. Sarah bought a new outfit for a dance. The total cost was $275.28. She paid $75.50 in cash and charged the remainder on her credit card. How much did she charge on her credit card?

   _____

20. Melanie left her house to do some errands. She drove 0.65 mile to her first stop. From there she drove 3.5 miles to her second stop. She then found that the shortest distance home was 4.1 miles. What was the total distance she drove?

   _____

21. Michael needs $84.69 to purchase a new pair of baseball shoes. He has saved $68.78. How much more money does he need to buy the shoes?

   _____

22. Wendy's rent is $525 per month. Last month she had these additional bills: telephone bill, $25.75; utility bill, $84.19; and cable television, $33.70. What was the total amount of her bills last month, including the rent?

   _____

3.4 MULTIPLYING DECIMAL NUMBERS

Multiplying decimal numbers is also very similar to multiplying whole numbers. The only difference is in the handling of the decimal point.

> **TO MULTIPLY DECIMAL NUMBERS**
> 1. Multiply the factors without regard for the decimal point(s).
> 2. Count the total number of decimal digits in the factors.
> 3. Place the decimal point in the product so that you will have the same number of decimal digits in the product that you counted in step 2.

The process of positioning the decimal point in the product is called **pointing off places.**

EXAMPLE 1 Multiply: 3.8 × 42

SOLUTION

$$\begin{array}{r} 3.8 \\ \times\ 4\,2 \\ \hline 7\,6 \\ 152\ \ \\ \hline 159\,6 \end{array}$$

Set the problem up just as you do with whole numbers. Multiply without regard for the decimal point.

$$\begin{array}{r} 3.8 \\ \times\ 4\,2 \\ \hline 7\,6 \\ 152\ \ \\ \hline 159.6 \end{array}$$

There is one decimal digit in the two factors, so you need one decimal digit in the product. Point off one place from the right. The decimal point goes to the left of the 6.

The product is 159.6.

The same procedure is used when the factors have more than one decimal digit.

EXAMPLE 2 Multiply: 7.6 × 8.02

SOLUTION

$$\begin{array}{r} 8.0\,2 \\ \times\ \ \ 7.6 \\ \hline 48\,1\,2 \\ 561\,4\ \ \ \\ \hline 609\,5\,2 \end{array}$$

Set up the problem as you would with whole numbers. Multiply without regard for the decimal point.

$$\begin{array}{r} 8.0\,2 \\ \times\ \ \ 7.6 \\ \hline 48\,1\,2 \\ 56\,1\,4\ \ \ \\ \hline 60.9\,5\,2 \end{array}$$

There are three decimal digits in the two factors, so point off three places in the product. The decimal point goes to the left of the 9.

The product is 60.952.

There may be times when there are not enough places in the product to point off the decimal point properly. When this happens, zeros must be added to the left of the product as placeholders.

EXAMPLE 3 Multiply: 0.005×0.43

SOLUTION

$$\begin{array}{r} 0.0\,05 \\ \times\ 0.43 \\ \hline 15 \\ 20 \\ \hline 2\,15 \end{array}$$

Set the problem up as you do with whole numbers. Multiply without regard for the decimal points. There are five decimal digits in the factors but only three places in the product. Add two zeros to the left of the product. This allows you to point off the correct number of places. Place the decimal point to the left of the zeros.

$$\begin{array}{r} 0.0\,05 \\ \times\ \ .43 \\ \hline 15 \\ 20 \\ \hline .002\,15 \end{array}$$

The product is 0.00215.

If either of the factors is a 1 followed by zeros (such as 10, 100, 1,000, etc.), the product may be obtained by moving the decimal point of the other factor to the right the same number of places as there are zeros in the factor.

EXAMPLE 4 Find the product in each problem:

a. 10×18.23 *b.* 6.75×100 *c.* $0.96 \times 10,000$

SOLUTIONS

a. $10 \times 18.23 = 182.3$. Move the decimal point one place since there is one zero in 10.
b. $6.75 \times 100 = 675$. Move the decimal point two places to the right since there are two zeros in 100.
c. $0.96 \times 10,000 = 9,600$. The decimal point should be moved four places. You must add two zeros to the right of the 6 in order to move the decimal point the correct number of places.

The factors made up of 1 followed by zeros in Example 4—10, 100, and 10,000—are called **powers** of 10. You learned in Section 1.5 that an exponential expression such as 3^4 is read "three to the fourth power." A power is the value of an exponential expression. You will learn more about working with powers of 10 in Section 3.6.

Use the Comprehension Checkpoint problems to measure your understanding of multiplying decimal numbers. If you have difficulty with these problems, review the material in this section before going on to the Practice Sets.

COMPREHENSION CHECKPOINT

Find the product in each problem.

1. 8.7 × 16 = _____

2. 4.13 × 17.32 = _____

3. 0.0034 × 12.83 = _____

4. 10 × 3.742 = _____

5. 1,000 × 28.64 = _____

3.4A PRACTICE SET

Find the product in each problem.

1. 5.9 × 6
2. 8.2 × 4
3. 9.7 × 3

4. 6.2 × 2.4
5. 7.5 × 4.7
6. 5.3 × 1.8

7. 0.57 × 0.4
8. 2.85 × 0.9
9. 6.26 × 3.4

10. 26.3 × 2.7
11. 53.95 × 4.25
12. 0.009 × 3.87

13. 3.56 × 100 = _____

14. 0.376 × 1,000 = _____

15. 845.27 × 10 = _____

16. 4.284 × 10,000 = _____

17. 0.072 × 10 = _____

Applications

18. Joel intends to buy four new shirts. The cost of each shirt is $25.95. What will be the total cost of the four shirts?

    _____

19. Juan bought a new car. His payments will be $195.27 each month for 36 months. What will be the total of the payments?

    _____

20. Polly needs eight pieces of ribbon, each 12.875 inches long, to make Christmas bows. What is the total length of ribbon Polly needs?

    _____

21. Chicken sells for $0.55 per pound. How much would you pay for a chicken that weighs 3.4 pounds?

_____

22. Concert tickets cost $25.75 each. How much will 1,000 tickets cost?

_____

3.4B PRACTICE SET

Find the product in each problem.

1. 2.6
   × 5

2. 9.3
   × 4

3. 5.4
   × 8

4. 8.4
   × 3.1

5. 2.5
   × 5.9

6. 4.7
   × 6.9

7. 1.96
   × 0.3

8. 5.28
   × 0.9

9. 7.55
   × 8.5

10. 89.3
    × 6.8

11. 10.62
    × 8.06

12. 0.016
    × 0.06

13. 76.28 × 10 = _____

14. 0.19 × 1,000 = _____

15. 9.2784 × 100 = _____

16. 0.004673 × 100,000 = _____

17. 290.985 × 10,000 = _____

Applications

18. A laser printer can print 6.5 pages per minute. The printer took 4.5 minutes to print Maria's term paper. How long was the term paper?

    _____

19. How much money would you save if you saved $24.75 every month for 12 months?

    _____

20. Courtney rode her bike 2.25 hours at an average speed of 7.45 miles per hour. How far did she travel?

    _____

21. A single computer diskette can be bought for $0.63. How much will 1,000 diskettes cost?

   _____

22. A gallon of premium unleaded gasoline costs $1.199. Stan's car holds 12.45 gallons of gas. How much will it cost to fill his car with the premium unleaded gasoline?

   _____

3.5 DIVIDING DECIMAL NUMBERS

The process of dividing decimal numbers is essentially the same as the process of dividing whole numbers. The only difference is positioning the decimal point in the quotient. There are two types of problems: the first involves a whole-number divisor, and the second involves a decimal-number divisor.

To refresh your memory, review the terminology of a division problem:

$$\text{divisor} \overline{)\text{dividend}}^{\text{quotient}}$$

Dividing by a Whole-Number Divisor

When the divisor is a whole number, set the problem up without regard for the decimal point. For instance, $86.4 \div 12$ would be

$$\text{divisor} \rightarrow 12\overline{)86.4} \leftarrow \text{dividend}$$

The decimal point in the quotient is placed directly above the decimal point in the dividend.

$$12\overline{)86.4}^{\,\cdot\,}$$

Next, complete the division. The position of the decimal point does not affect the division process.

$$\begin{array}{r} 7.2 \\ 12\overline{)86.4} \\ \underline{84} \\ 2\,4 \\ \underline{2\,4} \end{array}$$

The quotient is 7.2. Check by multiplication.

$$\begin{array}{r} 7.2 \\ \times\,1\,2 \\ \hline 14\,4 \\ 72 \\ \hline 86.4 \end{array}$$

EXAMPLE 1 Divide: $23.4 \div 18$

SOLUTION

$18\overline{)23.4}$ *Set up the problem.*

$18\overline{)23.4}^{\,\cdot\,}$ *Position the decimal point in the quotient.*

$\begin{array}{r} 1.3 \\ 18\overline{)23.4} \\ \underline{18} \\ 5\,4 \\ \underline{5\,4} \end{array}$ *Divide without concern for the decimal point. It is already in position in the quotient.*

The quotient is 1.3.

If the dividend happens to be a whole number as well, the procedure is the same. Place the decimal point in the quotient and add zeros to the right of the decimal point to continue the division process.

EXAMPLE 2 Divide: 30 ÷ 12

SOLUTION

$12\overline{)30}$ *Set up the problem.*

$12\overline{)30.00}$ *Position the decimal point in the quotient. Since 12 does not divide 30 evenly, add zeros to the dividend.*

$\begin{array}{r} 2.5 \\ 12\overline{)30.0} \\ \underline{24} \\ 6\,0 \\ \underline{6\,0} \end{array}$ *Divide without concern for the decimal point. Note that only one zero needed to be added in this problem. You can add as many as you need in a given problem.*

The quotient is 2.5.

Sometimes the quotient will need to be rounded to a particular place. You must carry the division out one place to the right of the place to be rounded to.

EXAMPLE 3 Divide 28.3 by 8. Round the answer to the nearest tenth.

SOLUTION

$8\overline{)28.3}$ *Set up the problem and place the decimal point in the quotient.*

$\begin{array}{r} 3.53 \\ 8\overline{)28.30} \\ \underline{24} \\ 4\,3 \\ \underline{4\,0} \\ 30 \\ \underline{24} \end{array}$ *Divide as usual. To round to the nearest tenth, you must know the digit in the hundredths place. You must divide until you have a digit in the hundredths place. Then you can round.*

It does not matter that the division did not come out even. It has been carried as far as necessary for this problem.

The quotient, 3.53, rounded to the nearest tenth is 3.5.

Dividing by a Decimal-Number Divisor

To understand dividing by a decimal number, you must understand the relationship between the divisor and the dividend. Consider the division problem

$$\textit{dividend} \rightarrow \underset{\underset{\textit{divisor}}{\uparrow}}{10 \div 5} = 2 \leftarrow \textit{quotient}$$

If you multiply both the dividend and the divisor by 10, you get

$$\textit{dividend} \rightarrow \underset{\underset{\textit{divisor}}{\uparrow}}{100 \div 50} = 2 \leftarrow \textit{quotient}$$

Notice that the relationship between the dividend and the divisor did not change; dividing one by the other yields the same quotient as before. Similarly, if you multiply both the dividend and the divisor (of the original problem, 10 ÷ 5 = 2) by 100, you get

Dividing Decimal Numbers

$$\text{dividend} \rightarrow \ 1{,}000 \div \underset{\underset{\textit{divisor}}{\uparrow}}{500} = 2 \ \leftarrow \textit{quotient}$$

Once again, the relationship between the dividend and the divisor remains the same.

If the divisor in a problem is a decimal number, you must change it to a whole number before dividing. Changing the divisor to a whole number requires moving the decimal point to the right end of the number. If you want to change 12.5 to a whole number, move the decimal point to the end of the number. This gives you 125. What you did was multiply 12.5 by 10. (Multiplying by a power of 10 was the last topic in Section 3.4.)

If you change the divisor, you must also change the dividend to maintain the same relationship and obtain the same quotient. If you move the decimal point one place in the divisor, you must move the decimal point one place in the dividend (multiplying each by 10). If you move the decimal point two places in the divisor, you must move the decimal point two places in the dividend (multiplying each by 100). However many places you move the decimal point in the divisor, you must move the decimal point in the dividend the same number of places. This is done to maintain the original relationship between the divisor and the dividend.

After you move the decimal points and have a whole-number divisor, follow the procedure for dividing by a whole number. Study the examples in this section carefully to make sure you understand dividing by a decimal number.

EXAMPLE 4 Divide 3.45 by 2.5. Round your answer to the nearest tenth.

SOLUTION

2.5)3.45   *Set up the problem.*

25.)34.5   *Move the decimal point one place in both the divisor and the dividend. Follow the procedures for dividing by a whole number.*

```
        1.38
  25.)34.50     Carry the division to the hundredths place since you are going to
      25        round to the nearest tenth.
      ──
       9 5
       7 5
       ──
       2 00
       2 00
```

The quotient, rounded to the nearest tenth, is 1.4.

EXAMPLE 5 Divide 89.5 by 5.38. Round your answer to the nearest tenth.

SOLUTION

5.38)89.5   *Set up the problem.*

538.)8950.00   *Move the decimal point two places in both the divisor and the dividend. Position the decimal point in the quotient and divide.*

```
        16.63
538)8950.00
     538
     ───
     3570
     3228
     ────
      342 0
      322 8
      ─────
       19 20
       16 14
```
Carry the division out to the hundredths place in order to round to the tenths place.

The quotient, rounded to the nearest tenth, is 16.6.

EXAMPLE 6   Divide 0.1284 by 0.063. Round your answer to the nearest hundredth.

SOLUTION

0.063)0.1284   *Set up the problem.*

063.)128.400   *Move the decimal point three places in both the divisor and the dividend. Place the decimal point in the quotient and divide.*

```
       2.038
63)128.400
   126
   ───
     2 40
     1 89
     ────
       510
       504
```
The division must be carried to the thousandths place to be able to round to the hundredths place. Add zeros to the end of the dividend.

The quotient, rounded to the nearest hundredth, is 2.04.

Dividing by a power of 10 (10, 100, 1,000, etc.) is the reverse of multiplying by a power of 10. To divide by a power of 10, move the decimal point to the left the same number of places as there are zeros in the power of 10.

EXAMPLE 7   Find the quotient in each of the following.

a. $16.8 \div 10 =$
b. $0.75 \div 100 =$
c. $928 \div 1{,}000 =$

SOLUTIONS

a. $16.8 \div 10 = 1.68$. Move the decimal point one place to the left since there is one zero in the divisor.
b. $0.75 \div 100 = 0.0075$. Move the decimal point two places to the left since there are two zeros in the divisor.
c. $928 \div 1{,}000 = 0.928$. Move the decimal point three places to the left since there are three zeros in the divisor.

Dividing Decimal Numbers

> **a + b = x**
> ## ALGEBRA CONNECTION
>
> In Chapter 2, the Algebra Connection suggested that mastery of common fractions is essential to success in algebra; the same can be said for decimal numbers. As you will learn in Chapter 6, common fractions and decimal fractions are sometimes interchangeable. Often it is more convenient to use a common fraction; at other times a decimal fraction may be preferable. If you have an equation like
>
> $$\frac{1}{2}x = 12$$
>
> it may be more convenient to use 0.5 in place of $\frac{1}{2}$, especially if you are using a calculator.
>
> You have now studied operations with whole numbers, common fractions, and decimal numbers. Every math problem that you encounter, whether in this text or in real-life situations, will involve one or more of these types of numbers. To be successful, not only in algebra but also in daily applications, you must know how to apply the four basic math operations (adding, subtracting, multiplying, and dividing) to whole numbers, common fractions, and decimal numbers. Learning these operations and applications now will avoid the necessity of reviewing them when they come up in algebra. You will have more time for learning new concepts and principles in algebra.

Work the Comprehension Checkpoint problems to test your understanding of dividing by decimal numbers. If you have difficulty with these problems, review the material in this section before going on to the Practice Sets.

COMPREHENSION CHECKPOINT

Find the quotient in each problem. In problems 1–4, round your answer to the nearest tenth.

1. 15 ÷ 8 = _____

2. 12.6 ÷ 3.5 = _____

3. 18 ÷ 4.7 = _____

4. 0.26 ÷ 0.14 = _____

5. 35.7 ÷ 100 = _____

6. 746.83 ÷ 10 = _____

3.5A PRACTICE SET

Find the quotient in each problem.

1. 5)4.5
2. 8)9.84
3. 15)2.25

4. 2.5)5.6
5. 3.8)17.1
6. 0.14)5.04

7. 36.89 ÷ 10 = _____

8. 0.46 ÷ 100 = _____

9. 356.28 ÷ 100 = _____

Find the quotient in each of the following problems. Round the answer to the nearest tenth.

10. 18.5 ÷ 8.6 = _____

11. 25.9 ÷ 1.25 = _____

Find the quotient in each of the following problems. Round the answer to the nearest hundredth.

12. 84 ÷ 12.8 = _____

13. 16.85 ÷ 3.6 = _____

Find the quotient in each of the following problems. Round the answer to the nearest whole cent.

14. $29.85 ÷ 4 = _____

15. $43 ÷ 7.25 = _____

Applications

16. How many half-dollars ($0.50) are in $25? _____

17. A ream of paper contains 500 sheets. How thick is each piece of paper if the ream is 2.6 inches thick?

_____

18. What is the cost of developing one picture if a roll of 24 pictures costs $8.88?

_____

19. Kelli drove her car 365.7 miles on a tank of gas. If she averaged 26.5 miles per gallon, how many gallons of gasoline does her car hold?

_____

20. After saving for 10 weeks, Patty had $153.50 in the bank. How much was she saving each week?

_____

3.5B PRACTICE SET

Find the quotient in each problem.

1. 8)9.6
2. 6)7.26
3. 14)1.96

4. 3.7)7.03
5. 4.2)22.68
6. 0.25)6.25

7. 64.92 ÷ 100 = _____

8. 0.28 ÷ 10 = _____

9. 25.986 ÷ 1,000 = _____

Find the quotient in each of the following problems. Round the answer to the nearest tenth.

10. 32.9 ÷ 0.57 = _____

11. 16.3 ÷ 1.85 = _____

Find the quotient in each of the following problems. Round the answer to the nearest hundredth.

12. 96 ÷ 25.5 = _____

13. 5.34 ÷ 9.55 = _____

Find the quotient in each of the following problems. Round the answer to the nearest whole cent.

14. $76.25 ÷ 15.25 = _____

15. $35 ÷ 8.9 = _____

Applications

16. Sam bought three lab books for his biology class. The total cost was $41.85. What was the cost of each book?

_____

17. If Carl drives at an average speed of 58.25 miles per hour, how many hours will it take him to go 426.875 miles? (Round to the nearest tenth.)

18. How many 3.5-foot-long boards could be cut from one board 24 feet long?

19. Marie, Tracey, and Nancy rent an apartment for the summer. The rent is $475 per month. What is each person's share per month?

20. Six people shared the expenses on a trip equally. Total expenses were $275.34. What was each person's share?

3.6 SCIENTIFIC NOTATION

Writing a Number in Scientific Notation

Recall from Section 3.4 that a power of 10 is a 1 followed by zeros. Examples of powers of 10 include:

$$10 = 10^1$$
$$100 = 10^2$$
$$1{,}000 = 10^3$$

Writing numbers in scientific notation means showing a number as the product of two factors: one is a power of 10, and the other is a decimal number. Examples of numbers written in scientific notation are:

$$2.9 \times 10^4$$
$$3.85 \times 10^{-3}$$

When you use scientific notation, the factor that is a decimal number has only one digit to the left of the decimal point, as shown in these two examples. Also note that one of the exponents in the examples is negative. Study the rules for writing a number in scientific notation.

TO WRITE A NUMBER IN SCIENTIFIC NOTATION

1. Move the decimal point in the original number until you have a single digit to the left of the decimal point.
2. Count the number of places you moved the decimal point. This number is the exponent on the base 10. The exponent is positive if the original number is more than 10 (or if the decimal point moved to the left). The exponent is negative* if the original number is less than 1 (or if the decimal point moved to the right).
3. The original number written in scientific notation is the factor obtained in step 1 multiplied by the exponential expression obtained in step 2.

---

*There will be more on negative numbers in Chapter 7. For now, a negative number is a number with a negative sign in front of it, such as -5 or -9.

Study the following examples carefully to see how the rules are applied.

EXAMPLE 1 Write each number in scientific notation:

a. 195,000   *b.* 0.00029

SOLUTIONS

a. 195,000   *Move the decimal point until there is one digit to the left of the decimal point. The decimal point moved five places to the left.*
   1.95

   10^5 *The exponent is 5, since you moved the decimal point five places.*

   1.95×10^5 is 195,000 written in scientific notation.

b. 0.00029   *Move the decimal point until there is one digit to the left of the decimal point. The decimal point moved four places to the right.*
   2.9

   10^{-4} *The exponent is -4, since you moved the decimal point to the right.*

   2.9×10^{-4} is 0.00029 written in scientific notation.

EXAMPLE 2 The sun is approximately 93,000,000 miles from the earth. Express this distance in scientific notation.

SOLUTION

93,000,000   *Move the decimal point until there is one digit to the left of the*
9.3   *decimal point. The decimal point moved seven places to the left.*

10^7 *The exponent is 7, since the decimal point moved to the left.*

9.3×10^7 is 93,000,000 written in scientific notation.

Writing a Number without Exponents

A number written in scientific notation can be written without exponents by reversing the process just described.

> **TO WRITE A NUMBER WITHOUT EXPONENTS**
> 1. Determine whether the exponent is positive or negative.
> 2. Move the decimal point the same number of places as the exponent number
>   a. to the left if the exponent is negative.
>   b. to the right if the exponent is positive.

Study the following examples.

EXAMPLE 3 Write each number without exponents:

a. 4.67×10^6 *b.* 5.2×10^{-4}

SOLUTIONS

a. 10^6 *The exponent is positive 6.*

     4,670,000   *Move the decimal point six places to the right.*

     $4.67 \times 10^6 = 4,670,000$

b. 10^{-4} *The exponent is negative 4.*

     0.00052   *Move the decimal point four places to the left.*

     $5.2 \times 10^{-4} = 0.00052$

EXAMPLE 4 The clearance for an engine piston is 1.5×10^{-3} inch. Express this number without exponents.

SOLUTION

10^{-3} *The exponent is negative 3.*

0.0015   *Move the decimal point three places to the left.*

The clearance for an engine piston is 0.0015 inch.

Work the Comprehension Checkpoint problems to measure your understanding of scientific notation. If you have difficulty with these problems, review the material in this section before going on to the Practice Sets.

Scientific Notation

COMPREHENSION CHECKPOINT

Write each number in scientific notation.

1. 37,000 _____

2. 0.000046 _____

Write each number without exponents.

3. 5.68×10^7 _____

4. 7.921×10^{-2} _____

3.6A PRACTICE SET

Write each number in scientific notation.

1. 5,100 _____
2. 63,000 _____
3. 248,000 _____
4. 47,300,000 _____
5. 95,000,000,000 _____
6. 0.021 _____
7. 0.000725 _____
8. 0.008412 _____
9. 0.000008 _____
10. 0.00004731 _____

Write each number without exponents.

11. 7.3×10^4 _____
12. 9.28×10^2 _____
13. 1.4×10^3 _____
14. 3.872×10^6 _____
15. 6.01×10^5 _____
16. 3.3×10^{-5} _____
17. 9.83×10^{-3} _____
18. 5.0065×10^{-7} _____
19. 8.154×10^{-4} _____
20. 2.18×10^{-9} _____

Applications

21. Light travels at a rate of approximately 186,000 miles per second. Express this number in scientific notation.

    _____

22. Some computers make calculations in 0.000000005 of a second. Express this time in scientific notation.

    _____

23. The thickness of a cell is approximately 4.3×10^{-6} inch. Express this measurement without exponents.

    _____

24. The distance around the earth is about 2.5×10^4 miles. Express this distance without exponents.

    _____

3.6B PRACTICE SET

Write each number in scientific notation.

1. 780 _____
2. 3,000 _____
3. 741,000 _____
4. 93,700,000 _____
5. 136,000,000,000 _____
6. 0.0075 _____
7. 0.092 _____
8. 0.0000067 _____
9. 0.000026 _____
10. 0.00000000013 _____

Write each number without exponents.

11. 2.5×10^3 _____
12. 8.29×10^5 _____
13. 4.74×10^2 _____
14. 7.191×10^8 _____
15. 6.73×10^9 _____
16. 1.93×10^{-6} _____
17. 5.38×10^{-4} _____
18. 3.0305×10^{-9} _____
19. 9.4×10^{-7} _____
20. 8.389×10^{-2} _____

Applications

21. A standard piece of paper is about 6.2×10^{-3} inch thick. Express the thickness without exponents.

    _____

22. The greatest known depth of the Pacific Ocean is 35,820 feet. Express this depth in scientific notation.

    _____

23. A nanosecond is 0.000000001 of a second. Express this unit of time in scientific notation.

    _____

24. A light-year is the distance light travels in one year. If light travels 5.88×10^{12} miles in one year, how many miles does light travel in one year?

    _____

3.7 ORDER OF OPERATIONS

Problems involving decimal numbers that require more than one operation are solved using the same procedures as problems involving whole numbers or fractions. Take a moment to review those procedures.

> **RULES FOR THE ORDER OF OPERATIONS**
> 1. Do all multiplication and division in order first, working from left to right.
> 2. After completing all multiplication and division, do addition and subtraction in order, from left to right.
> 3. If grouping symbols such as ( ) or [ ] are included, operations within these symbols should be performed first, using the same steps. After simplifying inside the grouping symbols, begin again using the first two steps for the operations that remain.

EXAMPLE 1 Simplify: $3.7 + 2 \times 5.9$

SOLUTION

$3.7 + 2 \times 5.9$ *multiply*

$3.7 + 11.8$ *add*

15.5

EXAMPLE 2 Simplify: $0.48 \div 0.12 - 3.8$

SOLUTION

$0.48 \div 0.12 - 3.8$ *divide*

$4 \quad - 3.8$ *subtract*

0.2

EXAMPLE 3 Simplify: $(4.75 + 2.56) \div 0.8$. Round your answer to the nearest tenth.

SOLUTION

$(4.75 + 2.56) \div 0.8$ *add*

$7.31 \quad \div 0.8$ *divide*

9.1375 *round*

9.1

Work the Comprehension Checkpoint problems to review the steps for solving multiple-operation problems. If you have difficulty with these problems, review the material in this section before going on to the Practice Sets.

COMPREHENSION CHECKPOINT

Simplify each expression.

1. $12 \div 2.4 + 0.95$ _____

2. $9.6 \times 3 - 4.7$ _____

3. $(18.2 - 3.7) \div 2.9$ _____

4. $1.2 \times 3 \div 0.4 - 6.2$ _____

3.7A PRACTICE SET

Simplify each expression.

1. 8.7 ÷ 3 − 1.8 _____

2. (4.6 + 0.4) ÷ 0.05 _____

3. (3.7 − 2.5) × 4.5 _____

4. 5.9 × 2.1 + 2.7 × 6 _____

5. 8.4 ÷ 6 − 0.02 × 3 _____

6. $100 − 6 × $7.50 _____

7. $50 + 3.4 × $2.75 − $49.95 _____

8. 2.85 × 16 ÷ (3.8 × 4) _____

Applications

9. Bill bought three cans of peaches at $1.29 per can and two cans of green beans at $0.79 per can. He gave the cashier a $10 bill. How much change did he receive?

   _____

10. Sara made a down payment of $1,500 and had 48 monthly payments of $169.64 for a new car. What was the total price of the car?

    _____

3.7B PRACTICE SET

Simplify each expression.

1. 2.4 + 3.7 × 8.1 _____

2. (6.2 + 3.8) ÷ 2.5 _____

3. (10 − 1.3) ÷ 2.9 _____

4. 9.5 × 5 ÷ 2.5 _____

5. 10.8 × 4.2 ÷ 1.8 _____

6. 12.6 ÷ 3 − 3.5 _____

7. (7.9 − 3.5 + 4.6) ÷ 2.25 _____

8. 93 ÷ 100 ÷ 0.31 + 0.7 _____

Applications

9. What would be the total cost of 100 flathead screws at $0.16 each, 100 nuts at $0.11 each, and 10 pounds of nails at $0.85 per pound?

   _____

10. Linda went shopping with $150. At the first store, she bought a purse for $37.50. She went to another store and spent $79.35 for a blouse. At the third store she found a watch for $49.50. Did she have enough to buy the watch? If yes, how much money did she have left after buying the watch? If no, how much more did she need to buy the watch?

    _____

SUMMARY OF KEY CONCEPTS

KEY TERMS

decimal numbers (3.1): Numbers that include a decimal point.

decimal point (3.1): A period that separates the whole number part from the fractional part of a decimal number.

decimal fraction (3.1): A fraction with a denominator of 10, 100, 1,000, and so on.

mixed decimal (3.1): A combination of a whole number and a decimal fraction.

decimal digits (3.1): The digits to the right of the decimal point.

pointing off places (3.4): The process of positioning the decimal point in the product of two numbers.

power (3.4): The value of an exponential expression.

KEY RULES

To round decimal numbers (3.2):

1. Determine which digit is in the rounding place and which digit is the determining digit.
2. The digit in the rounding place is
   a. unchanged if the determining digit is 4 or less.
   b. increased by 1 if the determining digit is 5 or more.
3. Drop all digits to the right of the rounding place.

To multiply decimal numbers (3.4):

1. Multiply the factors without regard for the decimal point(s).
2. Count the total number of decimal digits in the factors.
3. Place the decimal point in the product so that you will have the same number of decimal digits in the product that you counted in step 2.

Division may not be done with a decimal-number divisor (3.5):

The decimal point must be moved the same number of places in the dividend as in the divisor.

To write a number in scientific notation (3.6):

1. Move the decimal point in the original number until you have a single digit to the left of the decimal point.
2. Count the number of places you moved the decimal point. This number is the exponent on the base 10. The exponent is positive if the original number is more than 10 (meaning that the decimal point moved to the left). The exponent is negative if the original number is less than 1 (meaning that the decimal point moved to the right).
3. The original number written in scientific notation is the factor obtained in step 1 multiplied by the exponential expression obtained in step 2.

▬ To write a number without exponents (3.6)

1. Determine whether the exponent is positive or negative.
2. Move the decimal point the same number of places as the exponent number
   a. to the left if the exponent is negative.
   b. to the right if the exponent is positive.

PRACTICE TEST 3A

1. Write the place name for the underlined digit in each number.

    a. 0.43<u>8</u> _____

    b. 10.<u>7</u>5 _____

    c. 8.9321<u>0</u>6 _____

2. Write 94.063 in words.

    _____

3. Write six hundred forty-two and fifty-six hundredths in digits.

    _____

4. Round each decimal number to the indicated place.

    a. 0.435 to the nearest tenth _____

    b. 14.2864 to the nearest thousandth _____

    c. $65.485 to the nearest cent _____

In problems 5–15, perform the indicated operation.

5. 19.8
    6.47
    + 532.003

6. 16.3 + 0.25 + 9.712 = _____

7. 0.642
    − 0.589

8. 24 − 8.56 = _____

9. 6.4
    × 8

10. 0.035
    × 0.489

11. 0.37 × 10 = _____

12. 7)9.17

13. 18.7 ÷ 5.24 (round to the nearest tenth) = _____

14. 472.6 ÷ 100 = _____

15. Express each number in scientific notation:

   a. 952,000,000 _____

   b. 0.000038 _____

16. Write each number without exponents.

   a. 4.287×10^{10} _____

   b. 7.3×10^{-2} _____

17. $(2.4 + 3.7) \div 1.9$ (round your answer to the nearest tenth) = _____

CHAPTER 3

PRACTICE TEST 3B

1. Write the place name for the underlined digit in each number.

    a. 5.<u>6</u>9 _____

    b. 0.56<u>3</u>21 _____

    c. 75.60<u>7</u>594 _____

2. Write 10.75 in words.

    _____

3. Write four and eighteen thousandths in digits. _____

4. Round $76.473 as indicated:

    a. to the nearest cent _____

    b. to the nearest whole dollar _____

    c. to the nearest tenth of a dollar _____

In problems 5–15, perform the indicated operation.

5.   82.6
          5.902
    + 117.548

6. 5.64 + 0.2 + 12.976 = _____

7.   84.9
    − 38.17

8. 6.75 − 0.9 = _____

9.   7.9
    × 6

10.   0.042
    × 1.5

11. 0.842 × 100 = _____

12. 16)25.6

13. 14.2 ÷ 3.8 (round to the nearest tenth) = _____

14. 6.75 ÷ 1,000 = _____

243

15. Express each number in scientific notation:

   a. 63,800 _____

   b. 0.00000000502 _____

16. Write each number without exponents:

   a. 4.21×10^6 _____

   b. 8.4×10^{-5} _____

17. $12.3 \times 4.2 \div 8.4 =$ _____

CHAPTERS 1-3

SKILLS REVIEW

1. Write 85,423 in words.
   _____

2. Round 64,593 to the nearest ten thousand. _____

3. Divide 768 by 32. _____

4. Multiply 8,476 by 10. _____

5. Write the prime factorization of 50. _____

6. Convert $2\frac{5}{8}$ to an improper fraction. _____

7. Convert $\frac{35}{4}$ to a mixed number. _____

8. Reduce $\frac{30}{54}$ to lowest terms. _____

In problems 9–12, perform the indicated operation. Reduce your answers to lowest terms. If the answer is an improper fraction, convert it to a mixed number.

9. $\frac{2}{5} + \frac{3}{8} + \frac{7}{10} =$ _____

10. $\frac{4}{9} - \frac{2}{7} =$ _____

11. $\frac{2}{3} \times \frac{27}{16} =$ _____

12. $\frac{5}{16} \div \frac{25}{36} =$ _____

13. Write 64.95 in words.
    _____

14. Round 50.09563 to the nearest thousandth. _____

In problems 15–20, perform the indicated operation.

15. $19.5 + 32.76 + 0.003 =$ _____

16. $28.7 - 16.82 =$ _____

17. $6.3 \times 5.2 =$ _____

18. $0.42 \times 100 =$ _____

19. $16.5 \div 2.75 =$ _____

20. $8.43 \div 100 =$ _____

21. Simplify: $12.9 \div (2 + 4.45) =$ _____

CHAPTER 3 SOLUTIONS

Skills Preview

1. *a.* hundredths
   b. ten-thousandths
   c. tenths

2. two hundred forty-seven and six hundred forty-five thousandths

3. 16.24

4. *a.* 0.85
   b. 12.4
   c. $49

5. 354.276

6. 324.082

7. 5.915

8. 4.177

9. 220.5

10. 0.0274

11. 3,563

12. 1.4

13. 6.24

14. 0.649

15. *a.* 4.8×10^4
    b. 8.3×10^{-4}

16. *a.* 89,500,000
    b. 0.00000519

17. 40

Section 3.1 Comprehension Checkpoint

1. *a.* ten-thousandths
   b. hundred-thousandths

2. *a.* eight hundred seventy-two thousandths
   b. six hundred fifty-seven and ninety-one hundredths

3. *a.* 5.73
   b. 0.0214

3.1A Practice Set

1. ten-thousandths
2. hundredths
3. hundred-thousandths
4. tenths
5. millionths
6. thousandths
7. ten-millionths
8. hundred-millionths
9. six tenths
10. three thousandths
11. four and eighty-two hundredths
12. twelve and nine thousand five hundred thirty-two ten-thousandths
13. ninety-five and eighty-three thousandths
14. nine hundred sixty-seven and thirty-eight thousand two hundred twenty-one hundred-thousandths
15. six and forty-five hundred-thousandths
16. six hundred fifty-three and five hundred eighty-three thousand nine hundred eighty-two millionths
17. 0.17
18. 0.3074
19. 4.23
20. 15.095
21. 58.209
22. 36.004725
23. 926.0543
24. 71.06004

Section 3.2   Comprehension Checkpoint

1. 37.6
2. 4.08
3. 0.8277
4. $47.64
5. $95

3.2A   Practice Set

1. 0.5   0.48   0.483
2. 0.1   0.06   0.058
3. 0.5   0.53   0.531
4. 3.1   3.13   3.129
5. 5   4.96   4.965
6. 2.7   2.68   2.679
7. 17.4   17.35   17.353
8. 34.7   34.7   34.704
9. *a.* 0.7457
   b. 0.7457
   c. 0.745698
10. *a.* 0.4266
    b. 0.42659
    c. 0.426587
11. $8.50 $8.50 $8
12. $12.63 $12.60 $13
13. $35.51 $35.50 $36
14. $13.24 $13.20 $13
15. $50.16 $50.20 $50
16. $0.63
17. 8,900

Section 3.3   Comprehension Checkpoint

1. 8.57
2. 260.982
3. 6.25
4. 0.67
5. 5.11

3.3A   Practice Set

1. $8.95
    3.56
    + 2.38
    ———
    $14.89

2. 12.94
    462.731
    + 28.5
    ———
    504.171

3. 76.379
    0.7821
    + 253.53684
    ———
    330.69794

4. 18.54
    + 0.37
    ———
    18.91

5. 171.09
    2.986
    + 12.25
    ———
    186.326

6. $ 95.29
    46.87
    1.72
    + 163.68
    ———
    $307.56

7. 0.46
    0.709
    0.1812
    + 0.985723
    ———
    2.335923

8. $ 67.00
    91.50
    + 6.93
    ———
    $165.43

9. $9.65
    − 3.41
    ———
    $6.24

10. 18.47
    − 9.384
    ———
    9.086

11. $354.05
    − 65.85
    ———
    $288.20

12. $7.00
    − 3.67
    ———
    $3.33

13. 81
    − 65.49
    ———
    15.51

14. 54.9
    − 26.09
    ———
    28.81

15. 389.92485
    − 248.5563
    ———
    141.36855

16. 0.41823
    − 0.008471
    ———
    0.409759

17. 89.83
    − 65.27
    ———
    24.56

18. 42.7328
    − 42.65
    ———
    0.0828

Solutions

19.
$0.89
  0.67
+ 1.25
―――
$2.81

20.
  $5.00
− 2.81
―――
  $2.19

21.
  42.8
  64.25
+ 50
―――
157.05

22.
  $75.23
− 37.25
―――
  $37.98

Section 3.4 Comprehension Checkpoint

1. 139.2   **2.** 71.5316   **3.** 0.043622   **4.** 37.42   **5.** 28,640

3.4A Practice Set

1.
    5.9
×   6
―――
  35.4

2.
    8.2
×   4
―――
  32.8

3.
    9.7
×   3
―――
  29.1

4.
    6.2
× 2.4
―――
  2 4 8
12 4
―――
14.8 8

5.
    7.5
× 4.7
―――
  5 2 5
30 0
―――
35.2 5

6.
    5.3
× 1.8
―――
  4 2 4
  5 3
―――
  9.5 4

7.
  0.5 7
× 0.4
―――
0.2 2 8

8.
  2.8 5
× 0.9
―――
2.56 5

9.
  6.2 6
× 3.4
―――
  2 50 4
18 78
―――
21.28 4

10.
    26.3
×   2.7
―――
  18 4 1
52 6
―――
  71.0 1

11.
    53.95
×   4.25
―――
  2 69 75
10 79 0
215 80
―――
229.28 75

12.
    0.009
×   3.87
―――
      63
    7 2
  27
―――
0.034 83

13. 3.56 × 100 = 356   **14.** 0.376 × 1,000 = 376   **15.** 845.27 × 10 = 8,452.7

16. 4.284 × 10,000 = 42,840   **17.** 0.072 × 10 = 0.72

18.
   $25.95
×   4
―――
 $103.80

19.
    $195.27
×   36
―――
  1 171 62
5 858 1
―――
$7,029.72

20.
   12.875
×   8
―――
 103.000   103 inches

21.
    $ 0.5 5
×   3.4
―――
    2 2 0
  16 5
―――
$1.8 7 0 = $1.87

22. $25.75 × 1,000 = $25,750

Section 3.5 Comprehension Checkpoint

1. 1.9 (rounded) 2. 3.6 3. 3.8 (rounded) 4. 1.9 (rounded)
5. 0.357 6. 74.683

3.5A Practice Set

1. $5\overline{)4.5}^{.9}$ $\underline{4\,5}$

2. $8\overline{)9.84}^{1.23}$
   $\underline{8}$
   $1\,8$
   $\underline{1\,6}$
   24
   $\underline{24}$

3. $15\overline{)2.25}^{.15}$
   $\underline{1\,5}$
   75
   $\underline{75}$

4. $2.5\overline{)5.6} = 25\overline{)56.00}^{2.24}$
   $\underline{50}$
   $6\,0$
   $\underline{5\,0}$
   $1\,00$
   $\underline{1\,00}$

5. $3.8\overline{)17.1} = 38\overline{)171.0}^{4.5}$
   $\underline{152}$
   $19\,0$
   $\underline{19\,0}$

6. $0.14\overline{)5.04} = 14\overline{)504.}^{36.}$
   $\underline{42}$
   84
   $\underline{84}$

7. $36.89 \div 10 = 3.689$

8. $0.46 \div 100 = 0.0046$

9. $356.28 \div 100 = 3.5628$

10. $8.6\overline{)18.5} = 86\overline{)185.00}^{2.15} = 2.2$ rounded
    $\underline{172}$
    $13\,0$
    $\underline{8\,6}$
    $4\,40$
    $\underline{4\,30}$

11. $1.25\overline{)25.9} = 125\overline{)2590.00}^{20.72} = 20.7$ rounded
    $\underline{250}$
    $90\,0$
    $\underline{87\,5}$
    $2\,50$

12. $12.8\overline{)84} = 128\overline{)840.000}^{6.562} = 6.56$ rounded
    $\underline{768}$
    $72\,0$
    $\underline{64\,0}$
    $8\,00$
    $\underline{7\,68}$
    320
    $\underline{256}$
    64

13. $3.6\overline{)16.85} = 36\overline{)168.500}^{4.680} = 4.68$ rounded
    $\underline{144}$
    $24\,5$
    $\underline{21\,6}$
    $2\,90$
    $\underline{2\,88}$
    20

14. $4\overline{)29.850}^{7.462} = \$7.46$ rounded
    $\underline{28}$
    $1\,8$
    $\underline{1\,6}$
    25
    $\underline{24}$
    10

15. $7.25\overline{)43} = 725\overline{)4300.000}^{5.931} = \$5.93$
    $\underline{3625}$
    6750
    $\underline{6525}$
    2250
    $\underline{2175}$
    750
    $\underline{725}$

Solutions 251

16. $0.50\overline{)25} = 50\overline{)2500.}$ There are 50 half-dollars in $25.
$$\begin{array}{r} 50. \\ \underline{250} \\ 0 \end{array}$$

17. $500\overline{)2.6000}$ Each sheet of paper is 0.0052 inch thick.
$$\begin{array}{r} .0052 \\ \underline{2\ 500} \\ 1\ 000 \\ \underline{1\ 000} \end{array}$$

18. $24\overline{)8.88}$ It costs $0.37 to develop each picture.
$$\begin{array}{r} .37 \\ \underline{7\ 2} \\ 1\ 68 \\ \underline{1\ 68} \end{array}$$

19. $26.5\overline{)365.7} = 265\overline{)3657.0}$ Her car holds 13.8 gallons of gas.
$$\begin{array}{r} 13.8 \\ \underline{265} \\ 1007 \\ \underline{795} \\ 2120 \\ \underline{2120} \end{array}$$

20. $153.50 ÷ 10 = $15.35 savings per week

Section 3.6   Comprehension Checkpoint

1. 3.7×10^4 2. 4.6×10^{-5} 3. 56,800,000   4. 0.07921

3.6A   Practice Set

1. $5,100 = 5.1 \times 10^3$ 2. $63,000 = 6.3 \times 10^4$

3. $248,000 = 2.48 \times 10^5$ 4. $47,300,000 = 4.73 \times 10^7$

5. $95,000,000,000 = 9.5 \times 10^{10}$ 6. $0.021 = 2.1 \times 10^{-2}$

7. $0.000725 = 7.25 \times 10^{-4}$ 8. $0.008412 = 8.412 \times 10^{-3}$

9. $0.000008 = 8 \times 10^{-6}$ 10. $0.00004731 = 4.731 \times 10^{-5}$

11. $7.3 \times 10^4 = 73,000$ 12. $9.28 \times 10^2 = 928$

13. $1.4 \times 10^3 = 1,400$ 14. $3.872 \times 10^6 = 3,872,000$

15. $6.01 \times 10^5 = 601,000$ 16. $3.3 \times 10^{-5} = 0.000033$

17. $9.83 \times 10^{-3} = 0.00983$ 18. $5.0065 \times 10^{-7} = 0.00000050065$

19. $8.154 \times 10^{-4} = 0.0008154$ 20. $2.18 \times 10^{-9} = 0.00000000218$

21. $186,000 = 1.86 \times 10^5$ 22. $0.000000005 = 5 \times 10^{-9}$

23. $4.3 \times 10^{-6} = 0.0000043$ 24. $2.5 \times 10^4 = 25,000$

Section 3.7 Comprehension Checkpoint

1. 5.95   2. 24.1   3. 5   4. 2.8

3.7A Practice Set

1. 8.7 ÷ 3 − 1.8
   └─┬─┘ divide
   2.9 − 1.8
   └──┬──┘ subtract
   1.1

2. (4.6 + 0.4) ÷ 0.05
   └────┬────┘ add
   5 ÷ 0.05
   └──┬──┘ divide
   100

3. (3.7 − 2.5) × 4.5
   └────┬────┘ subtract
   1.2 × 4.5
   └──┬──┘ multiply
   5.4

4. 5.9 × 2.1 + 2.7 × 6
   multiply └─┬─┘   └─┬─┘ multiply
   12.39 + 16.2
   └──┬──┘ add
   28.59

5. 8.4 ÷ 6 − 0.02 × 3
   divide └─┬─┘   └─┬─┘ multiply
   1.4 − 0.06
   └──┬──┘ subtract
   1.34

6. $100 − 6 × $7.50
   └──┬──┘ multiply
   $100 − $45
   └──┬──┘ subtract
   $55

7. $50 + 3.4 × $2.75 − $49.95
   └──┬──┘ multiply
   $50 + $9.35   − $49.95
   └──┬──┘ add
   $59.35   − $49.95
   └──────┬──────┘ subtract
   $9.40

8. 2.58 × 16 ÷ (3.8 × 4)
   multiply └─┬─┘   └─┬─┘ multiply
   45.6 ÷ 15.2
   └──┬──┘ divide
   3

9. $10 − (3 × $1.29 + 2 × $0.79)
   └─┬─┘   └─┬─┘ multiply
   $10 − ($3.87 + $1.58)
   └──┬──┘ add
   $10 − $5.45
   └──┬──┘ subtract
   $4.55

10. $1,500 + 48 × $169.64
    └──┬──┘ multiply
    $1,500 + 8,142.72
    └──┬──┘ add
    $9,642.72

Practice Test 3A

1. *a.* thousandths
   b. tenths
   c. hundred-thousandths

2. ninety-four and sixty-three thousandths

3. 642.56

4. *a.* 0.4
   b. 14.286
   c. $65.49

Solutions

5. 558.273
6. 26.262
7. 0.053
8. 15.44
9. 51.2
10. 0.017115
11. 3.7
12. 1.31
13. 3.6 (rounded)
14. 4.726
15. a. 9.52×10^8
    b. 3.8×10^{-5}
16. a. 42,870,000,000
    b. 0.073
17. 3.2 (rounded)

Skills Review Chapters 1–3

1. eighty-five thousand, four hundred twenty-three
2. 60,000
3. 24
4. 84,760
5. $2 \times 5 \times 5$
6. $\dfrac{21}{8}$
7. $8\dfrac{3}{4}$
8. $\dfrac{5}{9}$
9. $1\dfrac{19}{40}$
10. $\dfrac{10}{63}$
11. $1\dfrac{1}{8}$
12. $\dfrac{9}{20}$
13. sixty-four and ninety-five hundredths
14. 50.096
15. 52.263
16. 11.88
17. 32.76
18. 42
19. 6
20. 0.0843
21. 2

4 RATIOS AND PROPORTIONS

OUTLINE

4.1 Ratios

4.2 Proportions

4.3 Solving Proportions

In the first three chapters, you learned how to apply the four basic math operations (addition, subtraction, multiplication, and division) to whole numbers, common fractions, and decimal fractions. In this chapter you will learn about ratios and proportions. A ratio is the comparison of two quantities, while a proportion is a statement that two ratios are equal. A true proportion shows that the relationship between two sets of numbers is the same.

SKILLS PREVIEW

1. Convert each ratio to colon form.

   a. 8 to 9   b. $\dfrac{13}{9}$

   a. _____   b. _____

2. Convert each ratio to fraction form.

   a. 5 to 18   b. 22 : 9

   a. _____   b. _____

3. Convert each ratio to verbal form.

   a. 10 : 3   b. $\dfrac{18}{1}$

   a. _____   b. _____

4. Reduce each ratio to lowest terms.

   a. 5 : 20   b. $\dfrac{51}{34}$

   a. _____   b. _____

5. A math book has 500 pages. An English book has 420 pages. Express the ratio of pages in the math book to pages in the English book as a fraction in lowest terms. _____

6. Is the proportion $\dfrac{10}{6} = \dfrac{30}{18}$ a true proportion? _____

7. In the proportion $\dfrac{9}{11} = \dfrac{27}{33}$, what are the means and what are the extremes?

   means: _____

   extremes: _____

8. Solve the proportion for the missing term:

   $\dfrac{n}{15} = \dfrac{25}{75}$ _____

9. One oil well produces 5 barrels of oil in two days. At the same rate, how long will it take the well to produce 30 barrels of oil? _____

4.1 RATIOS

A **ratio** is a comparison of two quantities. Suppose Bob's Burger Bonanza employs 10 people: 6 women and 4 men. The ratio of women to men could be expressed in three different ways:

$$6 \text{ to } 4 \quad \textit{verbal form}$$
$$6 : 4 \quad \textit{colon form}$$
$$\frac{6}{4} \quad \textit{fraction form}$$

Each of the forms is read "six to four." All of them have the same meaning.

It is important to note the order in a ratio. If you wanted to write the ratio of men to women at Bob's Burger Bonanza, it would be

$$4 \text{ to } 6 \quad \textit{verbal form}$$
$$4 : 6 \quad \textit{colon form}$$
$$\frac{4}{6} \quad \textit{fraction form}$$

Notice that in the ratio of women to men, the number of women is listed first in each of the three forms. In the ratio of men to women, the number of men is listed first in each of the three forms.

EXAMPLE 1 A car dealer sold three trucks and five cars one day. Write in colon form:

a. The ratio of trucks sold to cars sold for the day
b. The ratio of trucks sold to total vehicles sold

SOLUTION

a. 3 : 5   *The number of trucks sold will be listed first since you are comparing "trucks to cars."*

b. 3 : 8   *There were eight total vehicles sold. The number of trucks sold will be listed first since you are comparing "trucks sold to total vehicles sold."*

Ratios should be reduced to lowest terms. Write them as fractions and reduce them the way you reduce fractions.

EXAMPLE 2 Write the ratio of 4 to 12 in fraction form and reduce to lowest terms.

SOLUTION

$\frac{4}{12}$ *Write the ratio in fraction form with the numbers in the correct positions.*

$\frac{4 \div 4}{12 \div 4} = \frac{1}{3}$ *Divide the numerator and the denominator by a common factor—in this case, 4.*

The ratio $\frac{4}{12}$ reduces to $\frac{1}{3}$.

Notice that there is one significant difference between reducing fractions and reducing ratios. Consider the fraction $\frac{16}{8}$. This reduces to 2 (16 ÷ 8 = 2). If you are working with a ratio of $\frac{16}{8}$, though, and you want to reduce it to lowest terms, you must keep both a numerator and a denominator. Thus the ratio $\frac{16}{8}$ reduces to $\frac{2}{1}$. Remember that a ratio is a comparison of two quantities and therefore must always include two values.

Ratios will not always involve whole numbers. If common fractions or decimals are included in ratios, they can be reduced just as whole numbers can.

EXAMPLE 3 Express the ratio 24.6 to 6 in lowest terms.

SOLUTION

$\frac{24.6}{6}$ *Write the ratio in fraction form.*

$\frac{24.6}{6} = \frac{24.6 \div 6}{6 \div 6} = \frac{4.1}{1}$ *Divide numerator and denominator by a common factor—6, in this case.*

The ratio $\frac{24.6}{6}$ reduces to $\frac{4.1}{1}$.

Work the Comprehension Checkpoint problems to measure your understanding of ratios. If you have difficulty with these problems, review the material in this section before going on to the Practice Sets.

COMPREHENSION CHECKPOINT

1. Write each ratio in colon form:

   a.   4 to 6   *b.* $\frac{5}{3}$

   a. _____   *b.* _____

2. Write each ratio in fraction form:

   a.   5 to 9   *b.*   6 : 7

   a. _____   *b.* _____

3. Write each ratio in verbal form:

   a. $\frac{4}{8}$ *b.*   3 : 1

   a. _____   *b.* _____

4. Reduce each ratio to lowest terms. Write your answers in fraction form.

   a.   5 : 15   *b.*   27 : 3   *c.*   18 : 14

   a. _____   *b.* _____   *c.* _____

4.1A PRACTICE SET

Fill in each blank in problems 1–6 by converting each given ratio to the other forms.

	Verbal Form	Colon Form	Fraction Form
1.	1 to 4	_____	_____
2.	_____	5 : 9	_____
3.	_____	_____	$\frac{3}{10}$
4.	15 to 1	_____	_____
5.	_____	7 : 2	_____
6.	_____	_____	$\frac{12}{8}$

Reduce each ratio to lowest terms.

7. 18 to 12 _____

8. 8 to 32 _____

9. 28 : 7 _____

10. 13 : 65 _____

11. $\frac{36}{9}$ _____

12. $\frac{20}{44}$ _____

Applications

Express each ratio in the indicated form in lowest terms.

13. Sam works 8 hours each day. Karen works 6 hours each day. Write the ratio, in fraction form, of Sam's hours to Karen's hours.

_____

14. A carpenter has two boards, one 6 feet long and the other 2 feet long. Write the ratio, in verbal form, of the longer board to the shorter board.

15. A box of oranges weighs 40 pounds while a box of apples weighs 35 pounds. Express the ratio of the weight of the apples to the weight of the oranges in colon form.

16. Two glasses have a capacity of 8 ounces and 12 ounces, respectively. What is the ratio, in verbal form, of the volume of the larger glass to the volume of the smaller glass?

17. Sherry spends $650 each month on bills. Her rent is $250. Express the ratio (in colon form) of Sherry's rent to her total bills.

18. ComputerTime sold 20 boxes of 3.5-inch diskettes for every 4 boxes of 5.25-inch diskettes. What is the ratio, in fraction form, of 3.5-inch diskettes sold to 5.25-inch diskettes sold?

19. Melodious Music Madness sells 18 cassettes for every 10 compact disks. Express the ratio of cassettes sold to total sales in fraction form.

20. Kimberly makes 20 sales calls in an 8-hour workday. Write in verbal form the ratio of sales calls to hours worked.

4.1B PRACTICE SET

Fill in each blank in problems 1–6 by converting each given ratio to the other forms.

	Verbal Form	Colon Form	Fraction Form
1.	_____	_____	$\dfrac{3}{8}$
2.	2 to 1	_____	_____
3.	_____	_____	$\dfrac{5}{4}$
4.	_____	18 : 16	_____
5.	12 to 9	_____	_____
6.	_____	10 : 25	_____

Reduce each ratio to lowest terms.

7. 28 to 35 _____

8. 57 to 19 _____

9. 16 : 24 _____

10. 40 : 30 _____

11. $\dfrac{12}{60}$ _____

12. $\dfrac{75}{30}$ _____

Applications

Express each ratio in the indicated form in lowest terms.

13. Jill is taking 6 credit hours this semester; Sandra is taking 12 credit hours. Express, in fraction form, the ratio of the number of hours Sandra is taking to the number of hours Jill is taking.

_____

14. John drove 250 miles on vacation. Marie drove 100 miles. What is the ratio of John's miles to Marie's miles? Express the ratio in colon form.

15. David paid $48 for a dozen roses. Express the ratio for the cost of a single rose in colon form.

16. What is the ratio, in colon form, of the number of days in June to the number of days in one year (not a leap year)?

17. Clinton hits a golf ball 300 yards with his driver. Jason hits a golf ball 275 yards with his driver. Express the ratio, in fraction form, of the distance Clinton hits the ball to the distance Jason hits the ball.

18. Linda types 75 words per minute. Maria types 85 words per minute. Express the ratio of the number or words Linda types to the number of words Maria types in fraction form.

19. An apartment building has 48 one-bedroom apartments and 30 two-bedroom apartments. Express the ratio of the number of two-bedroom apartments to total apartments in colon form.

20. Ray's Market sells 600 pounds of chicken each week and 350 pounds of beef. Express the ratio of pounds of chicken sold to total pounds of meat sold each week in verbal form.

4.2 PROPORTIONS

A **proportion** is a statement that two ratios are equal. Proportions are usually written in fraction form such as

$$\frac{a}{b} = \frac{c}{d}$$

Proportions can also be written in colon form, as $a:b = c:d$; or in verbal form, as a to $b = c$ to d.

The **terms** of a proportion are the two numerators and two denominators of the ratios. The first and last terms, a and d, are called the **extremes**, while the middle terms, b and c, are called the *means*. In fraction form,

$$\begin{array}{r} extreme \leftarrow \\ mean \leftarrow \end{array} \frac{a}{b} = \frac{c}{d} \begin{array}{l} \rightarrow mean \\ \rightarrow extreme \end{array}$$

A proportion is read "a is to b as c is to d." This means that the relationship between a and b is the same as the relationship between c and d.

You can determine whether two ratios form a true proportion by using the rule of proportions.

> ### THE RULE OF PROPORTIONS
> The product of the extremes equals the product of the means.

The multiplication process used to determine whether proportions are true is called **cross multiplication.** Cross multiplication is multiplying the numerator of one fraction by the denominator of the other fraction. (To help you remember which terms to multiply, place a large X over the equal sign. One line of the X points toward the extremes and the other line of the X points toward the means. These are the terms to be multiplied.)

$$\begin{array}{r} extreme \\ mean \end{array} \frac{a}{b} \diagup\!\!\!\!\diagdown \frac{c}{d} \begin{array}{l} mean \\ extreme \end{array}$$

product of the extremes $\rightarrow ad = bc \leftarrow$ product of the means

The two products will be equal if the two ratios form a true proportion. To determine whether the two ratios

$$\frac{3}{6} \stackrel{?}{=} \frac{12}{24}$$

form a true proportion, use cross multiplication:

$$\begin{array}{cc} extremes & means \\ (3)(24) & \stackrel{?}{=} (6)(12) \\ 72 & = 72 \end{array}$$

If the result of cross-multiplying is an equality, such as

$$72 = 72$$

from the preceding problem, the two ratios are equal and the proportion is true. If the result of cross-multiplying is a statement that is not true, such as

$$12 = 15$$

then the two ratios are not equal and you do not have a true proportion.

EXAMPLE 1 Determine whether the following proportions are true proportions.

a. $\dfrac{3}{4} = \dfrac{9}{12}$ b. $\dfrac{5.9}{8.2} = \dfrac{20.6}{32.8}$ c. $\dfrac{1\frac{1}{2}}{2\frac{1}{2}} = \dfrac{4\frac{1}{2}}{7\frac{1}{2}}$

SOLUTIONS

a. $(3)(12) \stackrel{?}{=} (4)(9)$
$36 = 36$

This is a true proportion.

b. $(5.9)(32.8) \stackrel{?}{=} (8.2)(20.6)$
$193.52 = 168.92$

This is not a true proportion.

c. $\left(1\dfrac{1}{2}\right)\left(7\dfrac{1}{2}\right) \stackrel{?}{=} \left(4\dfrac{1}{2}\right)\left(2\dfrac{1}{2}\right)$

$11.25 = 11.25$

This is a true proportion.

Parts *b* and *c* in Example 1 illustrate that the terms of a proportion do not have to be whole numbers.

Use the Comprehension Checkpoint problems to test your understanding of proportions. If you have difficulty with these problems, review the material in this section before going on to the Practice Sets.

COMPREHENSION CHECKPOINT

1. Name the means and the extremes in each proportion.

   a. $3 : 4 = 6 : 8$ means: _____ extremes _____

   b. $\dfrac{2}{10} = \dfrac{6}{30}$ means: _____ extremes _____

2. Determine whether each proportion is a true proportion.

   a. $5 : 9 = 24 : 40$ _____

   b. $\dfrac{3}{12} = \dfrac{18}{72}$ _____

4.2A PRACTICE SET

In problems 1–3, identify the means and the extremes.

1. $4 : 5 = 16 : 20$ means: _____

    extremes: _____

2. $\dfrac{12}{8} = \dfrac{18}{12}$ means: _____

    extremes: _____

3. $10 : 1 = 20 : 2$ means: _____

    extremes: _____

In problems 4–16, determine whether each proportion is true.

4. $8 : 7 = 24 : 21$ _____

5. $5 : 8 = 25 : 35$ _____

6. $12 : 1 = 36 : 3$ _____

7. $15 : 9 = 45 : 18$ _____

8. $6 : 9 = 12 : 18$ _____

9. $14 : 7 = 42 : 21$ _____

10. $\dfrac{3}{2} = \dfrac{18}{14}$ _____

11. $\dfrac{20}{5} = \dfrac{100}{20}$ _____

12. $\dfrac{13}{39} = \dfrac{52}{156}$ _____

13. $\dfrac{24}{6} = \dfrac{144}{36}$ _____

14. $\dfrac{12}{25} = \dfrac{48}{100}$ _____

15. $\dfrac{33}{132} = \dfrac{1}{4}$ _____

16. $\dfrac{17}{1} = \dfrac{51}{4}$ _____

Applications

In each word problem, set up a proportion and check to see whether it is true.

17. One grocer loses 5 apples out of each 40 apples that he buys. Another loses 4 apples out of every 30. Is the ratio of loss to apples purchased the same for both grocers?

    _____

18. Handy Hardware sells nails for $3 for 6 pounds. Just Do It Hardware sells the same type of nail for $4 for 8 pounds. Are the ratios of price to pounds the same in both stores?

    _____

19. Jeremy runs 8 miles each day. Bryan runs 56 miles each week. Are the ratios of miles run the same for each runner?

    _____

20. The ABC Corporation employs 25 men and 36 women. The XYZ Corporation employs 15 men and 24 women. Is the ratio of men to women the same in both corporations?

    _____

4.2B PRACTICE SET

In problems 1–3, identify the means and the extremes.

1. $9 : 6 = 18 : 12$ means: _____

   extremes: _____

2. $\dfrac{3}{27} = \dfrac{9}{81}$ means: _____

   extremes: _____

3. $7 : 15 = 21 : 45$ means: _____

   extremes: _____

In problems 4–16, determine whether each proportion is true.

4. $8 : 24 = 16 : 48$ _____

5. $9 : 5 = 27 : 14$ _____

6. $13 : 15 = 39 : 45$ _____

7. $12 : 7 = 84 : 49$ _____

8. $25 : 100 = 4 : 1$ _____

9. $8 : 1 = 64 : 7$ _____

10. $\dfrac{8}{12} = \dfrac{32}{48}$ _____

11. $\dfrac{2}{9} = \dfrac{18}{81}$ _____

12. $\dfrac{7}{3} = \dfrac{28}{12}$ _____

13. $\dfrac{42}{126} = \dfrac{1}{3}$ _____

14. $\dfrac{40}{10} = \dfrac{10}{2}$ _____

15. $\dfrac{36}{16} = \dfrac{18}{6}$ _____

16. $\dfrac{45}{180} = \dfrac{90}{360}$ _____

Applications

In each word problem, set up a proportion and check to see whether it is true.

17. A new car goes 315 miles on a tank of 15 gallons of gasoline. An older model car can go 350 miles on 17 gallons of gas. Do the two cars get the same mileage per gallon?

_____

18. A math class has 27 students, 15 female and 12 male. A physics class has 20 women and 16 men. Is the ratio of women to total students in each class the same?

_____

19. The 10:00 AM tour group in a museum included 30 children and 18 adults. The 12:00 noon tour group had 25 children and 15 adults. Were the ratios of adults to children the same for both tour groups?

_____

20. Steve weighs 160 pounds, but he can bench-press 280 pounds. Allen weighs 240 pounds and can bench-press 450 pounds. Are the ratios of body weight to bench-press weight equal?

_____

4.3 SOLVING PROPORTIONS

Any time three of the four terms of a proportion are known, the fourth term can be computed. You learned previously that a letter can be used to represent an unknown quantity. The letter n will be used in the examples in this section to represent the missing term of the proportion. Remember that the n only stands in the place of a number until that number is determined.

Remember that a proportion written in fraction form looks like

$$\frac{a}{b} = \frac{c}{d}$$

Look at the following four proportions:

$$\frac{2}{4} = \frac{n}{8} \qquad \frac{3}{n} = \frac{6}{4} \qquad \frac{n}{2} = \frac{3}{6} \qquad \frac{2}{3} = \frac{10}{n}$$

In each of the proportions, the unknown term is in a different place. In each of the first two, one of the means is unknown. The last two are missing one of the extremes. Each of the unknowns can be found using the same method.

> **TO SOLVE A PROPORTION**
> 1. If a mean is unknown, multiply the extremes and divide the product by the given mean.
> 2. If an extreme is unknown, multiply the means and divide the product by the given extreme.

Study the examples to see how to solve any proportion.

EXAMPLE 1 Solve: $\dfrac{5}{8} = \dfrac{n}{40}$

SOLUTION

$\dfrac{(5)(40)}{8} = n$ *A mean is unknown. Multiply the extremes and divide by the given mean.*

$\dfrac{200}{8} = n$ *The product of the extremes is 200.*

$25 = n$ *Dividing 200 by 8 gives the missing mean term.*

The missing term in this proportion is 25.

CHECK

$\dfrac{5}{8} = \dfrac{25}{40}$ *Substitute 25 in the original proportion.*

$(5)(40) \stackrel{?}{=} (8)(25)$ *Cross-multiply.*

$200 = 200$

The check shows that the ratios are equal, so the proportion is true. The number represented by n in this problem is 25.

> **a + b = x**
>
> ## ALGEBRA CONNECTION
>
> You learned previously that as you develop more math skills, you will be able to solve the same problem in different ways. Any proportion can be solved by first setting up an equation and then solving the equation. An equation is a mathematical statement that two quantities are equal. Consider Example 1 again.
>
> $$\frac{5}{8} = \frac{n}{40}$$
>
> The proportion rule says that the product of the means equals the product of the extremes. If you cross-multiply the terms of this proportion, you get
>
> $$(8)(n) = (5)(40)$$
>
> or
>
> $$8n = (5)(40)$$
>
> When a number is multiplied by a letter, you can write them side by side to show multiplication, like the $8n$ in the illustration. In Chapter 8 you will learn different strategies for solving equations. For our purposes here, all you have to do is divide each side of the equation by 8.
>
> $$\frac{8n}{8} = \frac{(5)(40)}{8}$$
>
> This division gives you
>
> $$n = 25$$
>
> which is the same result you got by solving the proportion using another method. The ability to set up equations using a letter for an unknown quantity and then solving the equation will enable you to solve more problems more easily.

Another example illustrates that the same method can be used to solve a proportion if the missing term is an extreme.

EXAMPLE 2 Solve: $\frac{n}{0.5} = \frac{72}{24}$

SOLUTION

$\frac{(0.5)(72)}{24} = n$ *The unknown term is an extreme. Multiply the means and divide by the given extreme.*

$\frac{36}{24} = n$ *The product of the means is 36.*

$1.5 = n$ *Dividing the product of the means by the given extreme indicates that the unknown term is 1.5.*

The check is left for you.

Proportions are convenient to use to solve everyday problems. They are also flexible, as the next example shows. In solving word problems, it will often be

Solving Proportions

necessary to compare different quantities. The important thing is to set up the proportions correctly.

EXAMPLE 3 A florist sold eight plants for $40. How many plants must the florist sell in order to take in $75?

SOLUTION
Method 1

$\dfrac{\text{Plants}}{\text{Plants}} = \dfrac{\text{Dollars}}{\text{Dollars}}$ *You could compare like quantities.*

$\dfrac{8}{n} = \dfrac{40}{75}$ *Substitute the given values in the proportion.*

$\dfrac{(8)(75)}{40} = n$ *Solve the proportion.*

$15 = n$ *The missing term is 15.*

The florist must sell 15 plants to take in $75.

Method 2

$\dfrac{\text{Plants}}{\text{Dollars}} = \dfrac{\text{Plants}}{\text{Dollars}}$ *You can use ratios comparing different quantities.*

$\dfrac{8}{40} = \dfrac{n}{75}$ *Set up two ratios comparing plants to dollars.*

$\dfrac{(8)(75)}{40} = n$ *Solve the proportion.*

$15 = n$ *The missing term is 15.*

The florist would have to sell 15 plants to take in $75.

Work the Comprehension Checkpoint problems to check your understanding before going on to the Practice Sets.

COMPREHENSION CHECKPOINT

1. Solve for the missing term:

   $\dfrac{9}{15} = \dfrac{n}{60}$ _____

2. Solve for the missing term:

   $\dfrac{3}{n} = \dfrac{18}{4.5}$ _____

3. Solve for the missing term:

   $\dfrac{6\frac{1}{2}}{13} = \dfrac{19\frac{1}{2}}{n}$ _____

4.3A PRACTICE SET

Solve for the missing term.

1. $\dfrac{2}{3} = \dfrac{n}{9}$ _____

2. $\dfrac{5}{n} = \dfrac{20}{36}$ _____

3. $\dfrac{n}{16} = \dfrac{12}{24}$ _____

4. $\dfrac{3}{8} = \dfrac{15}{n}$ _____

5. $\dfrac{n}{13} = \dfrac{24}{52}$ _____

6. $\dfrac{18}{11} = \dfrac{n}{99}$ _____

7. $\dfrac{20}{n} = \dfrac{140}{7}$ _____

8. $\dfrac{24}{35} = \dfrac{n}{140}$ _____

9. $\dfrac{3\frac{1}{3}}{5} = \dfrac{n}{15}$ _____

10. $\dfrac{1\frac{1}{4}}{2\frac{1}{2}} = \dfrac{3\frac{3}{4}}{n}$ _____

11. $\dfrac{n}{3\frac{1}{4}} = \dfrac{6}{6\frac{1}{2}}$ _____

12. $\dfrac{5.75}{23} = \dfrac{28.75}{n}$ _____

13. $\dfrac{12}{0.6} = \dfrac{n}{3}$ _____

14. $\dfrac{1.6}{n} = \dfrac{9.6}{12}$ _____

Applications

15. An electronics store sells 10 television sets for every 3 radios sold. How many televisions did the store sell during a week in which it sold 9 radios?

16. On a map, one inch represents 25 miles. The distance between two cities is 175 miles. How many inches apart are the cities on the map?

17. The cost of a pound of margarine is $0.99. What would be the cost of 12 pounds of margarine?

18. Carol earned $40.25 for a 7-hour work shift. At the same rate, how much would she earn for an 8-hour shift?

19. Apples cost $0.89 for 3 pounds. How much would 12 pounds cost?

20. Bag N Wag grocery has 3 supervisors for every 18 clerks. How many supervisors does the company employ if it has a total of 126 clerks?

21. If a professional baseball player hit 3 home runs for every 44 at bats, how many home runs would he hit in a season if he batted 616 times?

22. Medication is given according to body weight. If a 180-pound adult's dosage is 9 grams, what would be the dosage for a 60-pound child?

23. An airplane travels from Dallas to Houston, a distance of 240 miles, in 45 minutes. At the same rate, how long would it take to fly from Dallas to Miami, a distance of 1,500 miles?

24. What is the cost per ounce of a 15.5-ounce can of green beans that sells for $0.55? (Round to the nearest thousandth of a cent.)

4.3B PRACTICE SET

Solve for the missing term.

1. $\dfrac{5}{9} = \dfrac{n}{45}$ _____

2. $\dfrac{n}{8} = \dfrac{9}{24}$ _____

3. $\dfrac{8}{12} = \dfrac{32}{n}$ _____

4. $\dfrac{19}{n} = \dfrac{1}{3}$ _____

5. $\dfrac{n}{51} = \dfrac{3}{17}$ _____

6. $\dfrac{15}{36} = \dfrac{n}{72}$ _____

7. $\dfrac{n}{12} = \dfrac{25}{60}$ _____

8. $\dfrac{40}{10} = \dfrac{120}{n}$ _____

9. $\dfrac{1\frac{1}{2}}{n} = \dfrac{3}{5}$ _____

10. $\dfrac{6\frac{3}{4}}{9} = \dfrac{2\frac{1}{4}}{n}$ _____

11. $\dfrac{6}{15\frac{9}{10}} = \dfrac{n}{5\frac{3}{10}}$ _____

12. $\dfrac{7.5}{n} = \dfrac{30}{22}$ _____

13. $\dfrac{4.25}{6.2} = \dfrac{21.25}{n}$ _____

14. $\dfrac{n}{0.75} = \dfrac{20}{6}$ _____

Applications

15. An auto dealer sells nine cars for every five trucks he sells. If he sold 81 cars one month, how many trucks did he sell?

    _____

16. Samuel charges $3 for a dozen doughnuts. How much would he charge for 8 dozen doughnuts?

    _____

17. Marianne runs 5 miles in 30 minutes. How many miles could she run in 72 minutes?

    _____

18. The temperature dropped 18° in $4\frac{1}{2}$ hours. By how much was the temperature dropping each hour?

    _____

19. Joanne earned $70 commission on sales of $500. How much would she earn if she sold $750 worth of merchandise?

   _____

20. Two cities are 5 inches apart on a map. The actual distance between them is 375 miles. How many miles does 1 inch represent on the map?

   _____

21. The cost of 1 pound of pork is $2.49. What would be the cost of a 3.25-pound pork roast?

   _____

22. What would be the tax on $50 if the tax on $35 is $2.45?

   _____

23. Jocelyn used 25 milliliters of solution to make 48 slides. How many milliliters of solution would she need to prepare 120 slides?

   _____

24. The speed limit is 55 miles per hour. How many hours would it take to go 467.5 miles?

   _____

SUMMARY OF KEY CONCEPTS

KEY TERMS

ratio (4.1): A comparison of two quantities.

proportion (4.2): A statement that two ratios are equal.

terms (of a proportion) (4.2): The two numerators and two denominators of the ratios; also called the *means* and the *extremes*.

extremes (4.2): The outside terms of a proportion.

means (4.2): The inside terms of a proportion.

cross-multiplying (4.2): The process of multiplying the top term of one ratio by the bottom term of another ratio.

KEY RULES

There are three ways to express a ratio (4.1):

a to b verbal form

$a : b$ colon form

$\dfrac{a}{b}$ fraction form

The rule of proportions (4.2):

The product of the extremes equals the product of the means.

To solve a proportion (4.3):

1. If a mean is unknown, multiply the extremes and divide the product by the given mean.
2. If an extreme is unknown, multiply the means and divide the product by the given extreme.

CHAPTER 4

PRACTICE TEST 4A

1. Convert each ratio to colon form.

   a. 3 to 7   b. $\dfrac{5}{9}$

   a. _____
   b. _____

2. Convert each ratio to fraction form.

   a. 6 : 11   b. 4 to 1

   a. _____
   b. _____

3. Convert each ratio to verbal form.

   a. $\dfrac{12}{5}$ b. 9 : 4

   a. _____
   b. _____

4. Reduce each ratio to lowest terms.

   a. $\dfrac{24}{6}$ b. 15 : 10

   a. _____
   b. _____

5. One concrete block at a construction site weighed 42 pounds. A second concrete block weighed 70 pounds. Express the ratio, in lowest terms in colon form, of the weight of the lighter block to the weight of the heavier block.

   _____

6. Determine whether $\dfrac{15}{51} = \dfrac{5}{17}$ is a true proportion. _____

7. In the proportion $\dfrac{3}{10} = \dfrac{45}{150}$ what are the means and what are the extremes?

   means: _____

   extremes: _____

8. Solve for the missing term in the proportion.

   $\dfrac{9}{14} = \dfrac{n}{70}$

   _____

9. Every $\dfrac{1}{2}$ inch on a map represents a distance of 50 miles. How far apart are two cities if they are $2\dfrac{1}{2}$ inches apart on the map?

   _____

PRACTICE TEST 4B

1. Convert each ratio to colon form.

   a. 8 to 17   b. $\dfrac{12}{5}$

2. Convert each ratio to fraction form.

   a. 22 to 15   b. 3 : 11

3. Convert each ratio to verbal form.

   a. $\dfrac{6}{23}$ b. 1 : 10

4. Reduce each ratio to lowest terms.

   a. $\dfrac{25}{40}$ b. 18 : 8

5. A factory employs 48 men and 25 women. Express the ratio, in lowest terms in fraction form, of women to the total number of employees.

6. Determine whether $\dfrac{13}{24} = \dfrac{50}{96}$ is a true proportion.

7. In the proportion $\dfrac{19}{15} = \dfrac{38}{30}$ what are the means and what are the extremes?

   means: _____

   extremes: _____

8. Solve for the missing term in the proportion.

   $\dfrac{n}{20} = \dfrac{25.5}{100}$

9. A car will go 300 miles on 15 gallons of gasoline. How far will the car go on 18 gallons of gasoline at the same rate?

CHAPTERS 1-4

SKILLS REVIEW

1. Write 476,094 in words.
   _____

2. Write thirty-seven thousand, eight hundred twelve in numbers.
   _____

3. Round 98,746 to the nearest hundred. _____

4. Round 364.0294 to the nearest hundredth. _____

Perform the indicated operation. Express answers in lowest terms.

5. $\dfrac{3}{5} + \dfrac{7}{8} + \dfrac{9}{10} =$ _____

6. $\dfrac{5}{7} - \dfrac{1}{3} =$ _____

7. $\dfrac{2}{5} \cdot \dfrac{15}{22} =$ _____

8. $\dfrac{7}{10} \div \dfrac{3}{4} =$ _____

9. $(17.6)(2.023) =$ _____

10. $37.5 \div 2.5 =$ _____

11. Convert $\dfrac{5}{16}$ to a decimal (four places). _____

12. Write the ratio of 12 to 15 in fraction form in lowest terms. _____

13. Is $\dfrac{13}{52} = \dfrac{39}{156}$ a true proportion? _____

14. Solve for the missing term:
    $\dfrac{8}{17} = \dfrac{n}{51}$ _____

CHAPTER 4   SOLUTIONS

Skills Preview

1. *a.* 8 : 9
   b. 13 : 9

2. *a.* $\dfrac{5}{18}$
   b. $\dfrac{22}{9}$

3. *a.* 10 to 3
   b. 18 to 1

4. *a.* 1 : 4
   b. 3 : 2

5. $\dfrac{25}{21}$

6. yes

7. means: 11, 27
   extremes: 9, 33

8. 5

9. 12

Section 4.1   Comprehension Checkpoint

1. *a.* 4 : 6
   b. 5 : 3

2. *a.* $\dfrac{5}{9}$
   b. $\dfrac{6}{7}$

3. *a.* 4 to 8
   b. 3 to 1

4. *a.* $\dfrac{1}{3}$
   b. $\dfrac{9}{1}$
   c. $\dfrac{9}{7}$

4.1A   Practice Set

	Verbal Form	Colon Form	Fraction Form			
1.	1 to 4	1 : 4	$\dfrac{1}{4}$			
2.	5 to 9	5 : 9	$\dfrac{5}{9}$			
3.	3 to 10	3 : 10	$\dfrac{3}{10}$			
4.	15 to 1	15 : 1	$\dfrac{15}{1}$			
5.	7 to 2	7 : 2	$\dfrac{7}{2}$			
6.	12 to 8	12 : 8	$\dfrac{12}{8}$			

7. 18 to 12
   $\dfrac{18}{12} = \dfrac{18 \div 6}{12 \div 6} = \dfrac{3}{2}$ 3 to 2

8. 8 to 32
   $\dfrac{8}{32} = \dfrac{8 \div 8}{32 \div 8} = \dfrac{1}{4}$ 1 to 4

9. 28 : 7
   $\dfrac{28}{7} = \dfrac{28 \div 7}{7 \div 7} = \dfrac{4}{1}$ 4 : 1

10. 13 : 65
    $\dfrac{13}{65} = \dfrac{13 \div 13}{65 \div 13} = \dfrac{1}{5}$ 1 : 5

11. $\dfrac{36}{9} = \dfrac{36 \div 9}{9 \div 9} = \dfrac{4}{1}$

12. $\dfrac{20}{44} = \dfrac{20 \div 4}{44 \div 4} = \dfrac{5}{11}$

286   CHAPTER 4   Ratios and Proportions

13. $\dfrac{8 \text{ (Sam)}}{6 \text{ (Karen)}} = \dfrac{8 \div 2}{6 \div 2} = \dfrac{4}{3}$

14. $\dfrac{6 \text{ feet (longer)}}{2 \text{ feet (shorter)}} = \dfrac{6 \div 2}{2 \div 2} = \dfrac{3}{1}$ 3 to 1

15. $\dfrac{35 \text{ (apples)}}{40 \text{ (oranges)}} = \dfrac{35 \div 5}{40 \div 5} = \dfrac{7}{8}$ 7 : 8

16. $\dfrac{12 \text{ (larger)}}{8 \text{ (smaller)}} = \dfrac{12 \div 4}{8 \div 4} = \dfrac{3}{2}$ 3 to 2

17. $\dfrac{\$250 \text{ rent}}{\$650 \text{ total bills}} = \dfrac{250 \div 50}{650 \div 50} = \dfrac{5}{13}$ 5 : 13

18. $\dfrac{20 \text{ (3.5-inch)}}{4 \text{ (5.25-inch)}} = \dfrac{20 \div 4}{4 \div 4} = \dfrac{5}{1}$

19. $\dfrac{18 \text{ cassettes}}{28 \text{ total sales}} = \dfrac{18 \div 2}{28 \div 2} = \dfrac{9}{14}$

20. $\dfrac{20 \text{ calls}}{8 \text{ hours}} = \dfrac{20 \div 8}{8 \div 8} = \dfrac{2.5}{1}$ 2.5 to 1

Section 4.2   Comprehension Checkpoint

1. **a.** means: 4, 6
extremes: 3, 8

   b. means: 10, 6
extremes: 2, 30

2. **a.** no
   b. yes

4.2A   Practice Set

1. means: 5, 16
extremes: 4, 20

2. means: 8, 18
extremes: 12, 12

3. means: 1, 20
extremes: 10, 2

4. $(7)(24) \stackrel{?}{=} (8)(21)$
$168 = 168$
true proportion

5. $(8)(25) \stackrel{?}{=} (5)(35)$
$200 \neq 175$
not a true proportion

6. $(1)(36) \stackrel{?}{=} (3)(12)$
$36 = 36$
true proportion

7. $(9)(45) \stackrel{?}{=} (15)(18)$
$405 \neq 270$
not a true proportion

8. $(9)(12) \stackrel{?}{=} (6)(18)$
$108 = 108$
true proportion

9. $(7)(42) \stackrel{?}{=} (14)(21)$
$294 = 294$
true proportion

10. $(3)(14) \stackrel{?}{=} (2)(18)$
$42 \neq 36$
not a true proportion

11. $(20)(20) \stackrel{?}{=} (100)(5)$
$400 \neq 500$
not a true proportion

12. $(13)(156) \stackrel{?}{=} (39)(52)$
$2{,}028 = 2{,}028$
true proportion

13. $(24)(36) \stackrel{?}{=} (6)(144)$
$864 = 864$
true proportion

14. $(12)(100) \stackrel{?}{=} (25)(48)$
$1{,}200 = 1{,}200$
true proportion

15. $(33)(4) \stackrel{?}{=} (1)(132)$
$132 = 132$
true proportion

16. $(17)(4) \stackrel{?}{=} (1)(51)$
$68 \neq 51$
not a true proportion

17. $\dfrac{5}{40} \stackrel{?}{=} \dfrac{4}{30}$
$(5)(30) \stackrel{?}{=} (4)(40)$
$150 \neq 160$
The ratios are not the same.

18. $\dfrac{3}{6} \stackrel{?}{=} \dfrac{4}{8}$
$(3)(8) \stackrel{?}{=} (6)(4)$
$24 = 24$
The ratios are the same.

19. $\dfrac{8}{1} \stackrel{?}{=} \dfrac{56}{7}$
$(8)(7) \stackrel{?}{=} (1)(56)$
$56 = 56$
The ratios are equal.

20. $\dfrac{25}{36} \stackrel{?}{=} \dfrac{15}{24}$
$(25)(24) \stackrel{?}{=} (36)(15)$
$600 \neq 540$
The ratios are not the same.

Section 4.3   Comprehension Checkpoint

1. 36   2. 0.75   3. 39

4.3A Practice Set

1. $\dfrac{(2)(9)}{3} = n$
   $6 = n$

2. $\dfrac{(5)(36)}{20} = n$
   $9 = n$

3. $\dfrac{(16)(12)}{24} = n$
   $8 = n$

4. $\dfrac{(8)(15)}{3} = n$
   $40 = n$

5. $\dfrac{(13)(24)}{52} = n$
   $6 = n$

6. $\dfrac{(18)(99)}{11} = n$
   $162 = n$

7. $\dfrac{(20)(7)}{140} = n$
   $1 = n$

8. $\dfrac{(24)(140)}{35} = n$
   $96 = n$

9. $\dfrac{\left(3\frac{1}{3}\right)(15)}{5} = n$
   $10 = n$

10. $\dfrac{\left(2\frac{1}{2}\right)\left(3\frac{3}{4}\right)}{1\frac{1}{4}} = n$
    $7\frac{1}{2} = n$

11. $\dfrac{(6)\left(3\frac{1}{4}\right)}{6\frac{1}{2}} = n$
    $3 = n$

12. $\dfrac{(23)(28.75)}{5.75} = n$
    $115 = n$

13. $\dfrac{(12)(3)}{0.6} = n$
    $60 = n$

14. $\dfrac{(1.6)(12)}{9.6} = n$
    $2 = n$

15. $\dfrac{10}{3} = \dfrac{n}{9}$
    $\dfrac{(10)(9)}{3} = n$
    $30 = n$

16. $\dfrac{1}{25} = \dfrac{n}{175}$
    $\dfrac{(175)(1)}{25} = n$
    $7 = n$

17. $\dfrac{1}{0.99} = \dfrac{12}{n}$
    $\dfrac{(0.99)(12)}{1} = n$
    $\$11.88 = n$

18. $\dfrac{7}{40.25} = \dfrac{8}{n}$
    $\dfrac{(8)(40.25)}{7} = n$
    $\$46 = n$

19. $\dfrac{3}{0.89} = \dfrac{12}{n}$
    $\dfrac{(0.89)(12)}{3} = n$
    $\$3.56 = n$

20. $\dfrac{3}{18} = \dfrac{n}{126}$
    $\dfrac{(3)(126)}{18} = n$
    $21 = n$

21. $\dfrac{3}{44} = \dfrac{n}{616}$
    $\dfrac{(3)(616)}{44} = n$
    $42 = n$

22. $\dfrac{9}{180} = \dfrac{n}{60}$
    $\dfrac{(9)(60)}{180} = n$
    $3 = n$

23. $\dfrac{45}{240} = \dfrac{n}{1{,}500}$
    $\dfrac{(45)(1{,}500)}{240} = n$
    $281.25 = n$

24. $\dfrac{1}{n} = \dfrac{15.5}{0.55}$
    $\dfrac{(1)(0.55)}{15.5} = n$
    $\$0.035 = n$

Practice Test 4A

1. **a.** 3 : 7
   b. 5 : 9

2. **a.** $\dfrac{6}{11}$
   b. $\dfrac{4}{1}$

3. **a.** 12 to 5
   b. 9 to 4

4. **a.** $\dfrac{4}{1}$
   b. 3 : 2

5. 3 : 5

6. yes

7. means: 10, 45
   extremes: 3, 150

8. 45

9. 250 miles

Skills Review   Chapters 1–4

1. four hundred seventy-six thousand, ninety-four
2. 37,812
3. 98,700
4. 364.03
5. $2\frac{3}{8}$
6. $\frac{8}{21}$
7. $\frac{3}{11}$
8. $\frac{14}{15}$
9. 35.6048
10. 15
11. 0.3125
12. $\frac{4}{5}$
13. yes
14. 24

PERCENTS

OUTLINE

5.1 Introduction to Percents

5.2 Calculating the Amount

5.3 Calculating the Rate

5.4 Calculating the Base

5.5 Percent Increase and Decrease

5.6 Percent Applications

Percents are another special type of fraction that can be used to solve many word problems. Working with percents provides you with an opportunity to use whole numbers, decimal numbers, and fractions to solve problems. Learning to use percents successfully will make everyday life easier. You will be able to compute a 9% increase in tuition. You can compare amounts using percents. A grade of 86% is a comparison of the number of correct answers with the total number of answers on a test.

SKILLS PREVIEW

1. Convert each percent to its decimal or whole-number equivalent.

    a. 28% _____   b. 0.5% _____   c. 250% _____

2. Convert each decimal or whole number to a percent.

    a. 0.42 _____   b. 0.06 _____   c. 3 _____

3. Identify the base, rate, and amount:

    8 is 20% of 40

    rate: _____   base: _____   amount: _____

4. What is 30% of 75? _____

5. What percent of 40 is 12? _____

6. 42% of what number is 63? _____

7. What is 48 increased by 12.5%? _____

8. What is 36 decreased by 15%? _____

9. The price of a dress is marked down for a sale from $89 to $69. What is the rate of decrease?
_____

10. A cruise ship has a capacity of 350 passengers. On one cruise, the ship was 92% full. How many passengers were on the ship? _____

5.1 INTRODUCTION TO PERCENTS

Percent is a frequently used math application. Percent means "per hundred." A 6% sales tax means that for every dollar spent (100 cents), 6 cents will be paid in taxes. The symbol used to indicate percent is %.

Remember that a decimal fraction is a fraction with a denominator of 10, 100, 1,000, and so on. A percent is a fraction with a denominator of 100. Thus,

$$6\% = \frac{6}{100} = 0.06$$

You can see that a percent can be converted to a common fraction or a decimal fraction.*

Converting a Percent to a Decimal

There are many applications using percent. Examples include a painter finishing 25% of a job before lunch or a credit card member getting a 1% discount for paying on time. Using percent is a convenient way to show comparisons. An

---

*Converting percents to common fractions will be covered in Chapter 6.

Introduction to Percents

example of a comparison is a firm announcing that sales this year are 9% more than last year. *To solve a problem involving percent, the percent must be converted to its decimal (or whole number) equivalent.*

TO CONVERT A PERCENT TO A DECIMAL

1. Move the decimal point two places to the left. (This is the same as dividing by 100.)
2. Drop the percent sign.

EXAMPLE 1 Convert each percent to its decimal equivalent.

a. 58%   *b.* 125%
c. 0.9%   *d.* 400%

SOLUTIONS

a. 58% = 0.58 The decimal point in 58 is to the right of the 8.
b. 125% = 1.25 The decimal point in 125 is to the right of the 5. Always move it two places.
c. 0.9% = 0.009 The decimal point always moves two places regardless of its location in the original number.
d. 400% = 4 Move the decimal point two places to the left and drop the % sign. This conversion results in a whole number.

Converting a Decimal to a Percent

Converting a decimal number (or whole number) to a percent is the reverse of converting a percent to a decimal number.

TO CONVERT A DECIMAL TO A PERCENT

1. Move the decimal point two places to the right. (This is the same as multiplying by 100.)
2. Add the percent symbol.

EXAMPLE 2 Convert each number to a percent.

a. 0.75   *b.* 0.023
c. 0.4   *d.* 5

SOLUTIONS

a. 0.75 = 75%. Move the decimal point two places to the right and add the symbol %. The 0 to the left of the 7 is dropped.
b. 0.023 = 2.3%. Always move the decimal point two places to the right. Drop both zeros to the left of the 2. Add the symbol %.
c. 0.4 = 40%. Always move the decimal point two places to the right, regardless of its original position. Add one zero so that the decimal point can be moved two places, and add the symbol %.
d. 5 = 500%. The decimal point in a whole number is to the right of the units digit. Add two zeros so that the decimal point can be moved two places, and add the symbol %.

The Basic Percent Formula

Consider the problem

$$40\% \text{ of } 60 \text{ is what number?}$$

The rate is always the number with the symbol % or before the word *percent*.

The base is the number after the phrase *percent of*.

The amount usually comes immediately before or after the word *is*.

Terms that you need to know to be able to work with percents effectively are **base, rate,** and **amount.** The rate is always the number with the percent sign attached. The rate shows the relationship between the base and the amount. The base represents the whole or the original number. The amount is the part of the base found by calculating a percent. The amount can be less than, equal to, or greater than the base.

A **formula** is a set of symbols expressing a mathematical principle. The basic percent formula shows the relationship among the three parts of a percent problem.

$$\boxed{\text{base} \times \text{rate} = \text{amount}}$$

The same formula could be written with letters to represent each element:

$$b \times r = a$$

This formula shows that the product of the base and the rate is the amount.

The original problem

$$40\% \text{ of } 60 \text{ is what number?}$$

could be substituted into the formula

$$b \times r = a$$

$$60 \times 40\% = a$$

and you can see that the amount is the unknown quantity.

Next look at the problem

$$25\% \text{ of what number is } 30?$$

The rate is 25%, since the symbol % is attached to 25. The amount is 30, because it follows the word *is*. The base is the unknown quantity in this problem.

Finally, consider the problem

$$\text{What percent of } 80 \text{ is } 40?$$

The number after the phrase *percent of* is the base, and the number after the word *is* is the amount. There is no number before the word *percent* (and there is no symbol %). This indicates that the rate is the unknown quantity in this problem.

In any percent problem, you will be given two of the three parts. Your task will be to determine the third part. To accomplish this, you must be able to identify what part you are trying to find.

EXAMPLE 3 Identify the unknown part in each statement.

a. 42% of 200 is what number?
b. What percent of 60 is 6?
c. 54% of what number is 20?

Introduction to Percents

SOLUTIONS

a. Amount: You are given the rate (42%) and the base (200 follows *percent of*).

b. Rate: You are given the base (60 follows *percent of*) and the amount (6 follows the word *is*).

c. Base: You are given the rate (54%) and the amount (20 follows the word *is*).

The next three sections in this chapter will show you how to solve percent problems for each of the three parts. Now try the Comprehension Checkpoint problems to measure your understanding of percents. If you have difficulty with these problems, review the material in this section before going on to the Practice Sets.

COMPREHENSION CHECKPOINT

1. Convert each percent to its decimal or whole-number equivalent.

    a. 47% _____ *b.* 0.05% _____

    c. 6.3% _____ *d.* 400% _____

2. Convert each decimal or whole number to a percent.

    a. 0.64 _____ *b.* 0.072 _____

    c. 5.9 _____ *d.* 8 _____

3. Identify the missing part in each statement.

    a. 65% of 300 is what number? _____

    b. What percent of 25 is 20? _____

    c. 15% of what number is 60? _____

5.1A PRACTICE SET

Convert each percent to its decimal or whole-number equivalent.

1. 82% _____
2. 70% _____
3. 37% _____
4. 42.5% _____
5. 91.37% _____
6. 0.4% _____
7. 6% _____
8. 150% _____
9. 200% _____
10. 1,000% _____

Convert each decimal or whole number to a percent.

11. 0.38 _____
12. 0.50 _____
13. 0.84 _____
14. 0.008 _____
15. 0.6 _____
16. 0.625 _____
17. 0.05 _____
18. 4.2 _____
19. 3 _____
20. 1 _____

Identify the missing part (rate, base, or amount) in each statement.

21. What percent of 60 is 45? _____
22. 800 is what percent of 1,000? _____
23. What number is 15% of 45? _____
24. 110 is 25% of what number? _____
25. What percent of 50 is 16? _____
26. 250 is 20% of what number? _____
27. 25% of 64 is what number? _____

28. 12 is what percent of 36? _____

29. 30 is 18% of what number? _____

30. 12% of 42 is what number? _____

5.1B PRACTICE SET

Convert each percent to its decimal or whole-number equivalent.

1. 41% _____
2. 92% _____
3. 79% _____
4. 62.6% _____
5. 28.75% _____
6. 0.9% _____
7. 8% _____
8. 175% _____
9. 600% _____
10. 2,000% _____

Convert each decimal or whole number to a percent.

11. 0.65 _____
12. 0.18 _____
13. 0.53 _____
14. 0.065 _____
15. 0.2 _____
16. 0.375 _____
17. 0.01 _____
18. 6.9 _____
19. 7 _____
20. 12 _____

Identify the missing part (rate, base, or amount) in each statement.

21. What percent of 30 is 10? _____
22. 600 is what percent of 950? _____
23. What number is 25% of 70? _____
24. 96 is 40% of what number? _____
25. What percent of 100 is 65? _____
26. 37 is 30% of what number? _____
27. 50% of 90 is what number? _____

28. 60 is what percent of 48? _____

29. 310 is 40% of what number? _____

30. 15% of 38 is what number? _____

5.2 CALCULATING THE AMOUNT

If you are given the base and the rate in a percent problem, you can determine the amount. Substituting the given numbers in the basic percent equation and multiplying them will produce the amount. Recall from Section 5.1 that the basic percent formula is

$$b \times r = a$$

EXAMPLE 1 What is 65% of 120?

SOLUTION

$b \times r = a$	*Write the basic percent equation.*
$120 \times 65\% = a$	*Substitute the given values.*
$120 \times 0.65 = a$	*Change the rate to its decimal equivalent. Multiply the base and the rate.*
$78 = a$	*The amount is 78.*

65% of 120 is 78.

The amount can be equal to the base. If the rate is 100%, the amount will be the same as the base.

EXAMPLE 2 What is 100% of 12?

SOLUTION

$b \times r = a$	*Write the basic percent equation.*
$12 \times 100\% = a$	*Substitute the given values.*
$12 \times 1 = a$	*Convert the rate to its whole-number equivalent. Multiply the base and the rate.*
$12 = a$	*The amount is 12, the same as the base.*

100% of 12 is 12.

Example 2 demonstrates that the base is always 100% of itself.

The amount can also be more than the base. If the rate is more than 100%, the amount will be more than the base.

EXAMPLE 3 130% of 40 is what number?

SOLUTION

$b \times r = a$	*Write the basic percent equation.*
$40 \times 130\% = a$	*Substitute the given values.*
$40 \times 1.3 = a$	*Convert the rate to its decimal equivalent. Multiply the base and the rate.*
$52 = a$	*The amount is 52, which is more than the base.*

130% of 40 is 52.

CHAPTER 5 Percents

Each example in this section illustrates that if you are given the base and the rate, you should use the basic percent formula to find the amount. The basic percent formula is "solved" for the amount. You will learn in Sections 5.3 and 5.4 that this formula can be adjusted and used to solve for the rate or the base.

Work the Comprehension Checkpoint problems. If you have difficulty with these problems, review the material in this section before going on to the Practice Sets.

COMPREHENSION CHECKPOINT

Determine the amount in each problem.

1. 42% of 140 is what number? _____

2. What is 8% of $35? _____

3. 100% of 246 is what number? _____

4. 112% of 315 is what number? _____

5. What is 300% of 9? _____

5.2A PRACTICE SET

Determine the amount in each problem. Round each answer to the nearest tenth (or the nearest cent if you are working with dollars).

1. $b = \$180$ $r = 20\%$ _____
2. $b = 43$ $r = 6\%$ _____
3. $b = \$162$ $r = 48\%$ _____
4. $b = 86$ $r = 3.8\%$ _____
5. $b = 48$ $r = 225\%$ _____
6. $b = 23$ $r = 600\%$ _____
7. $b = 76$ $r = 0.5\%$ _____
8. $b = 1{,}250$ $r = 0.08\%$ _____
9. $b = \$470$ $r = 98\%$ _____
10. $b = \$38$ $r = 12.6\%$ _____
11. What is 450% of $1? _____
12. What is 100% of 12.7? _____
13. 7.5% of 224 is what number? _____
14. 14.8% of 346 is what number? _____
15. What is 109% of 280? _____
16. What is 50% of $436? _____
17. What is 20.5% of 802? _____
18. 32% of $40,000 is what dollar amount? _____
19. 8.5% of $55,600 is what dollar amount? _____
20. 175% of $6,000 is what dollar amount? _____

Applications

21. Texas has a sales tax of 7.25%. How much tax would be charged on a sweater that cost $50?

22. Amy must correctly answer 70% of the questions on her English final exam to pass. If there are 120 questions on the exam, how many must she answer correctly to pass?

23. A family that recycles gets a 5% discount on its utility bill. What would be the amount of the discount if the utility bill is $115?

5.2B PRACTICE SET

Determine the amount in each problem. Round each answer to the nearest tenth (or the nearest cent if you are working with dollars).

1. $b = \$260$ $r = 25\%$ _____

2. $b = 84$ $r = 9\%$ _____

3. $b = \$125$ $r = 15\%$ _____

4. $b = 106$ $r = 5.4\%$ _____

5. $b = 80$ $r = 375\%$ _____

6. $b = 57$ $r = 500\%$ _____

7. $b = 92$ $r = 0.8\%$ _____

8. $b = 4,210$ $r = 0.02\%$ _____

9. $b = \$590$ $r = 24\%$ _____

10. $b = \$76$ $r = 30.6\%$ _____

11. What is 250% of 3? _____

12. What is 100% of 9,452? _____

13. 3.1% of 182 is what number? _____

14. 78.3% of 281 is what number? _____

15. What is 185% of $310? _____

16. What is 60% of $800? _____

17. What is 40.2% of 327? _____

18. 75% of $80,000 is what dollar amount? _____

19. 6.25% of $29,200 is what dollar amount? _____

20. 325% of $1,500 is what dollar amount? _____

Applications

21. During the lunch hour, a fast-food restaurant made 150 burgers. Only 96% of the burgers were sold. How many burgers were sold during the lunch hour?

    _____

22. A car loses about 30% of its original value during its first year. How much of the original value of $9,500 will a car lose in the first year?

    _____

23. Chris sells cosmetics. She earns 6% commission on everything she sells. What would her commission be on a sale of $680?

    _____

5.3 CALCULATING THE RATE

A variation of the basic percent formula is used to calculate the rate. If you take the formula

$$\text{base} \times \text{rate} = \text{amount}$$

and divide each side by the base:

$$\frac{\text{base} \times \text{rate}}{\text{base}} = \frac{\text{amount}}{\text{base}}$$

you get

$$\boxed{\text{rate} = \frac{\text{amount}}{\text{base}} \quad \text{or} \quad r = \frac{a}{b}}$$

This new formula is "solved" for the rate. Substituting the given values for the base and amount into this formula and dividing will give the rate in decimal form. You must then convert it to a rate form with the symbol %.

EXAMPLE 1 What percent of 16 is 4?

SOLUTION

$r = \dfrac{a}{b}$ Write the proper formula.

$r = \dfrac{4}{16}$ Substitute the given values into the formula.

$r = 0.25$ Divide the amount by the base.

$r = 25\%$ Convert the decimal to its percent equivalent.

4 is 25% of 16.

If the amount is more than the base, the rate will be more than 100%.

EXAMPLE 2 What percent of 24 is 60?

SOLUTION

$r = \dfrac{a}{b}$ Write the proper formula.

$r = \dfrac{60}{24}$ Substitute the given values into the formula.

$r = 2.5$ Divide the amount by the base.

$r = 250\%$ Convert the decimal number to its percent equivalent.

60 is 250% of 24.

If the amount and the base are the same, the rate will be 100%.

EXAMPLE 3 What percent of 25 is 25?

SOLUTION

$r = \dfrac{a}{b}$ Write the proper formula.

$r = \dfrac{25}{25}$ Substitute the given values into the formula.

$r = 1$ Divide the amount by the base.

$r = 100\%$ Convert the number to its percent equivalent.

100% of 25 is 25.

Whenever you determine that you are looking for the rate, use the formula $r = \dfrac{a}{b}$. Remember to make the conversion to rate form after dividing.

Try the Comprehension Checkpoint problems. If you have difficulty with these problems, review the material in this section before going on to the Practice Sets.

COMPREHENSION CHECKPOINT

Calculate the rate in each problem. Round to the nearest tenth if necessary.

1. What percent of 40 is 26? _____

2. What percent of 48 is 6? _____

3. 13.5 is what percent of 225? _____

4. 50 is what percent of 10? _____

5. What percent of 30 is 75? _____

6. What percent of 6 is 6? _____

5.3A PRACTICE SET

Calculate the rate in each problem. Round each answer to the nearest tenth if necessary.

1. $b = 60$ $a = 6$ _____
2. $b = 90$ $a = 75$ _____
3. $b = 25$ $a = 12$ _____
4. $b = 20$ $a = 17$ _____
5. $b = 100$ $a = 47$ _____
6. $b = 7$ $a = 21$ _____
7. $b = 12$ $a = 66$ _____
8. $b = 25$ $a = 90$ _____
9. $b = 35$ $a = 140$ _____
10. $b = 150$ $a = 120$ _____
11. What percent of 82 is 82? _____
12. What percent of 35 is 14? _____
13. What percent of 45 is 180? _____
14. What percent of 2.9 is 8.7? _____
15. What percent of 25 is 625? _____
16. 30 is what percent of 240? _____
17. 70 is what percent of 16? _____
18. 84 is what percent of 300? _____
19. 280 is what percent of 28? _____
20. 9 is what percent of 1,500? _____

Applications

21. A carpenter cut a 6-inch-long section off a board that was 48 inches long. What percent of the board did he cut off?

    _____

22. If you pay $22.50 tax on a purchase of $300, what is the tax rate?

    _____

23. Mr. Harrison has a total of 240 students. One day 72 of his students were absent. What percent of his students were absent?

    _____

5.3B PRACTICE SET

Calculate the rate in each problem. Round each answer to the nearest tenth if necessary.

1. $b = 20$ $a = 5$ _____

2. $b = 80$ $a = 60$ _____

3. $b = 36$ $a = 9$ _____

4. $b = 95$ $a = 19$ _____

5. $b = 180$ $a = 135$ _____

6. $b = 12$ $a = 48$ _____

7. $b = 15$ $a = 37.5$ _____

8. $b = 47$ $a = 47$ _____

9. $b = 81$ $a = 108$ _____

10. $b = 300$ $a = 125$ _____

11. What percent of 64 is 48? _____

12. What percent of 25 is 9? _____

13. What percent of 52 is 78? _____

14. What percent of 2.5 is 12.5? _____

15. What percent of 40 is 35? _____

16. 28 is what percent of 98? _____

17. 65 is what percent of 13? _____

18. 75 is what percent of 500? _____

19. 425 is what percent of 42.5? _____

20. 6 is what percent of 3,000? _____

Applications

21. Lori borrowed $1,500 for one year. Interest on the loan was $223.50. What was the interest rate?

    _____

22. The original price of a dress was $80, but the dress was on sale for $10 off. What was the discount rate?

    _____

23. In a local wet/dry election, 720 of the 960 eligible voters actually voted. What percent of the eligible voters voted?

    _____

5.4 CALCULATING THE BASE

Calculating the base requires another variation of the basic percent formula. Divide each side of the formula by the rate to produce a different formula to be solved for the base.

$$\frac{\text{base} \times \text{rate}}{\text{rate}} = \frac{\text{amount}}{\text{rate}}$$

$$\boxed{\text{base} = \frac{\text{amount}}{\text{rate}} \quad \text{or} \quad b = \frac{a}{r}}$$

Dividing the amount by the rate gives the base. Remember that the rate must be converted to a decimal before dividing.

EXAMPLE 1   40% of what number is 20?

SOLUTION

$b = \dfrac{a}{r}$ *Write the proper formula.*

$b = \dfrac{20}{40\%}$ *Substitute the given values into the formula.*

$b = \dfrac{20}{0.4}$ *Convert the rate to its decimal equivalent. Divide the amount by the rate.*

$b = 50$ *The base is 50.*

40% of 50 is 20.

If the rate is more than 100%, the base will be less than the amount.

EXAMPLE 2   125% of what number is 40?

SOLUTION

$b = \dfrac{a}{r}$ *Write the proper formula.*

$b = \dfrac{40}{125\%}$ *Substitute the given values into the formula.*

$b = \dfrac{40}{1.25}$ *Convert the rate to its decimal equivalent. Divide the amount by the rate.*

$b = 32$ *The base is 32.*

125% of 32 is 40.

When the rate is 100%, the base will be the same as the amount.

ALGEBRA CONNECTION

`a + b = x`

In Section 5.1, you learned about the basic percent formula, $b \times r = a$. This formula shows a fundamental relationship between the elements of a percent problem. The essence of mathematics is the study of relationships between numbers.

If you are given any formula, you can adjust it to solve for any of the variables. A formula that you will learn in Chapter 10, Introduction to Geometry, is

$$A = \frac{1}{2}bh$$

This is the formula for the area of a triangle. In the formula, A stands for area, b stands for base, and h stands for height. If you know the base and height of a triangle, you can compute the area by using this formula. What would you do if you knew the area and the base and you wanted to find the height? Solve the formula for h by multiplying each side by 2 and then dividing each side by b. You get

$$\frac{2A}{b} = h$$

and the formula is solved for the height. The same formula could be solved for b if you needed to find the base.

$$\frac{2A}{h} = b$$

As you learned in this chapter, being able to adjust a formula gives you much more flexibility in solving problems. As you learn more math, you will find that there will be more ways to solve the same problems.

EXAMPLE 3   100% of what number is 10?

SOLUTION

$b = \dfrac{a}{r}$ *Write the proper formula.*

$b = \dfrac{10}{100\%}$ *Substitute the given values into the formula.*

$b = \dfrac{10}{1}$ *Convert the rate to its decimal equivalent (a whole number in this case). Divide the amount by the rate.*

$b = 10$ *The base is 10.*

100% of 10 is 10.

Try the Comprehension Checkpoint problems to measure your understanding of how to calculate a base number. If you have difficulty with these problems, review the material in this section before going on to the Practice Sets.

COMPREHENSION CHECKPOINT

Calculate the base in each problem.

1. 20% of what number is 9? _____

2. 35% of what number is 105? _____

3. 5% of what number is 3? _____

4. 150% of what number is 18? _____

5. 300% of what number is 21? _____

6. 100% of what number is 42? _____

5.4A PRACTICE SET

Calculate the base in each problem.

1. $r = 20\%$ $a = 60$ _____

2. $r = 15\%$ $a = 12$ _____

3. $r = 8\%$ $a = 32$ _____

4. $r = 22\%$ $a = 33$ _____

5. $r = 9\%$ $a = 3.6$ _____

6. $r = 135\%$ $a = 27$ _____

7. $r = 160\%$ $a = 20$ _____

8. $r = 250\%$ $a = 75$ _____

9. $r = 125\%$ $a = 100$ _____

10. $r = 150\%$ $a = 48$ _____

11. 40% of what number is 24? _____

12. 65% of what number is 156? _____

13. 24% of what number is 16.8? _____

14. 100% of what number is 12.8? _____

15. 65 is 130% of what number? _____

16. 48 is 120% of what number? _____

17. 84 is 140% of what number? _____

18. 95 is 125% of what number? _____

19. 63 is 112.5% of what number? _____

20. 198 is 300% of what number? _____

Applications

21. Ismail sells encyclopedias. One week he sold 12 sets. This represented 24% of his monthly quota. What was his monthly quota?

_____

22. A section of a new highway is 25 miles long. The new highway is 40% complete. How long will the highway be when construction is finished?

_____

23. There were 20 rainy days last April. This was 125% of the number of rainy days during March. How many rainy days did March have?

_____

5.4B PRACTICE SET

Calculate the base in each problem.

1. $r = 30\%$ $a = 36$ _____

2. $r = 25\%$ $a = 9$ _____

3. $r = 7\%$ $a = 10.5$ _____

4. $r = 28\%$ $a = 126$ _____

5. $r = 14\%$ $a = 45.5$ _____

6. $r = 120\%$ $a = 96$ _____

7. $r = 175\%$ $a = 350$ _____

8. $r = 300\%$ $a = 42$ _____

9. $r = 260\%$ $a = 78$ _____

10. $r = 110\%$ $a = 440$ _____

11. 35% of what number is 28? _____

12. 90% of what number is 54? _____

13. 23% of what number is 9.2? _____

14. 50% of what number is 3.4? _____

15. 12 is 150% of what number? _____

16. 58 is 145% of what number? _____

17. 21 is 200% of what number? _____

18. 8.8 is 110% of what number? _____

19. 250 is 125% of what number? _____

20. 20 is 1,000% of what number? _____

Applications

21. Marquis delivers pizzas. One night he delivered six pizzas the first hour of his shift. If this was 40% of the pizzas he delivered that night, how many pizzas did he deliver?

_____

22. A plane is 75% full. There are 141 passengers on board. How many passengers will the plane hold?

_____

23. The Good Times Company decided to have a picnic for its employees and their guests. The total number of people attending the picnic was 120, which was equivalent to 80% of the workforce. How many people did the Good Times Company employ?

_____

5.5 PERCENT INCREASE AND DECREASE

Percent is often used as a method to compare figures over time. A university may announce that tuition will be 8% more this year than last year, for example. The decrease in the price of a house, from the original asking price to the actual selling price, could also be expressed as a percent.

In increase or decrease problems, the original value is the base. The increased or decreased number is the new total. The percent increase or decrease is always given as a rate applied to the base.

> **TO FIND THE PERCENT INCREASE OR DECREASE**
> 1. Determine the amount of increase or decrease.
> 2. Divide the amount from step 1 by the base.

In formula form, it looks like this:

$$\text{percent increase (or decrease)} = \frac{\text{amount of change}}{\text{base}}$$

EXAMPLE 1 What is the percent increase from $30 to $36?

SOLUTION

$36   *new total*
− 30   *base*
$ 6   *amount of change*

$$\text{Percent increase} = \frac{\text{amount of change}}{\text{base}}$$

$$\text{Percent increase} = \frac{6}{30} = 0.20 = 20\%$$

The procedure is essentially the same for percent decrease problems.

EXAMPLE 2 What is the percent decrease from $0.99 to $0.69?

SOLUTION

$0.99   *base*
− 0.69   *new total*
$0.30   *amount of change*

$$\text{Percent decrease} = \frac{\text{amount of change}}{\text{base}}$$

$$\text{Percent decrease} = \frac{0.30}{0.99} = 0.303 = 30.3\%$$

If you know the base and the rate of increase, you can compute the new total.

EXAMPLE 3   What is 80 increased by 12%?

SOLUTION

Either of two methods can be used to solve this problem.

Method 1

$b \times r = a$	Use the basic percent formula to find the amount of increase.
$80 \times 12\% = a$	
$80 \times 0.12 = a$	
$9.6 = a$	This is the amount of increase.
$80 + 9.6 =$ new total	Add the increase amount to the base.
$89.6 =$ new total	

Method 2

$100\% + 12\% = 112\%$	The new total will be 112% of the base, since the base is 100% and the increase amount is 12%.
$b \times r = a$	Use the basic percent formula.
$80 \times 112\% = a$	
$80 \times 1.12 = a$	
$89.6 = a$	The result of this step is the new total.

Notice that the new total is the same using either method to solve the problem.

If you are given the base and the rate of change in a decrease problem, you can find the new total.

EXAMPLE 4   What is 35 decreased by 18%?

SOLUTION

There are two methods for solving a decrease problem as well.

Method 1

$b \times r = a$	Use the basic percent formula to find the amount of decrease.
$35 \times 18\% = a$	
$35 \times 0.18 = a$	
$6.3 = a$	This is the amount of decrease.
$35 - 6.3 =$ new total	Subtract the amount of change from the base to find the new total.
$28.7 =$ new total	

Method 2

$100\% - 18\% = 82\%$	Subtract the rate of change from 100%. The new total will be the base minus the amount of change.
$b \times r = a$	Use the basic percent formula with the rate from the first step to compute the new total.
$35 \times 82\% = a$	
$35 \times 0.82 = a$	
$28.7 = a$	This is the new total.

As with increase problems, the answer is the same using either method.

A significant difference between increase or decrease problems and problems where you are just finding the amount or rate is that increase and decrease problems require two steps to solve.

Work the Comprehension Checkpoint problems to assess your understanding of increase and decrease problems. If you have difficulty with these problems, review the material in this section before going on to the Practice Sets.

COMPREHENSION CHECKPOINT

Determine the rate of increase or decrease.

1. $48 increased to $60 _____

2. $90 decreased to $72 _____

Calculate the new total.

3. 96 increased by 12.5% _____

4. 42 decreased by 5% _____

5.5A PRACTICE SET

Calculate the percent increase.

1. 50 increased to 54 _____

2. 75 increased to 90 _____

3. 150 increased to 195 _____

4. $400 increased to $464 _____

5. $225 increased to $247.50 _____

6. $1,200 increased to $1,302 _____

Calculate the percent decrease.

7. 80 decreased to 64 _____

8. 95 decreased to 57 _____

9. $150 decreased to $138 _____

10. $125 decreased to $62.50 _____

11. $56 decreased to $49 _____

12. $500 decreased to $462.50 _____

Calculate the new total.

13. $64 increased by 5% _____

14. 40 increased by 20% _____

15. 30 increased by 50% _____

16. $88 increased by 12.5% _____

17. $120 increased by 15% _____

18. $240 decreased by 6.25% _____

19. $1,550 decreased by 2% _____

20. 40 decreased by 9% _____

21. 65 decreased by 18% _____

22. 110 decreased by 6.5% _____

Applications

23. A manufacturer decreased the wholesale price of a $250 table by 10%. What was the new wholesale price of the table?

    _____

24. Jennifer invested $375 in savings bonds. Her investment increased by 12% in one year. What was the value of her investment after one year?

    _____

25. Round-trip airfare overseas was $1,080. The airline decreased it by 15% during fare wars. What was the new airfare?

    _____

5.5B PRACTICE SET

Calculate the percent increase.

1. 35 increased to 38.5 _____

2. 60 increased to 69 _____

3. 140 increased to 175 _____

4. $200 increased to $212.50 _____

5. $625 increased to $650 _____

6. $2,400 increased to $2,520 _____

Calculate the percent decrease.

7. 40 decreased to 32 _____

8. 75 decreased to 67.5 _____

9. 160 decreased to 140 _____

10. $210 decreased to $199.50 _____

11. $340 decreased to $311.10 _____

12. $495 decreased to $297 _____

Calculate the new total.

13. $25 increased by 6% _____

14. 70 increased by 11% _____

15. 120 increased by 12.5% _____

16. $85 increased by 20% _____

17. $112 increased by 8.5% _____

18. $380 decreased by 7.25% _____

19. $650 decreased by 14.5% _____

20. 72 decreased by 25% _____

21. 150 decreased by 3.5% _____

22. 290 decreased by 4% _____

Applications

23. Jamie's rent decreased from $200 to $181.40 for a summer special. What was the rate of decrease?

_____

24. Calvin earned $424 per week until he received a raise of 12.5%. How much did he earn per week after the raise?

_____

25. Lisa typed 90 words per minute using her manual typewriter. She bought a new word processor and was able to type 20% faster. How many words per minute can she type using the new word processor?

_____

5.6 PERCENT APPLICATIONS

In each of the previous four sections in this chapter, you learned how to solve a particular type of percent problem. To find the amount, you used the basic percent formula

$$\text{base} \times \text{rate} = \text{amount}$$

To find either the base or the rate, you used a variation of the formula. You were told what to solve for in each problem.

Mathematics is a participant sport rather than a spectator sport. You must be able to use what you have learned to solve everyday problems. In this section, you will be given problems and you will have to determine yourself what element is missing and then solve for that element using one of the methods described in the previous sections.

EXAMPLE 1   What percent of 14 is 3.5?

SOLUTION

First you must determine what information you are given. Remember that the number after the phrase *percent of* is the base. The number after the word *is* is the amount. You are given the base and the amount. You need to solve for the rate.

$r = \dfrac{a}{b}$ *Write the formula that you will use to solve for the rate.*

$r = \dfrac{3.5}{14}$ *Substitute the given amounts into the equation.*

$r = 0.25$ *Divide.*

$r = 25\%$ *Convert the decimal to a rate.*

3.5 is 25% of 14.

To solve Example 1, you merely had to look back to find a problem like it in one of the previous sections. With word problems, you must read the problem carefully. Analyze the wording of the problem and determine what it is that you are trying to find. Once you know what you are looking for, select the proper formula and substitute the given values. Work the formula to find the solution to the problem.

EXAMPLE 2   Patricia's shorthand speed is 120 words per minute. Her goal is to improve her speed by 25%. How many words per minute will she take shorthand if she reaches her goal?

SOLUTION

You are trying to determine Patricia's new shorthand speed. You know her current speed, 120 words per minute. You also know the rate she is trying to improve by. This is a percent increase problem. Use the method for finding the increase amount.

$b \times r = a$ $r = 100\% + 25\% = 125\%$

$120 \times 125\% = a$

$120 \times 1.25 = a$

$150 = a$

Patricia's new shorthand speed will be 150 words per minute if she reaches her goal.

EXAMPLE 3 The cost of a textbook is $32. The sales tax is 6.25% of the cost. How much sales tax will be paid on the book?

SOLUTION

You are looking for an amount. The cost of the book, $32, is the base that the tax will be a part of. The rate is 6.25%. Use the basic percent formula for finding the amount.

$b \times r = a$

$32 \times 6.25\% = a$

$32 \times 0.0625 = a$

$\$2 = a$

The tax on the book will be $2.

Notice that the wording in the two word-problem examples is similar to the wording in the problems in the previous sections. Solving a word problem should be a thought-provoking process. You must analyze the problem to determine what it is that you are trying to find. Once you have made that determination, analyze the given information to identify what it is that you already know. Then select a formula that will allow you to solve for the missing element and answer the question. *Be sure that the answer you give answers the question that is asked.*

Try the Comprehension Checkpoint problems. Make sure you know what is being asked for in each problem before you try to solve it. If you have difficulty with these problems, review the material in this section before going on to the Practice Sets.

COMPREHENSION CHECKPOINT

1. What is 60 increased by 15%? _____

2. 120 is what percent of 200? _____

3. The cost of a stamp increased from $0.25 to $0.29. What was the rate of increase?

   _____

4. How many of 4,500 students are freshmen if the freshman class includes 28% of the student body of a college?

   _____

5.6A   PRACTICE SET

Solve each problem.

1. 20% of 50 is what number? _____

2. 12 is 25% of what number? _____

3. 60 increased by 10% is what number? _____

4. What percent of 75 is 45? _____

5. What is 130 decreased by 40%? _____

6. What is 125 increased by 5.5%? _____

7. What is 37% of 250? _____

8. Decreasing 90 by 8% gives what number? _____

9. 28 is what percent of 700? _____

10. 5% of what number is 17? _____

Applications

11. Mario borrowed $450 for one year. The interest rate was 9%. How much interest will he pay at the end of the year?
_____

12. The sales tax is $2.25 on a shirt that costs $30. What is the sales tax rate?
_____

13. Steffi paid $700 tuition last year. This year the tuition will be $750 for the same number of credits. What is the percent increase (rounded to the nearest tenth of a percent) from last year to this year?
_____

14. A road crew is resurfacing a 25-mile segment of highway. How many miles have been completed if the job is 45% finished?
_____

15. Kim's utility bill last month was $86. This month her bill is $78. What is the rate of decrease from last month to this month? (Round to the nearest tenth of a percent.)
_____

16. A hotel had accepted reservations for 136 rooms for the July 4 holiday. This number was 85% of the rooms available. How many rooms total were available?

_____

17. A cart of groceries at store A costs $36.50. The same groceries at store B cost 8% more. What is the cost of the groceries at store B?

_____

18. ACE Management Group occupies offices that encompass 5,000 square feet of floor space. The copy and mail room has 125 square feet of floor space. What percent of the total floor space is in the copy and mail room?

_____

19. The Great Outdoors reduced summer merchandise at the end of the season. A patio set that normally sold for $190 was reduced by 15%. What was the new selling price of the patio set?

_____

20. Ivan planted 405 acres of corn. This amount was only 90% of the number of acres he planted the previous year. How many acres of corn did he plant the previous year?

_____

21. April correctly answered 32 questions on a test. The test included 40 questions. What percent of the questions did she answer correctly?

_____

22. Carmen earns $6.50 per hour. She will receive a 6% raise on her next paycheck. What will her new hourly rate be?

_____

23. Gilbert deposited $1,500 in a savings account. At the end of one year he had $1,552.50 in the account. The new total included the original investment and interest earned. What was the interest rate?

_____

24. Zachary's assignment for English literature included reading a 472-page novel. He planned to read 25% of the book each evening. How many pages will he need to read each evening?

_____

25. A beaker containing 30 ounces of saline solution is 75% full. How many ounces does it take to fill the beaker?

_____

5.6B PRACTICE SET

Solve each problem.

1. What is 12 increased by 20%? _____

2. What number is 40% of 70? _____

3. What percent of 65 is 26? _____

4. 25 decreased by 10% is what number? _____

5. 48 is 8% of what number? _____

6. 18% of 500 is what number? _____

7. 35 decreased by 5% is what number? _____

8. 6 is 15% of what number? _____

9. 24 increased by 30% is what number? _____

10. 16 is what percent of 128? _____

Applications

11. Home Furnishings decreased the price of an entertainment center from $625 to $575 for a sale. What was the percent decrease?

    _____

12. Rainfall in a west Texas county averaged 13.8 inches per month in 1991. During 1992, the average monthly rainfall fell by 8.5%. What was the average monthly rainfall for the county in 1992?

    _____

13. Tonia scored 78% on her final exam. There were 250 questions on the exam. How many did she answer correctly?

    _____

14. Enrollment in freshman English classes at a community college was 1,428 in 1990. This was 112% of the enrollment for 1989. What was the enrollment for 1989?

    _____

15. The sales tax on a $240 computer monitor is $15. What is the sales tax rate?

16. Rafael's hourly rate of pay went from $8.40 to $10.50 over a three-year period. What was the rate of increase?

17. The workforce at Owens Office Supply includes 18 female and 12 male employees. What percent of the employees is female?

18. The Mendozas' house payment each month is $525. This represents 48% of their total monthly bills. What is the total of their monthly bills?

19. The cost of a leather jacket was 22% more in 1992 than in 1991. If the jacket cost $295 in 1991, how much did it cost in 1992?

20. At the end of a trip, a truck driver had 16% of his fuel remaining. His fuel tanks hold a total of 540 gallons. How many gallons of fuel did he have remaining?

21. Babe Ruth hit 714 home runs during his career. Hank Aaron hit 5.74% more home runs than Ruth did. How many home runs did Aaron hit?

22. Drivers who successfully complete a defensive driving course are eligible for a 10% reduction in liability insurance premiums. Michelle paid $142 for her insurance before taking the course. How much was her premium after she took the course?

23. What is the percent decrease in the price of an automobile that listed for $12,000 and sold for $10,800?

24. What percent of an hour is used by a class that is scheduled for 50 minutes?

25. A full-time employee is one who works 40 hours per week. If the average full-time employee misses 3.5% of his or her scheduled hours, how many hours will the employee miss during a year? (There are 52 weeks per year.)

SUMMARY OF KEY CONCEPTS

KEY TERMS

percent (5.1): One hundredth of a quantity.

base (5.1): The whole amount or original amount on which a percent is calculated.

rate (5.1): A figure that shows the relationship between the base and the amount.

amount (5.1): The part of a base found by calculating a percent. It can be less than, equal to, or greater than the base.

formula (5.1): A set of symbols expressing a mathematical principle.

KEY RULES

To convert a percent to a decimal (5.1):

1. Move the decimal point two places to the left.
2. Drop the percent symbol.

To convert a decimal to a percent (5.1):

1. Move the decimal point two places to the right.
2. Add the percent symbol.

To find the percent increase or decrease (5.5):

1. Determine the amount of increase or decrease.
2. Divide the amount from step 1 by the base.

Formulas used to solve percent problems include:

(5.1) base × rate = amount $b \times r = a$

(5.3) rate = $\dfrac{\text{amount}}{\text{base}}$ $r = \dfrac{a}{b}$

(5.4) base = $\dfrac{\text{amount}}{\text{rate}}$ $b = \dfrac{a}{r}$

(5.5) percent increase or decrease = $\dfrac{\text{amount of change}}{\text{base}}$

PRACTICE TEST 5A

1. Convert each percent to its decimal or whole-number equivalent.

    35% _____   0.12% _____   400% _____

2. Convert each decimal or whole number to a percent.

    0.78 _____   0.089 _____   10 _____

3. Identify the base, rate, and amount:

    60 times 40% is 24

    rate: _____   base: _____   amount: _____

4. 75% of 44 is what number? _____

5. 14 is what percent of 112? _____

6. 5% of what number is 6? _____

7. What is 50 increased by 16%? _____

8. What is 120 decreased by 20%? _____

9. How many seniors will graduate with honors if 16% of the 250 graduating seniors have earned honors?

    _____

10. A bus ticket costs $80. Airfare to the same destination is 120% of the bus fare. What is the cost of the airline ticket?

    _____

11. Jimmy bought a software package when it was reduced by 20%. The original cost was $240. How much did Jimmy pay for the software package?

    _____

12. In a coed softball league, 40% of the participants are women. There are 88 women in the league. How many players are in the league altogether?

    _____

PRACTICE TEST 5B

1. Convert each percent to its decimal or whole-number equivalent.

   83% _____   0.92% _____   825% _____

2. Convert each decimal or whole number to a percent.

   0.13 _____   0.005 _____   6 _____

3. Identify the base, rate, and amount:

   42 is 12% of 350

   rate: _____   base: _____   amount: _____

4. 28% of 300 is what number? _____

5. 62 is what percent of 248? _____

6. 80% of what number is 10? _____

7. What is 75 increased by 20%? _____

8. What is 180 decreased by 30%? _____

9. Interest on a $500 loan was $45. What was the interest rate on the loan?

   _____

10. The price of a reserved-seat ticket to a baseball game increased by 20% in one year. If the original price of a ticket was $12, what was the new price?

    _____

11. Jennifer's utility bill decreased by 15% from the August billing period to the September billing period. The August bill was $150. What was the amount of her bill in September?

    _____

12. Gilbert and Roberta took a math test. There were 40 problems on the test. Gilbert answered 75% of the questions correctly while Roberta answered 85% correctly. How many more problems did Roberta answer correctly than Gilbert?

    _____

SKILLS REVIEW

CHAPTERS 1-5

1. Write 489,212 in words. _____

2. Write thirty-seven million, eight hundred sixty thousand, four hundred ten in numbers. _____

3. Round 2,629 to the nearest hundred. _____

4. Round 89,521 to the nearest thousand. _____

Perform the indicated operation. Express each answer in lowest terms.

5. $\frac{5}{6} + \frac{9}{16} + \frac{7}{12} =$ _____

6. $\frac{8}{9} - \frac{11}{20} =$ _____

7. $\frac{4}{5} \times \frac{2}{3} \times \frac{15}{16} =$ _____

8. $\frac{3}{2} \div \frac{1}{8} =$ _____

9. $4.72 \times 8.006 =$ _____

10. $16.75 \div 0.75 =$ _____

11. Convert $\frac{7}{20}$ to a decimal. _____

12. Convert 0.06 to a percent. _____

13. What is 40% of 130? _____

14. 75% of what number is 96? _____

15. 20 is what percent of 1,000? _____

16. What is 60 increased by 12%? _____

17. What is 120 decreased by 6%? _____

18. Write 0.27 in words.

_____

19. Write 57.065 in words.

_____

20. Is the following a true proportion?

$\dfrac{18}{20} = \dfrac{72}{80}$ _____

21. Find the missing element in the proportion.

$\dfrac{7}{8} = \dfrac{35}{n}$ _____

CHAPTER 5 SOLUTIONS

Skills Preview

1. *a.* 0.28
   b. 0.005
   c. 2.5
2. *a.* 42%
   b. 6%
   c. 300%
3. rate = 20%
   base = 40
   amount = 8
4. 22.5
5. 30%
6. 150
7. 54
8. 30.6
9. 22.5%
10. 322

Section 5.1 Comprehension Checkpoint

1. *a.* 0.47
   b. 0.0005
   c. 0.063
   d. 4
2. *a.* 64%
   b. 7.2%
   c. 590%
   d. 800%
3. *a.* amount
   b. rate
   c. base

5.1A Practice Set

1. 82% = 0.82
2. 70% = 0.7
3. 37% = 0.37
4. 42.5% = 0.425
5. 91.37% = 0.9137
6. 0.4% = 0.004
7. 6% = 0.06
8. 150% = 1.5
9. 200% = 2
10. 1,000% = 10
11. 0.38 = 38%
12. 0.50 = 50%
13. 0.84 = 84%
14. 0.008 = 0.8%
15. 0.6 = 60%
16. 0.625 = 62.5%
17. 0.05 = 5%
18. 4.2 = 420%
19. 3 = 300%
20. 1 = 100%
21. rate
22. rate
23. amount
24. base
25. rate
26. base
27. amount
28. rate
29. base
30. amount

Section 5.2 Comprehension Checkpoint

1. 58.8
2. $2.80
3. 246
4. 352.8
5. 27

5.2A Practice Set

1. $b \times r = a$
   $180 \times 20\% = a$
   $180 \times 0.20 = a$
   $\$36 = a$

2. $b \times r = a$
   $43 \times 6\% = a$
   $43 \times 0.06 = a$
   $2.58 = a$
   $a = 2.6$ (rounded)

3. $b \times r = a$
   $162 \times 48\% = a$
   $162 \times 0.48 = a$
   $\$77.76 = a$

4. $b \times r = a$
   $86 \times 3.8\% = a$
   $86 \times 0.038 = a$
   $3.268 = a$
   $a = 3.3$ (rounded)

5. $b \times r = a$
   $48 \times 225\% = a$
   $48 \times 2.25 = a$
   $108 = a$

6. $b \times r = a$
   $23 \times 600\% = a$
   $23 \times 6 = a$
   $138 = a$

7. $b \times r = a$
   $76 \times 0.5\% = a$
   $76 \times 0.005 = a$
   $0.38 = a$
   $a = 0.4$ (rounded)

8. $b \times r = a$
   $1{,}250 \times 0.08\% = a$
   $1{,}250 \times 0.0008 = a$
   $1 = a$

9. $b \times r = a$
   $470 \times 98\% = a$
   $470 \times 0.98 = a$
   $\$460.60 = a$

10. $b \times r = a$
    $38 \times 12.6\% = a$
    $38 \times 0.126 = a$
    $\$4.788 = a$
    $\$4.79$ (rounded) $= a$

11. $b \times r = a$
    $1 \times 450\% = a$
    $1 \times 4.5 = a$
    $\$4.50 = a$

12. $b \times r = a$
    $12.7 \times 100\% = a$
    $12.7 \times 1 = a$
    $12.7 = a$

13. $b \times r = a$
    $224 \times 7.5\% = a$
    $224 \times 0.075 = a$
    $16.8 = a$

14. $b \times r = a$
    $346 \times 14.8\% = a$
    $346 \times 0.148 = a$
    $51.208 = a$
    51.2 (rounded) $= a$

15. $b \times r = a$
    $280 \times 109\% = a$
    $280 \times 1.09 = a$
    $305.2 = a$

16. $b \times r = a$
    $436 \times 50\% = a$
    $436 \times 0.5 = a$
    $\$218 = a$

17. $b \times r = a$
    $802 \times 20.5\% = a$
    $802 \times 0.205 = a$
    $164.41 = a$
    164.4 (rounded) $= a$

18. $b \times r = a$
    $40{,}000 \times 32\% = a$
    $40{,}000 \times 0.32 = a$
    $\$12{,}800 = a$

19. $b \times r = a$
    $55{,}600 \times 8.5\% = a$
    $55{,}600 \times 0.085 = a$
    $\$4{,}726 = a$

20. $b \times r = a$
    $6{,}000 \times 175\% = a$
    $6{,}000 \times 1.75 = a$
    $\$10{,}500 = a$

21. $b \times r = a$
    $50 \times 7.25\% = a$
    $50 \times 0.0725 = a$
    $3.625 = a$
    $\$3.63 = a$

22. $b \times r = a$
    $120 \times 70\% = a$
    $120 \times 0.7 = a$
    $84 = a$

23. $b \times r = a$
    $115 \times 5\% = a$
    $115 \times 0.05 = a$
    $\$5.75 = a$

Section 5.3 Comprehension Checkpoint

1. 65%   2. 12.5%   3. 6%   4. 500%   5. 250%   6. 100%

5.3A Practice Set

1. $r = \dfrac{a}{b}$ $r = \dfrac{6}{60} = 0.1 = 10\%$

2. $r = \dfrac{a}{b}$ $r = \dfrac{75}{90} = 0.8333 = 83.3\%$

3. $r = \dfrac{a}{b}$ $r = \dfrac{12}{25} = 0.48 = 48\%$

4. $r = \dfrac{a}{b}$ $r = \dfrac{17}{20} = 0.85 = 85\%$

5. $r = \dfrac{a}{b}$ $r = \dfrac{47}{100} = 0.47 = 47\%$

6. $r = \dfrac{a}{b}$ $r = \dfrac{21}{7} = 3 = 300\%$

7. $r = \dfrac{a}{b}$ $r = \dfrac{66}{12} = 5.5 = 550\%$

8. $r = \dfrac{a}{b}$ $r = \dfrac{90}{25} = 3.6 = 360\%$

9. $r = \dfrac{a}{b}$ $r = \dfrac{140}{35} = 4 = 400\%$

10. $r = \dfrac{a}{b}$ $r = \dfrac{120}{150} = 0.8 = 80\%$

11. $r = \dfrac{a}{b}$ $r = \dfrac{82}{82} = 1 = 100\%$

12. $r = \dfrac{a}{b}$ $r = \dfrac{14}{35} = 0.4 = 40\%$

Solutions

13. $r = \dfrac{a}{b}$ $r = \dfrac{180}{45} = 4 = 400\%$

14. $r = \dfrac{a}{b}$ $r = \dfrac{8.7}{2.9} = 3 = 300\%$

15. $r = \dfrac{a}{b}$ $r = \dfrac{625}{25} = 25 = 2{,}500\%$

16. $r = \dfrac{a}{b}$ $r = \dfrac{30}{240} = 0.125 = 12.5\%$

17. $r = \dfrac{a}{b}$ $r = \dfrac{70}{16} = 4.375 = 437.5\%$

18. $r = \dfrac{a}{b}$ $r = \dfrac{84}{300} = 0.28 = 28\%$

19. $r = \dfrac{a}{b}$ $r = \dfrac{280}{28} = 10 = 1{,}000\%$

20. $r = \dfrac{a}{b}$ $r = \dfrac{9}{1{,}500} = 0.006 = 0.6\%$

21. $r = \dfrac{a}{b}$ $r = \dfrac{6}{48} = 0.125 = 12.5\%$

22. $r = \dfrac{a}{b}$ $r = \dfrac{22.5}{300} = 0.075 = 7.5\%$

23. $r = \dfrac{a}{b}$ $r = \dfrac{72}{240} = 0.3 = 30\%$

Section 5.4   Comprehension Checkpoint

1. 45   2. 300   3. 60   4. 12   5. 7   6. 42

5.4A   Practice Set

1. $b = \dfrac{a}{r}$ $b = \dfrac{60}{20\%} = \dfrac{60}{0.20} = 300$

2. $b = \dfrac{a}{r}$ $b = \dfrac{12}{15\%} = \dfrac{12}{0.15} = 80$

3. $b = \dfrac{a}{r}$ $b = \dfrac{32}{8\%} = \dfrac{32}{0.08} = 400$

4. $b = \dfrac{a}{r}$ $b = \dfrac{33}{22\%} = \dfrac{33}{0.22} = 150$

5. $b = \dfrac{a}{r}$ $b = \dfrac{3.6}{9\%} = \dfrac{3.6}{0.09} = 40$

6. $b = \dfrac{a}{r}$ $b = \dfrac{27}{135\%} = \dfrac{27}{1.35} = 20$

7. $b = \dfrac{a}{r}$ $b = \dfrac{20}{160\%} = \dfrac{20}{1.6} = 12.5$

8. $b = \dfrac{a}{r}$ $b = \dfrac{75}{250\%} = \dfrac{75}{2.5} = 30$

9. $b = \dfrac{a}{r}$ $b = \dfrac{100}{125\%} = \dfrac{100}{1.25} = 80$

10. $b = \dfrac{a}{r}$ $b = \dfrac{48}{150\%} = \dfrac{48}{1.5} = 32$

11. $b = \dfrac{a}{r}$ $b = \dfrac{24}{40\%} = \dfrac{24}{0.4} = 60$

12. $b = \dfrac{a}{r}$ $b = \dfrac{156}{65\%} = \dfrac{156}{0.65} = 240$

13. $b = \dfrac{a}{r}$ $b = \dfrac{16.8}{24\%} = \dfrac{16.8}{0.24} = 70$

14. $b = \dfrac{a}{r}$ $b = \dfrac{12.8}{100\%} = \dfrac{12.8}{1} = 12.8$

15. $b = \dfrac{a}{r}$ $b = \dfrac{65}{130\%} = \dfrac{65}{1.3} = 50$

16. $b = \dfrac{a}{r}$ $b = \dfrac{48}{120\%} = \dfrac{48}{1.2} = 40$

17. $b = \dfrac{a}{r}$ $b = \dfrac{84}{140\%} = \dfrac{84}{1.4} = 60$

18. $b = \dfrac{a}{r}$ $b = \dfrac{95}{125\%} = \dfrac{95}{1.25} = 76$

19. $b = \dfrac{a}{r}$ $b = \dfrac{63}{112.5\%} = \dfrac{63}{1.125} = 56$

20. $b = \dfrac{a}{r}$ $b = \dfrac{198}{300\%} = \dfrac{198}{3} = 66$

21. $b = \dfrac{a}{r}$ $b = \dfrac{12}{24\%} = \dfrac{12}{0.24} = 50$

22. $b = \dfrac{a}{r}$ $b = \dfrac{25}{40\%} = \dfrac{25}{0.4} = 62.5$

23. $b = \dfrac{a}{r}$ $b = \dfrac{20}{125\%} = \dfrac{20}{1.25} = 16$

Section 5.5 Comprehension Checkpoint

1. 25%   2. 20%   3. 108   4. 39.9

5.5A Practice Set

1. $54 - 50 = 4 =$ amount of increase
   $4 \div 50 = 0.08 = 8\%$ increase

2. $90 - 75 = 15 =$ amount of increase
   $15 \div 75 = 0.2 = 20\%$ increase

3. $195 - 150 = 45 =$ amount of increase
   $45 \div 150 = 0.3 = 30\%$ increase

4. $\$464 - \$400 = \$64 =$ amount of increase
   $\$64 \div \$400 = 0.16 = 16\%$ increase

5. $\$247.50 - \$225 = \$22.50 =$ amount of increase
   $\$22.50 \div \$225 = 0.1 = 10\%$ increase

6. $\$1{,}302 - \$1{,}200 = \$102 =$ amount of increase
   $\$102 \div \$1{,}200 = 0.085 = 8.5\%$ increase

7. $80 - 64 = 16 =$ amount of decrease
   $16 \div 80 = 0.2 = 20\%$ decrease

8. $95 - 57 = 38 =$ amount of decrease
   $38 \div 95 = 0.4 = 40\%$ decrease

9. $\$150 - \$138 = \$12 =$ amount of decrease
   $\$12 \div \$150 = 0.08 = 8\%$ decrease

10. $\$125 - \$62.50 = \$62.50 =$ amount of decrease
    $\$62.50 \div \$125 = 0.5 = 50\%$ decrease

11. $\$56 - \$49 = \$7 =$ amount of decrease
    $\$7 \div \$56 = 0.125 = 12.5\%$ decrease

12. $\$500 - \$462.50 = \$37.50 =$ amount of decrease
    $\$37.50 \div \$500 = 0.075 = 7.5\%$

13. $b \times r = a$ $r = 100\% + 5\% = 105\%$
    $64 \times 105\% = a$
    $64 \times 1.05 = a$
    $\$67.20 = a$

14. $b \times r = a$ $r = 100\% + 20\% = 120\%$
    $40 \times 120\% = a$
    $40 \times 1.2 = a$
    $48 = a$

15. $b \times r = a$ $r = 100\% + 50\% = 150\%$
    $30 \times 150\% = a$
    $30 \times 1.5 = a$
    $45 = a$

16. $b \times r = a$ $r = 100\% + 12.5\% = 112.5\%$
    $88 \times 112.5\% = a$
    $88 \times 1.125 = a$
    $\$99 = a$

17. $b \times r = a$ $r = 100\% + 15\% = 115\%$
    $120 \times 115\% = a$
    $120 \times 1.15 = a$
    $\$138 = a$

18. $b \times r = a$ $r = 100\% - 6.25\% = 93.75\%$
    $\$240 \times 93.75\% = a$
    $\$240 \times 0.9375 = a$
    $\$225 = a$

19. $b \times r = a$ $r = 100\% - 2\% = 98\%$
    $\$1{,}550 \times 98\% = a$
    $\$1{,}550 \times 0.98 = a$
    $\$1{,}519 = a$

20. $b \times r = a$ $r = 100\% - 9\% = 91\%$
    $40 \times 91\% = a$
    $40 \times 0.91 = a$
    $36.4 = a$

21. $b \times r = a$ $r = 100\% - 18\% = 82\%$
    $65 \times 82\% = a$
    $65 \times 0.82 = a$
    $53.3 = a$

22. $b \times r = a$ $r = 100\% - 6.5\% = 93.5\%$
    $110 \times 93.5\% = a$
    $110 \times 0.935 = a$
    $102.85 = a$

23. $b \times r = a$ $r = 100\% - 10\% = 90\%$
    $\$250 \times 90\% = a$
    $\$250 \times 0.9 = a$
    $\$225 = a$

24. $b \times r = a$ $r = 100\% + 12\% = 112\%$
    $\$375 \times 112\% = a$
    $\$375 \times 1.12 = a$
    $\$420 = a$

25. $b \times r = a$ $r = 100\% - 15\% = 85\%$
    $\$1{,}080 \times 85\% = a$
    $\$1{,}080 \times 0.85 = a$
    $\$918 = a$

Solutions

Section 5.6 Comprehension Checkpoint

1. 69   **2.** 60%   **3.** 16%   **4.** 1,260

5.6A Practice Set

1. $b \times r = a$
$50 \times 20\% = a$
$50 \times 0.2 = a$
$10 = a$

2. $b = \dfrac{a}{r}$ $b = \dfrac{12}{25\%} = \dfrac{12}{0.25} = 48$

3. $b \times r = a$ $r = 100\% + 10\% = 110\%$
$60 \times 110\% = a$
$60 \times 1.1 = a$
$66 = a$

4. $r = \dfrac{a}{b}$ $r = \dfrac{45}{75} = 0.60 = 60\%$

5. $b \times r = a$ $r = 100\% - 40\% = 60\%$
$130 \times 60\% = a$
$130 \times 0.60 = a$
$78 = a$

6. $b \times r = a$ $r = 100\% + 5.5\% = 105.5\%$
$125 \times 105.5\% = a$
$125 \times 1.055 = a$
$131.875 = a$

7. $b \times r = a$
$250 \times 37\% = a$
$250 \times 0.37 = a$
$92.5 = a$

8. $b \times r = a$ $r = 100\% - 8\% = 92\%$
$90 \times 92\% = a$
$90 \times 0.92 = a$
$82.8 = a$

9. $r = \dfrac{a}{b}$ $r = \dfrac{28}{700} = 0.04 = 4\%$

10. $b = \dfrac{a}{r}$ $b = \dfrac{17}{5\%} = \dfrac{17}{0.05} = 340$

11. $b \times r = a$
$450 \times 9\% = a$
$450 \times 0.09 = a$
$\$40.50 = a$

12. $r = \dfrac{a}{b}$ $r = \dfrac{2.25}{30} = 0.075 = 7.5\%$

13. $\$750 - \$700 = \$50 =$ amount of increase
$\$50 \div \$700 = 0.0714 = 7.1\%$ percent increase

14. $b \times r = a$
$25 \times 45\% = a$
$25 \times 0.45 = a$
11.25 miles $= a$

15. $\$86 - \$78 = \$8 =$ amount of decrease
$\$8 \div \$86 = 0.093 = 9.3\%$ rate of decrease

16. $b = \dfrac{a}{r}$ $b = \dfrac{136}{85\%} = \dfrac{136}{0.85} = 160$

17. $b \times r = a$ $r = 100\% + 8\% = 108\%$
$36.50 \times 108\% = a$
$36.50 \times 1.08 = a$
$\$39.42 = a$

18. $r = \dfrac{a}{b}$ $r = \dfrac{125}{5{,}000} = 0.025 = 2.5\%$

19. $b \times r = a$ $r = 100\% - 15\% = 85\%$
$190 \times 85\% = a$
$190 \times 0.85 = a$
$\$161.50 = a$

20. $b = \dfrac{a}{r}$ $b = \dfrac{405}{90\%} = \dfrac{405}{0.9} = 450$

21. $r = \dfrac{a}{b}$ $r = \dfrac{32}{40} = 0.8 = 80\%$

22. $b \times r = a$ $r = 100\% + 6\% = 106\%$
$6.50 \times 106\% = a$
$6.50 \times 1.06 = a$
$\$6.89 = a$

23. $1,552.50 − $1,500 = $52.50 = amount of increase
$52.50 ÷ $1,500 = 0.035 = 3.5%

24. $b \times r = a$
472 × 25% = a
472 × 0.25 = a
118 = a

25. $b = \dfrac{a}{r}$ $b = \dfrac{30}{75\%} = \dfrac{30}{0.75} = 40$

Practice Test   5A

1. **a.** 0.35
   b. 0.0012
   c. 4
2. **a.** 78%
   b. 8.9%
   c. 1,000%
3. rate = 40%
   base = 60
   amount = 24
4. 33
5. 12.5%
6. 120
7. 58
8. 96
9. 40
10. $96
11. $192
12. 220

Skills Review

1. four hundred eighty-nine thousand, two hundred twelve
2. 37,860,410
3. 2,600
4. 90,000
5. $1\dfrac{47}{48}$
6. $\dfrac{61}{180}$
7. $\dfrac{1}{2}$
8. 12
9. 37.78832
10. 22.333 . . .
11. 0.35
12. 6%
13. 52
14. 128
15. 2%
16. 67.2
17. 112.8
18. twenty-seven hundredths
19. fifty-seven and sixty-five thousandths
20. yes
21. 40

6 CONVERTING FRACTIONS

OUTLINE

6.1 Equivalent Common Fractions, Decimal Fractions, and Percents

6.2 Terminating and Repeating Decimals

6.3 Using Conversions to Solve Problems

In previous chapters, you learned about common fractions, decimal fractions, and percents. Each type of fraction represents a part of a whole amount. For example, $\frac{1}{2}$, 0.5, and 50% all represent the same amount. Since that is true, $\frac{1}{2} = 0.5 = 50\%$. In this chapter you will learn how to convert from one type of fraction to another, equivalent fraction. You will also learn the advantages of using one type of fraction rather than another.

SKILLS PREVIEW

Fill in the chart for problems 1–6 by converting each given fraction to its two equivalent fractions. Reduce common fractions to lowest terms.

	Common Fraction	Decimal Fraction	Percent
1.	$\frac{4}{5}$	_____	_____
2.	$\frac{7}{12}$	_____	_____
3.	_____	0.75	_____
4.	_____	0.42	_____
5.	_____	_____	95%
6.	_____	_____	$6\frac{1}{4}\%$

7. Convert each of the following common fractions to decimals and identify each as terminating or repeating.

		Decimal	Terminating or Repeating
a.	$\frac{4}{10} =$	_____	_____
b.	$\frac{2}{9} =$	_____	_____
c.	$\frac{5}{7} =$	_____	_____
d.	$\frac{6}{12} =$	_____	_____

Solve each of the following problems.

8. A shirt is reduced by $37\frac{1}{2}\%$ for a sale. If the shirt sold for $20 originally, what is the amount of the discount? _____

9. Sheila's telephone bill increased by $66\frac{2}{3}\%$ as a result of a computer error. Her bill was $54 before the error. What was her bill after the error? _____

10. In a recent election, $\frac{8}{25}$ of those voting voted against public improvement bonds. What percent of the votes were against the bonds? _____

11. Carlos worked $\frac{5}{11}$ of his 44-hour workweek during the first two days of the week. How many hours did he work during those two days? _____

12. Kim ordered a megalarge square pizza. It was cut into 24 pieces. She ate $\frac{1}{6}$ of the pizza when it arrived and saved the remainder. How many pieces of pizza did she eat? _____

6.1 EQUIVALENT COMMON FRACTIONS, DECIMAL FRACTIONS, AND PERCENTS

You learned about common fractions in Chapter 2, decimal fractions in Chapter 3, and percents in Chapter 5. Each type of fraction can be converted to the other two types. Being able to make these conversions should enable you to solve problems more easily.

Converting Common Fractions to Decimal Fractions

> To convert a common fraction to a decimal fraction, divide the numerator by the denominator.

EXAMPLE 1 Convert each common fraction to its decimal-fraction equivalent.

a. $\frac{1}{2}$ b. $\frac{5}{16}$

SOLUTIONS

a. $\frac{1}{2} = 2\overline{)1.0}^{\,.5} = 0.5$ $\frac{1}{2}$ and 0.5 have the same value.

b. $\frac{5}{16} = 16\overline{)5.0000}^{\,.3125} = 0.3125$ $\frac{5}{16}$ and 0.3125 have the same value.

Any common fraction can be converted to a decimal fraction using the method described in Example 1.

Converting a Decimal Fraction to a Common Fraction

> **TO CONVERT A DECIMAL FRACTION TO A COMMON FRACTION**
> 1. Write the decimal fraction as the numerator of a common fraction. Drop the decimal point.
> 2. Place a 0 in the denominator under each digit in the numerator and a 1 to the left of the zeros.
> 3. Reduce to lowest terms.

EXAMPLE 2 Convert each decimal fraction to a common fraction.

a. 0.4   *b.* 0.87   *c.* 0.0625

SOLUTIONS

a. $0.4 = \dfrac{4}{10}$ *Write the 4 as the numerator. The denominator will have one zero under the 4. Place a 1 to the left of the zero. (Remember that 0.4 is read "four tenths.")*

$\dfrac{4}{10} = \dfrac{2}{5}$ *Reduce to lowest terms.*

0.4 has the same value as $\dfrac{2}{5}$.

b. $0.87 = \dfrac{87}{100}$ *The numerator will be 87. Place a zero under each digit in the numerator and a 1 to the left of the zeros. (Remember that 0.87 is read "eighty-seven hundredths.")*

$\dfrac{87}{100}$ will not reduce.

0.87 and $\dfrac{87}{100}$ have the same value.

c. $0.0625 = \dfrac{0625}{10000}$ *Notice that the zero is included in the numerator. This is to make sure that you place enough zeros in the denominator.*

$\dfrac{625}{10000} = \dfrac{1}{16}$ *Drop the zero in the numerator and reduce.*

0.0625 and $\dfrac{1}{16}$ have the same value.

Every decimal fraction (terminating and repeating, which you will study in Section 6.2) has a common-fraction equivalent. Reading the decimal fraction properly tells you what the denominator of the common fraction should be.

Converting Common Fractions to Percents

TO CONVERT COMMON FRACTIONS TO PERCENTS
1. Convert the common fraction to its decimal equivalent.
2. Move the decimal point two places to the right and add the symbol %.

EXAMPLE 3 Convert each common fraction or mixed number to a percent.

a. $\dfrac{3}{8}$ *b.* $1\dfrac{1}{5}$

SOLUTIONS

a. $\dfrac{3}{8} = 0.375$ *Convert the common fraction to its decimal equivalent.*

37.5%   *Move the decimal point two places to the right and add the symbol %.*

b. $1\frac{1}{5} = 1.2$ *Convert the fraction part of the mixed number to a decimal.*

120%   *Move the decimal point two places to the right and add the symbol %. Notice that you must add one zero to be able to move the decimal point two places.*

Converting Percents to Common Fractions

You learned in Chapter 5 that percents must be converted to decimal fractions to solve problems. Percents can also be converted to common-fraction equivalents in order to solve problems.

TO CONVERT A PERCENT TO A COMMON FRACTION

1. Multiply the number preceding the percent sign by $\frac{1}{100}$.
2. Reduce the answer to lowest terms.

EXAMPLE 4   Convert each percent to a common fraction (or mixed number).

a. 45%   *b.* $16\frac{2}{3}\%$ *c.* $\frac{1}{4}\%$ *d.* 250%

SOLUTIONS

a. $45 \times \frac{1}{100} = \frac{45}{1} \times \frac{1}{100} = \frac{9}{20}$

45% and $\frac{9}{20}$ are equivalent values.

b. $16\frac{2}{3} \times \frac{1}{100} = \frac{50}{3} \times \frac{1}{100} = \frac{50}{300} = \frac{1}{6}$

$16\frac{2}{3}\%$ and $\frac{1}{6}$ are equivalent values.

c. $\frac{1}{4} \times \frac{1}{100} = \frac{1}{400}$

$\frac{1}{4}\%$ and $\frac{1}{400}$ are equivalent values.

d. $250 \times \frac{1}{100} = \frac{250}{1} \times \frac{1}{100} = \frac{250}{100} = \frac{5}{2} = 2\frac{1}{2}$

250% and $2\frac{1}{2}$ are equivalent values.

Any percent can be converted to a common-fraction equivalent using the method described in Example 4.

You learned in Chapter 5 how to convert between percents and decimal fractions. As a reminder, to convert from a percent to a decimal fraction, move the decimal point two places to the left and drop the percent symbol (%). Thus,

$$42\% = 0.42$$

To convert a decimal fraction to a percent, move the decimal point two places to the right and add a percent symbol (%):

$$0.65 = 65\%$$

Work the Comprehension Checkpoint problems to measure your understanding of the conversion processes discussed in this section. If you have difficulty, review the material in this section before going on to the Practice Sets.

COMPREHENSION CHECKPOINT

Convert the given fraction to each of the other two types of fractions.

	Common Fraction	Decimal Fraction	Percent
1.	$\dfrac{3}{8}$	_____	_____
2.	_____	0.28	_____
3.	_____	_____	70%
4.	_____	_____	$\dfrac{3}{5}\%$

6.1A PRACTICE SET

Convert each common fraction to its decimal-fraction and percent equivalents.

	Decimal Fraction	Percent
1. $\dfrac{3}{4}$	_____	_____
2. $\dfrac{9}{16}$	_____	_____
3. $\dfrac{12}{25}$	_____	_____
4. $\dfrac{1}{8}$	_____	_____
5. $\dfrac{3}{5}$	_____	_____
6. $\dfrac{7}{10}$	_____	_____

Convert each decimal fraction to its percent and common-fraction (in lowest terms) equivalents.

	Percent	Common Fraction
7. 0.3	_____	_____
8. 0.85	_____	_____
9. 0.24	_____	_____
10. 0.375	_____	_____
11. 0.0625	_____	_____
12. 0.3125	_____	_____

Convert each percent to its decimal-fraction and common-fraction (in lowest terms) equivalents.

	Decimal Fraction	Common Fraction
13. 90%	_____	_____
14. 58%	_____	_____
15. 1%	_____	_____
16. $62\frac{1}{2}\%$	_____	_____
17. $43\frac{3}{4}\%$	_____	_____
18. $\frac{1}{5}\%$	_____	_____

Applications

19. Last Monday, $\frac{3}{20}$ of Ms. Johnson's second-grade class were absent. What percent of the students were absent?

_____

20. A piston has a tolerance of 0.0015 centimeter. Express this tolerance as a common fraction.

_____

21. A sightseeing tour bus was 98% full. What fractional part of the bus was full?

_____

22. Armando is a running back on his high school football team. On one play he carried the ball 18 yards. If the field is 100 yards long, express as a decimal fraction the portion of the field he covered on that one play.

_____

23. The workforce at an office is $12\frac{1}{2}\%$ male. What fractional part of the workforce is female?

_____

6.1B PRACTICE SET

Convert each common fraction to its decimal-fraction and percent equivalents.

	Decimal Fraction	Percent
1. $\frac{2}{5}$	_____	_____
2. $\frac{7}{8}$	_____	_____
3. $\frac{11}{16}$	_____	_____
4. $\frac{18}{25}$	_____	_____
5. $\frac{3}{10}$	_____	_____
6. $\frac{3}{100}$	_____	_____

Convert each decimal fraction to its percent and common-fraction (in lowest terms) equivalents.

	Percent	Common Fraction
7. 0.75	_____	_____
8. 0.4	_____	_____
9. 0.09	_____	_____
10. 0.67	_____	_____
11. 0.375	_____	_____
12. 0.0025	_____	_____

Convert each percent to its decimal-fraction and common-fraction (in lowest terms) equivalents.

	Decimal Fraction	Common Fraction
13. 10%	_____	_____
14. 17%	_____	_____
15. 96%	_____	_____
16. $12\frac{1}{2}\%$	_____	_____
17. $38\frac{1}{2}\%$	_____	_____
18. $\frac{1}{8}\%$	_____	_____

Applications

19. The first chapter in a physics book is $\frac{4}{25}$ of the entire book. What percent of the total book is the first chapter?

   _____

20. A blueprint has a line that measures 0.45 inch. Express the length of the line as a common fraction.

   _____

21. Kerensa made a down payment of $37\frac{1}{2}\%$ on a new refrigerator. Express the down payment as a decimal fraction.

   _____

22. The price of a car is expected to increase by $\frac{1}{10}$ in one year. Express the increase as a percent.

   _____

23. Melissa completed 80% of her homework before leaving school. What fractional part of her homework had she finished?

   _____

6.2 TERMINATING AND REPEATING DECIMALS

Each conversion from a common fraction to a decimal fraction in the previous section resulted in a **terminating decimal.** Terminating decimals occur when the denominator divides the numerator evenly. For example, 0.5 has no digits to the right of the 5. Zeros could be added, as in 0.50 or 0.500, but the value of the decimal would not change. Other examples of terminating decimals include 0.3, 0.47, 0.125, and 0.1439. Notice that terminating decimals can have more than one decimal digit.

Not all decimals terminate. In some decimal fractions, one or more digits repeat. These decimals are called **repeating decimals.** Examples of repeating decimals are 0.333... and 0.0909... The dots, called an *ellipsis,* indicate that the digits repeat indefinitely. Once you have determined that a decimal is a repeating decimal, you can indicate this by placing a bar (‾) over the digits that repeat. Thus, 0.333... could be written $0.\overline{3}$. The bar over the 3 indicates that the 3 repeats indefinitely. Similarly, 0.0909... could be written $0.\overline{09}$ to show that the 09 repeats indefinitely. Any number of digits can repeat. All of the digits that repeat must be written under the bar.

EXAMPLE 1 Identify each decimal fraction as either terminating or repeating.

a. $0.0\overline{6}$ b. 0.625   c. $0.\overline{1}$ d. 0.72

SOLUTIONS

a. repeating
b. terminating
c. repeating
d. terminating

EXAMPLE 2 Convert each common fraction to a decimal fraction. State whether each is a terminating or a repeating decimal.

a. $\dfrac{4}{7}$ b. $\dfrac{1}{5}$ c. $\dfrac{2}{3}$ d. $\dfrac{5}{9}$

SOLUTIONS

a. $\dfrac{4}{7} = 4 \div 7 = 0.\overline{571428}$ repeating decimal

b. $\dfrac{1}{5} = 1 \div 5 = 0.2$ terminating decimal

c. $\dfrac{2}{3} = 2 \div 3 = 0.\overline{6}$ repeating decimal

d. $\dfrac{5}{9} = 5 \div 9 = 0.\overline{5}$ repeating decimal

In order to work with repeating decimals, you must round them to a predetermined place.

Finally, it should be noted that some decimals neither terminate nor repeat. One that is frequently used is pi, symbolized π. The numerical value of π, carried to 12 places, is approximately 3.141592653589. You will learn some applications using π in Chapter 10, Introduction to Geometry. If you are working with decimals of this type, you will have to round. Any time you round a decimal for computation purposes, your answer will be an approximation at best (though reasonably accurate).

Work the Comprehension Checkpoint problems. If you have difficulty with these problems, review the material in this section before attempting the Practice Sets.

COMPREHENSION CHECKPOINT

Identify each of the following decimal fractions as either terminating or repeating.

1. 0.45 _____

2. $0.08\overline{3}$ _____

Convert each of the following fractions to decimals. Identify the decimals as terminating or repeating.

3. $\dfrac{5}{6} =$ _____   _____

4. $\dfrac{5}{8} =$ _____   _____

6.2A PRACTICE SET

Convert each common fraction to a decimal fraction. If the decimal is a repeating decimal, use a bar above the repeating digits.

1. $\dfrac{2}{3} =$ _____
2. $\dfrac{4}{5} =$ _____
3. $\dfrac{5}{12} =$ _____

4. $\dfrac{4}{9} =$ _____
5. $\dfrac{3}{4} =$ _____
6. $\dfrac{4}{25} =$ _____

7. $\dfrac{2}{15} =$ _____
8. $\dfrac{5}{6} =$ _____
9. $\dfrac{9}{32} =$ _____

10. $\dfrac{19}{50} =$ _____
11. $\dfrac{11}{20} =$ _____
12. $\dfrac{3}{7} =$ _____

13. $\dfrac{9}{11} =$ _____
14. $\dfrac{7}{8} =$ _____

6.2B PRACTICE SET

Convert each common fraction to a decimal fraction. If the decimal is a repeating decimal, use a bar above the repeating digits.

1. $\dfrac{6}{10} =$ _____

2. $\dfrac{1}{3} =$ _____

3. $\dfrac{5}{7} =$ _____

4. $\dfrac{1}{5} =$ _____

5. $\dfrac{2}{11} =$ _____

6. $\dfrac{1}{6} =$ _____

7. $\dfrac{3}{8} =$ _____

8. $\dfrac{15}{16} =$ _____

9. $\dfrac{8}{9} =$ _____

10. $\dfrac{9}{12} =$ _____

11. $\dfrac{17}{32} =$ _____

12. $\dfrac{10}{15} =$ _____

13. $\dfrac{19}{25} =$ _____

14. $\dfrac{31}{64} =$ _____

6.3 USING CONVERSIONS TO SOLVE PROBLEMS

Converting a percent to a decimal or common fraction is more than just an academic exercise. Making the proper conversion is a tool that can be used to solve problems more quickly and more accurately.

EXAMPLE 1 During the off-season, the Plush Hotel offers a discount of $12\frac{1}{2}\%$ off the regular room rate of $80 per night. What is the amount of the discount?

SOLUTION

Method 1

$12\frac{1}{2}\% = 0.125$ You can convert the rate to its decimal equivalent.

$r \times b = a$ Use the percent formula.

$0.125 \times \$80 = \$10$ Multiply the rate by the base to find the discount (amount).

Method 2

$12\frac{1}{2} \times \frac{1}{100} = \frac{1}{8}$ You can convert the rate to its common-fraction equivalent.

$\frac{1}{8} \times \$80 = \$10$ Then multiply the rate by the base to find the discount (amount).

The amount of the discount is $10 using either method.

Notice in Example 1 that converting the rate to either its decimal or common-fraction equivalent produces *exactly* the same result. This is because the decimal is a terminating decimal. Study Example 2 carefully to see what happens if the decimal does not terminate.

EXAMPLE 2 Emilio wants to buy a new car. The cost is $15,000. The dealer offers him an $8\frac{1}{3}\%$ discount. What is the amount of the discount?

SOLUTION

Method 1

$8\frac{1}{3}\% = 0.08\overline{3}$ Convert the rate to its decimal equivalent.

$0.083 \times \$15,000 = \$1,245 =$ discount   Multiply the rate by the base to get the discount.

Method 2

$8\frac{1}{3} \times \frac{1}{100} = \frac{1}{12}$ Convert the rate to its common-fraction equivalent.

$\frac{1}{12} \times \$15,000 = \$1,250 =$ discount   Multiply the rate by the base to get the discount.

> **a + b = x**
> ## ALGEBRA CONNECTION
>
> One of the most fundamental algebraic concepts is the idea of equivalent values. When you know that
>
> $$50\% = \frac{1}{2} = 0.5$$
>
> then all three forms of the fraction are equivalent. They can be interchanged as needed to solve problems.
>
> When studying equations, you may see something like
>
> $$x + 5 = 7$$
>
> It is obvious in this equation that x must have a value of 2, or
>
> $$x = 2$$
>
> These are equivalent values, so the 2 can be substituted for the x in the equation:
>
> $$2 + 5 = 7$$
> $$7 = 7$$
>
> This gives you a true statement. Equivalent values can be interchanged and used as needed.

Notice the difference in discount amounts using the two different methods. When you use the common-fraction equivalent of the rate (percent), the answer is more accurate. If the decimal had been rounded to the nearest hundredth, the discount would have been

$$0.08 \times \$15,000 = \$1,200$$

Or, if the decimal had been rounded to the nearest ten-thousandth, the discount would have been

$$0.0833 \times \$15,000 = \$1,249.50$$

The more decimal places you add, the closer you will get to an exact answer.

The rule of thumb is to convert to a common fraction rather than a repeating decimal. This will give you greater accuracy. If the conversion results in a terminating decimal, as in Example 1, it does not make any difference whether you use the decimal or the common fraction. The answer will be the same. (However, you should not round a terminating decimal for purposes of multiplication.)

One more thought on using equivalent values may be in order here. Many students do not like to work with common fractions and so will use decimals and sacrifice some accuracy. When you multiply a whole number by a fraction with a numerator of 1, all you have to do is divide the whole number by the fraction's reciprocal. In Example 1, you saw that

$$\frac{1}{8} \times \$80 = \frac{1}{8} \times \frac{80}{1} = \frac{80}{8} = \$10$$

The same was true in Example 2 when you multiplied $\frac{1}{12}$ by $15,000.

EXAMPLE 3 The senior class at Roosevelt High School included 240 students. Of these, $\frac{1}{6}$ graduated with honors. How many students graduated with honors?

SOLUTION

$$\frac{1}{6} \times 240 = 240 \div 6 = 40$$

40 students graduated with honors.

If you tried to work this problem with decimals, you would get:

$\frac{1}{6} = 0.1\overline{6}$ *Convert the fraction to a decimal.*

$0.17 \times 240 = 40.8$ *Multiply the rate by the base to get the number of honor graduates.*

Whatever place you round the decimal to, you will not get as accurate an answer as you do with the common fraction. Also, if you want an exact number of people, this answer would not make any sense.

The same principle can be used to multiply a fraction with a numerator other than 1. Remember that the denominator of a fraction tells you how many parts the whole is divided into. The numerator tells you how many of those parts are being considered. Study Example 4 for a slightly different approach to multiplying fractions.

EXAMPLE 4 Carmen invited 140 people to a party. Only $\frac{6}{7}$ of those invited attended. How many guests were at the party?

SOLUTION

The fraction $\frac{6}{7}$ converts to a repeating decimal, so use it as it is.

$\frac{6}{7} \times 140 = ?$ *Since the denominator tells you how many parts the whole is divided into, divide the whole by the denominator to see how many are in each part.*

$140 \div 7 = 20$

$20 \times 6 = 120$ *Since 6 parts are being considered (that's how many came to the party), multiply the number in each part by 6.*

There were 120 guests at the party.

This section is not meant to imply that working with decimals is incorrect. Some people think of math as being an exact science, but you can see that there will be some errors caused by rounding. The purpose here is to avoid round-off errors as much as possible. When you are working with nonterminating and nonrepeating decimals, rounding will be a necessity. After some practice, you will learn how much error is acceptable for your needs.

Work the Comprehension Checkpoint problems to determine your understanding of the material in this section. If you have difficulty with these problems, review the material in this section before going on to the Practice Sets.

COMPREHENSION CHECKPOINT

Solve each problem.

1. Monica increased her reading speed by $33\frac{1}{3}$% as the result of a speed-reading course. She could read 24 pages per hour before taking the course. By how many pages per hour did she increase her speed?

   _____

2. The cost of bananas will increase by $16\frac{2}{3}$% as a result of a poor harvest. What will the amount of increase be if the current price is $0.48 per pound?

   _____

3. Of all the vehicles registered in Grayson County, $\frac{1}{9}$ are pickups. There are 18,945 vehicles registered in the county. How many pickups are registered in the county?

   _____

4. Maria has paid $\frac{5}{6}$ of the total cost of her house. If the house cost $54,000, how much has she paid?

   _____

6.3A PRACTICE SET

Solve each problem using the methods discussed in this section.

1. What is $66\frac{2}{3}\%$ of $900? _____

2. What is $14\frac{2}{7}\%$ of 196? _____

3. What is $41\frac{2}{3}\%$ of 48? _____

4. What is $22\frac{2}{9}\%$ of $540? _____

5. What is $83\frac{1}{3}\%$ of 150? _____

6. What is $\frac{1}{5}$ of $800? _____

7. What is $\frac{3}{4}$ of $200? _____

8. What is $\frac{5}{8}$ of $320? _____

Applications

9. Jennifer needed to address 144 wedding invitations. She completed $66\frac{2}{3}\%$ of them before going to work. How many did she finish before going to work?

   _____

10. There are 48 students in a defensive driving class. Of these, $37\frac{1}{2}\%$ are female. How many female students are in the class?

    _____

11. If there are 60 minutes in an hour, how many minutes are in $33\frac{1}{3}\%$ of an hour?

    _____

12. Josh earns $4.75 per hour. What would his hourly rate be if he received a raise equal to $\frac{1}{5}$ of his current rate?

   _____

13. If $\frac{3}{4}$ of the mail received every week is junk mail, how many pieces of junk mail would be included in 44 pieces of mail?

   _____

14. Jeremy spends 6 hours studying each week. His sister, Amanda, spends only $83\frac{1}{3}$% as much time studying. How many hours does Amanda study each week?

   _____

15. Scott, a delivery driver, delivered $\frac{5}{9}$ of his 180 packages before taking a lunch break. How many packages did he have to deliver after lunch?

   _____

16. Stephanie used only $\frac{2}{15}$ of the time allocated for her math final exam. If the exam was scheduled for 135 minutes, how long did it take her to finish the exam?

   _____

17. After a trip, the gas gauge in Carla's car indicated that the tank was $\frac{1}{8}$ full. Her tank holds 18 gallons of gas, and it was full when she left. If she averages 20.5 miles per gallon, how many miles did she travel? Round to the nearest tenth.

   _____

18. To get a grade of A on an exam, Sherry must answer $\frac{9}{10}$ of the questions correctly. There were 140 questions on the exam. What is the maximum number of problems that she can miss and still get an A on the exam?

   _____

19. Voters cast 24,000 votes in a city election. One write-in candidate received only $\frac{1}{3}$% of the votes. How many votes did he get?

   _____

20. Rob earns $18,655 per year. How much does he have available to spend (disposable income) if all his deductions total $28\frac{4}{7}$%?

   _____

6.3B PRACTICE SET

Solve each problem using the methods discussed in this section.

1. What is $16\frac{2}{3}$% of $240? _____

2. What is $44\frac{4}{9}$% of 198? _____

3. What is $33\frac{1}{3}$% of 48? _____

4. What is $18\frac{2}{11}$% of $550? _____

5. What is $42\frac{6}{7}$% of 161? _____

6. What is $\frac{1}{12}$ of $720? _____

7. What is $\frac{10}{15}$ of 165? _____

8. What is $\frac{7}{16}$ of $320? _____

Applications

9. In a survey of 3,600 prospective voters, $83\frac{1}{3}$% said they would vote against a proposition to sell bonds. How many voters would vote against the bonds?

   _____

10. A new house costs $72,000. A veteran gets a discount equal to $11\frac{1}{9}$%. What is the amount of the discount?

    _____

11. Tonia bought a new television set. It cost $480. She made a $33\frac{1}{3}$% down payment. How much does she still owe on the television?

    _____

12. Sales tax and luxury tax are charged on new boats. The total of taxes is $16\frac{2}{3}$% of the price. What would the total tax be on a boat that cost $18,690?

   _____

13. For an experiment, a scientist needs a beaker to be 45% full of a gas. What common fraction would express this same amount?

   _____

14. The manager at SuperSaver Grocery bought 600 packages of hot dog buns for the July 4 holiday. He sold all but $\frac{1}{2}$%. How many packages did he sell?

   _____

15. Sharon bought 36 lottery tickets for $1 each. One-ninth of the tickets were $2 winners and one-fourth were $1 winners. How much money did Sharon gain or lose by buying the 36 tickets?

   _____

16. A mixture for killing weeds is $\frac{1}{320}$ weed killer; the remainder is water. If the total mixture is 960 ounces, how many ounces of weed killer are used?

   _____

17. A large city sanitation department buys trucks at a fleet price that is $42\frac{6}{7}$% off the regular price. What would the department pay for a truck that cost $49,000?

   _____

18. The cost of a new table is $180. To buy it without interest you must pay $\frac{1}{9}$ down and $\frac{1}{4}$ of the balance each month for four months. What is the amount of each monthly payment?

   _____

19. Enrique found $6\frac{1}{4}$% of 320 computer diskettes inspected to be defective. How many diskettes were defective?

   _____

20. Westlake Community College had 7,400 students (full- and part-time) during the fall semester. That number increased by $\frac{1}{2}$% for the spring. How many students enrolled for the spring semester?

   _____

SUMMARY OF KEY CONCEPTS

KEY TERMS

terminating decimal (6.2): Decimal fractions that end when 0 begins repeating.

repeating decimal (6.2): Decimal fractions in which one or more digits repeat indefinitely.

KEY RULES

To convert a common fraction to a decimal fraction (6.1):

Divide the numerator by the denominator.

To convert a decimal fraction to a common fraction (6.1):

1. Write the decimal fraction as the numerator of a common fraction. Drop the decimal point.
2. Place a 0 in the denominator under each digit in the numerator and a 1 to the left of the zeros.
3. Reduce to lowest terms.

To convert a common fraction to a percent (6.1):

1. Convert the common fraction to its decimal equivalent.
2. Move the decimal point two places to the right and add the symbol %.

To convert a percent to a common fraction (6.1):

1. Multiply the number preceding the percent sign by $\frac{1}{100}$.
2. Reduce the answer to lowest terms.

PRACTICE TEST 6A

Fill in the chart for problems 1–6 by converting each given fraction to its two equivalent fractions. Reduce common fractions to lowest terms.

	Common Fraction	Decimal Fraction	Percent
1.	$\frac{3}{8}$	_____	_____
2.	$\frac{2}{15}$	_____	_____
3.	_____	0.65	_____
4.	_____	0.08	_____
5.	_____	_____	24%
6.	_____	_____	$\frac{1}{5}$%

7. Convert each of the following common fractions to a decimal and identify each as terminating or repeating.

	Decimal	Terminating or Repeating
a. $\frac{5}{8} =$	_____	_____
b. $\frac{3}{11} =$	_____	_____
c. $\frac{10}{25} =$	_____	_____
d. $\frac{15}{32} =$	_____	_____

Solve each of the following problems.

8. What is $28\frac{4}{7}$% of 140? _____

9. A caterer used 50 pounds of fruit for a luncheon for 75 people. Only $\frac{1}{20}$ of the fruit was not eaten. How many pounds of fruit were not eaten?

_____

10. Only $\frac{29}{32}$ of the ticket holders to a football game attended the game. What percent of those holding tickets went to the game? Round to the nearest tenth of a percent.

_____

11. Of the 144 junior golfers who entered a tournament, $\frac{15}{16}$ actually played. How many golfers played in the tournament?

_____

12. Jennifer spent $33\frac{1}{3}$% of her paycheck each week on bills. If she earned $372 per week, how much did she spend on bills?

_____

PRACTICE TEST 6B

Fill in the chart for problems 1–6 by converting each given fraction to its two equivalent fractions. Reduce common fractions to lowest terms.

	Common Fraction	Decimal Fraction	Percent
1.	$\frac{1}{3}$	_____	_____
2.	$\frac{7}{16}$	_____	_____
3.	_____	0.35	_____
4.	_____	0.0625	_____
5.	_____	_____	20%
6.	_____	_____	$83\frac{1}{3}\%$

7. Convert each of the following common fractions to a decimal and identify each as terminating or repeating.

		Decimal	Terminating or Repeating
a.	$\frac{1}{7} =$	_____	_____
b.	$\frac{3}{8} =$	_____	_____
c.	$\frac{4}{16} =$	_____	_____
d.	$\frac{9}{64} =$	_____	_____

Solve each of the following problems.

8. What is $44\frac{4}{9}\%$ of 720? _____

9. The metropolitan section of a large newspaper is 18 pages each Monday. The entire newspaper is 90 pages. What percent of the total pages is the metropolitan section?

   _____

10. A political candidate sent out 8,000 letters seeking contributions. He estimated that he would receive responses from $\frac{1}{25}$ of the recipients. How many responses did he anticipate?

    _____

11. Gloria attends college in a small town 540 miles from home. When she drives home, she likes to stop at a rest area that is $\frac{7}{12}$ of the way from college to home. How many miles has she driven when she gets to the rest area?

    _____

12. What is the cost of a new dinette set if the original price of $198 is reduced by $33\frac{1}{3}$%?

    _____

CHAPTERS 1-6

SKILLS REVIEW

1. Write 32,609 in words.
   _____

2. Round 3,550 to the nearest thousand. _____

3. Round 54,963,072 to the nearest million. _____

Perform the indicated operation. Express each answer in lowest terms.

4. $\dfrac{3}{5} + \dfrac{1}{3} + \dfrac{11}{20} =$ _____

5. $\dfrac{9}{16} - \dfrac{5}{12} =$ _____

6. $\dfrac{2}{3} \cdot \dfrac{6}{7} =$ _____

7. $\dfrac{5}{9} \div \dfrac{1}{3} =$ _____

8. $(9.3)(0.02) =$ _____

9. $18 \div 0.03 =$ _____

10. Convert $\dfrac{7}{20}$ to a decimal fraction. _____

11. Convert $\dfrac{18}{25}$ to a percent. _____

12. Convert 0.46 to a percent. _____

13. Convert 0.06 to a common fraction. _____

14. What is 28% of 350? _____

15. 40% of what number is 24? _____

16. 50 is what percent of 400? _____

17. 70 decreased by 10% is what amount? _____

377

18. Write 0.03 in words.
    _____

19. Write 9.265 in words.
    _____

20. Is the following a true proportion?
    $\frac{5}{6} \stackrel{?}{=} \frac{55}{72}$ _____

21. Find the missing element in the proportion
    $\frac{16}{x} = \frac{48}{15}$ _____

22. What is $\frac{1}{5}$% of 2,000? _____

23. What is $83\frac{1}{3}$% of 420? _____

24. $\frac{1}{8}$ of 600 is what amount? _____

25. What is $\frac{3}{11}$ of 242? _____

CHAPTER 6   SOLUTIONS

Skills Preview

1. 0.8; 80%
2. $0.58\bar{3}$; 58.3%
3. $\frac{3}{4}$; 75%
4. $\frac{21}{50}$; 42%
5. $\frac{19}{20}$; 0.95
6. $\frac{1}{16}$; 0.0625
7. **a.** $0.\bar{4}$; terminating
   b. $0.\bar{2}$; repeating
   c. $0.\overline{714285}$; repeating
   d. 0.5; terminating
8. $7.50
9. $90
10. 32%
11. 20
12. 4

Section 6.1   Comprehension Checkpoint

1. 0.375; 37.5%
2. $\frac{7}{25}$; 28%
3. $\frac{7}{10}$; 0.7
4. $\frac{3}{500}$; 0.006

6.1A   Practice Set

1. $\frac{3}{4} = 3 \div 4 = 0.75$; $0.75 = 75\%$

2. $\frac{9}{16} = 9 \div 16 = 0.5625$; $0.5625 = 56.25\%$ or $56\frac{1}{4}\%$

3. $\frac{12}{25} = 12 \div 25 = 0.48$; $0.48 = 48\%$

4. $\frac{1}{8} = 1 \div 8 = 0.125$; $0.125 = 12.5\%$ or $12\frac{1}{2}\%$

5. $\frac{3}{5} = 3 \div 5 = 0.6$; $0.6 = 60\%$

6. $\frac{7}{10} = 7 \div 10 = 0.7$; $0.7 = 70\%$

7. $0.3 = 30\%$; $\frac{3}{10}$

8. $0.85 = 85\%$; $\frac{85}{100} = \frac{17}{20}$

9. $0.24 = 24\%$; $\frac{24}{100} = \frac{6}{25}$

10. $0.375 = 37.5\%$ or $37\frac{1}{2}\%$; $\frac{375}{1000} = \frac{3}{8}$

11. $0.0625 = 6.25\%$ or $6\frac{1}{4}\%$; $\frac{0625}{10000} = \frac{1}{16}$

12. $0.3125 = 31.25\%$ or $31\frac{1}{4}\%$; $\frac{3125}{10000} = \frac{5}{16}$

379

13. $90\% = 0.9$; $90 \times \dfrac{1}{100} = \dfrac{9}{10}$

14. $58\% = 0.58$; $58 \times \dfrac{1}{100} = \dfrac{58}{100} = \dfrac{29}{50}$

15. $1\% = 0.01$; $1 \times \dfrac{1}{100} = \dfrac{1}{100}$

16. $62\dfrac{1}{2}\% = 62.5\% = 0.625$; $62\dfrac{1}{2} \times \dfrac{1}{100} = \dfrac{125}{2} \times \dfrac{1}{100} = \dfrac{125}{200} = \dfrac{5}{8}$

17. $43\dfrac{3}{4}\% = 43.75\% = 0.4375$; $43\dfrac{3}{4} \times \dfrac{1}{100} = \dfrac{175}{4} \times \dfrac{1}{100} = \dfrac{175}{400} = \dfrac{7}{16}$

18. $\dfrac{1}{5}\% = 0.2\% = 0.002$; $\dfrac{1}{5} \times \dfrac{1}{100} = \dfrac{1}{500}$

19. $\dfrac{3}{20} = 3 \div 20 = 0.15$; $0.15 = 15\%$

20. $0.0015 = \dfrac{0015}{10,000} = \dfrac{15}{10,000} = \dfrac{3}{2,000}$

21. $98 \times \dfrac{1}{100} = \dfrac{98}{100} = \dfrac{49}{50}$

22. $\dfrac{18}{100} = 0.18$

23. $12\dfrac{1}{2} \times \dfrac{1}{100} = \dfrac{25}{2} \times \dfrac{1}{100} = \dfrac{1}{8} =$ fractional part of the workforce that is men

$1 - \dfrac{1}{8} = \dfrac{8}{8} - \dfrac{1}{8} = \dfrac{7}{8} =$ fractional part of the workforce that is women

Section 6.2 Comprehension Checkpoint

1. terminating 2. repeating 3. $0.8\overline{3}$; repeating 4. 0.625; terminating

6.2A Practice Set

1. $\dfrac{2}{3} = 2 \div 3 = 0.\overline{6}$

2. $\dfrac{4}{5} = 4 \div 5 = 0.8$ (terminating)

3. $\dfrac{5}{12} = 5 \div 12 = 0.41\overline{6}$

4. $\dfrac{4}{9} = 4 \div 9 = 0.\overline{4}$

5. $\dfrac{3}{4} = 3 \div 4 = 0.75$ (terminating)

6. $\dfrac{4}{25} = 4 \div 25 = 0.16$ (terminating)

7. $\dfrac{2}{15} = 2 \div 15 = 0.1\overline{3}$

8. $\dfrac{5}{6} = 5 \div 6 = 0.8\overline{3}$

9. $\dfrac{9}{32} = 9 \div 32 = 0.28125$ (terminating)

10. $\dfrac{19}{50} = 19 \div 50 = 0.38$ (terminating)

Solutions

11. $\dfrac{11}{20} = 11 \div 20 = 0.55$ (terminating)

12. $\dfrac{3}{7} = 3 \div 7 = 0.\overline{428571}$

13. $\dfrac{9}{11} = 9 \div 11 = 0.\overline{81}$

14. $\dfrac{7}{8} = 7 \div 8 = 0.875$ (terminating)

Section 6.3   Comprehension Checkpoint

1. 8   **2.** $0.08   **3.** 2,105   **4.** $45,000

6.3A   Practice Set

1. $66\dfrac{2}{3} \times \dfrac{1}{100} = \dfrac{200}{3} \times \dfrac{1}{100} = \dfrac{2}{3}$

$\dfrac{2}{3} \times \$900 = (\$900 \div 3) \times 2 = \$600$

2. $14\dfrac{2}{7} \times \dfrac{1}{100} = \dfrac{100}{7} \times \dfrac{1}{100} = \dfrac{1}{7}$

$\dfrac{1}{7} \times 196 = 196 \div 7 = 28$

3. $41\dfrac{2}{3} \times \dfrac{1}{100} = \dfrac{125}{3} \times \dfrac{1}{100} = \dfrac{5}{12}$

$\dfrac{5}{12} \times 48 = (48 \div 12) \times 5 = 20$

4. $22\dfrac{2}{9} \times \dfrac{1}{100} = \dfrac{200}{9} \times \dfrac{1}{100} = \dfrac{2}{9}$

$\dfrac{2}{9} \times \$540 = (\$540 \div 9) \times 2 = \$120$

5. $83\dfrac{1}{3} \times \dfrac{1}{100} = \dfrac{250}{3} \times \dfrac{1}{100} = \dfrac{5}{6}$

$\dfrac{5}{6} \times 150 = (150 \div 6) \times 5 = 125$

6. $\dfrac{1}{5} \times \$800 = \$800 \div 5 = \$160$

7. $\dfrac{3}{4} \times \$200 = (\$200 \div 4) \times 3 = \$150$

8. $\dfrac{5}{8} \times \$320 = (\$320 \div 8) \times 5 = \$200$

9. $66\dfrac{2}{3} \times \dfrac{1}{100} = \dfrac{200}{3} \times \dfrac{1}{100} = \dfrac{2}{3}$

$\dfrac{2}{3} \times \dfrac{144}{1} = 96$

10. $37\dfrac{1}{2}\% = 0.375;\ 0.375 \times 48 = 18$

11. $33\dfrac{1}{3} \times \dfrac{1}{100} = \dfrac{100}{3} \times \dfrac{1}{100} = \dfrac{1}{3}$

$\dfrac{1}{3} \times \dfrac{60}{1} = 20$

12. $\dfrac{1}{5} \times \$4.75 = \$4.75 \div 5 = \$0.95;$

$\$4.75 + \$0.95 = \$5.70$

13. $\dfrac{3}{4} \times 44 = (44 \div 4) \times 3 = 33$

14. $83\dfrac{1}{3} \times \dfrac{1}{100} = \dfrac{250}{3} \times \dfrac{1}{100} = \dfrac{5}{6}$

$\dfrac{5}{6} \times 6 = (6 \div 6) \times 5 = 5$

15. $\dfrac{5}{9} \times 180 = (180 \div 9) \times 5 = 100$ delivered before lunch

$180 - 100 = 80$ still to deliver after lunch

16. $\dfrac{2}{15} \times 135 = (135 \div 15) \times 2 = 18$ minutes

17. $\dfrac{1}{8} \times 18 = 18 \div 8 = 2.25$ gallons of gas in the tank

$18 - 2.25 = 15.75$ gallons of gas used on the trip

$15.75 \times 20.5 = 322.9$ miles traveled on the trip (rounded to the nearest tenth)

18. $\dfrac{9}{10} \times 140 = (140 \div 10) \times 9 = 126$

problems she must answer correctly
$140 - 126 = 14 =$ maximum number she can miss and still get an A

19. $\frac{1}{3} \times \frac{1}{100} = \frac{1}{300}$; $\frac{1}{300} \times 24{,}000 =$
    $24{,}000 \div 300 = 80$

20. $28\frac{4}{7} \times \frac{1}{100} = \frac{200}{7} \times \frac{1}{100} = \frac{2}{7}$
    $\frac{2}{7} \times \$18{,}655 = (\$18{,}655 \div 7) \times 2 =$
    $\$5{,}330 =$ all of his deductions
    $\$18{,}655 - \$5{,}330 = \$13{,}325$ disposable income

Practice Test 6A

1. 0.375; 37.5%
2. $0.1\overline{3}$; 13.3%
3. $\frac{13}{20}$; 65%
4. $\frac{2}{25}$; 8%
5. $\frac{6}{25}$; 0.24
6. $\frac{1}{500}$; 0.002
7. **a.** 0.625; terminating
   b. $0.\overline{27}$; repeating
   c. 0.4; terminating
   d. 0.46875; terminating
8. 40
9. 2.5 pounds
10. 90.6%
11. 135
12. $124

Skills Review Chapters 1–6

1. thirty-two thousand, six hundred nine
2. 4,000
3. 55,000,000
4. $1\frac{29}{60}$
5. $\frac{7}{48}$
6. $\frac{4}{7}$
7. $1\frac{2}{3}$
8. 0.186
9. 600
10. 0.35
11. 72%
12. 46%
13. $\frac{3}{50}$
14. 98
15. 60
16. $12\frac{1}{2}\%$
17. 63
18. three hundredths
19. nine and two hundred sixty-five thousandths
20. no
21. 5
22. 4
23. 350
24. 75
25. 66

SIGNED NUMBERS

OUTLINE

7.1 Introduction to Signed Numbers

7.2 Adding Signed Numbers

7.3 Subtracting Signed Numbers

7.4 Multiplying Signed Numbers

7.5 Dividing Signed Numbers

7.6 Order of Operations

In the first six chapters, all of the examples and problems were worked with numbers greater than zero. These are called *positive numbers*. Not all numbers that you will encounter will be positive. There are many occasions when you will need to use a number less than zero. You can use numbers less than zero to describe the temperature, an overdraft situation in a checking account, and the yardage a quarterback loses when tackled behind the line of scrimmage.

The objective of this chapter is to introduce you to numbers less than zero. You will learn how to apply the four basic math operations to these numbers.

SKILLS PREVIEW

In problems 1–4, insert the appropriate symbol, either < or >.

1. 3 _____ 9
2. $-\frac{1}{2}$ _____ $+\frac{1}{2}$
3. 4.23 _____ −3.7
4. 4 _____ −5

In problems 5 and 6, determine the absolute value.

5. $|-9|$ = _____
6. $-|6.8|$ = _____

In problems 7 and 8, write the additive inverse of the value given.

7. +15 _____
8. $-3\frac{1}{2}$ _____

Add:

9. $(+6) + (-3) =$ _____
10. $(-8) + (-18) =$ _____
11. $(+15) + (-9) + (-4) + (-7) =$ _____

Subtract:

12. $(+7) - (+3) =$ _____
13. $(-20) - (-6) =$ _____
14. $(-14) - (+15) =$ _____

Multiply:

15. $(+2)(+5) =$ _____
16. $(-8)(+9) =$ _____
17. $(-4)(-13) =$ _____
18. $(-10)(-5)(+2) =$ _____

19. $(-15)(-7)(-8) =$ _____

Divide:

20. $(+14) \div (+7) =$ _____

21. $(+35) \div (-5) =$ _____

22. $(-48) \div (-12) =$ _____

Simplify:

23. $(+6)(-2) \div (-3) + (-5) - (+2) =$ _____

24. $[(+12) \div (-6)] - (+1) + (+4)(-8) =$ _____

25. $(+3)(-12) - (+18) \div [(-72) \div (-8)] =$ _____

7.1 INTRODUCTION TO SIGNED NUMBERS

The concept of **signed numbers** expands the number system to include negative as well as positive numbers. Negative numbers are numbers less than 0; they allow us to discuss such things as elevation below sea level, a decrease in output by a company, or a debt owed to another person.

Positive and negative numbers can be represented pictorially on a number line as in Figure 7.1.

FIGURE 7.1

The point separating positive from negative numbers on the line is 0. All numbers to the right of 0 are positive, while all numbers to the left of 0 are negative. It is important to notice that 0 is neither positive nor negative. All numbers, including fractions, are represented on the line. Notice that a negative sign (−) is used to designate a negative number. Numbers without a sign are understood to be positive. As you move to the right from any point on the line, the movement is in a positive direction. Movement from any point on the line to the left is in a negative direction. The arrows at either end of the line indicate that the line extends indefinitely in each direction.

Comparing Signed Numbers

Numbers along the number line may be compared by use of **inequality symbols.** A < (less than) or > (greater than) between two numbers indicates that one number is larger than the other number. Whenever two numbers are compared, the inequality symbol points toward the lesser value; it also opens toward the greater value.

FIGURE 7.2

Notice in Figure 7.2 that -3 is to the left of $+2$ on the number line.

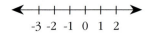

To compare these two numbers, we would write

$$-3 < 2$$

This is read "negative three is less than two." This same statement could be turned around to appear as

$$2 > -3$$

This would be read "two is greater than negative three." Either way, the meaning is the same: $+2$ represents a greater amount than -3 does. Notice also that the inequality symbol points toward the -3 in both instances because -3 is less than 2.

> **RELATIONSHIPS ON THE NUMBER LINE**
> 1. Any number on the line is greater than any other number to its left.
> 2. Any number on the line is less than any other number to its right.

EXAMPLE 1 Insert either $<$ or $>$ in each blank to show the proper relationship between the pair of numbers.

a. 4 _____ 2
b. -3 _____ 1
c. -2.7 _____ -2.9

SOLUTIONS

a. $4 > 2$

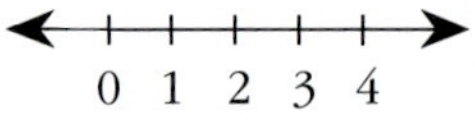

The inequality symbol must open toward the 4 since 4 is greater than 2. Notice that 4 is to the right of 2 on the number line.

b. $-3 < 1$

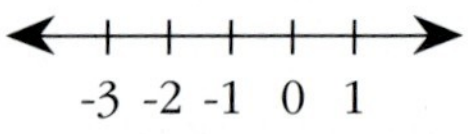

The symbol must open toward the 1 since 1 is greater than -3. On the number line, 1 is to the right of -3.

c. $-2.7 > -2.9$

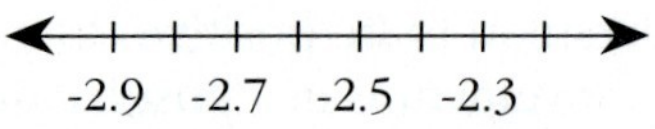

The symbol should open toward -2.7 since -2.7 is greater than -2.9. On the number line, -2.7 is to the right of -2.9.

Introduction to Signed Numbers

Absolute Value

Absolute value may be easily explained using the number line. The absolute value of any number is the distance that number is from 0 on the number line. The symbol to indicate absolute value is | |.

Figure 7.3 demonstrates that the absolute value of 4 is 4 since it is four units from 0.

FIGURE 7.3

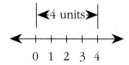

It is important to note that negative numbers also have absolute values. Figure 7.4 shows that −4 is also four units from 0, so that its absolute value is 4.

FIGURE 7.4

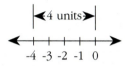

In symbols, |4| = 4 and |−4| = 4.

EXAMPLE 2 State the indicated absolute value.

a. |6| b. |−8| c. −|5.3| d. −|−7|

SOLUTIONS

a. |6| = 6   6 is 6 units from 0.

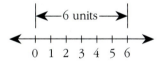

b. |−8| = 8   −8 is 8 units from 0.

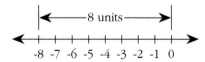

c. |5.3| = 5.3   The absolute value of 5.3 is 5.3.

   −|5.3| = −5.3   The negative sign outside the symbol | | indicates that after you determine |5.3|, you then need to make it negative.

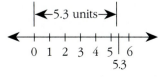

d. $|-7| = 7$ *The absolute value of -7 is 7.*

  $-|-7| = -7$ *The negative sign outside the symbol $|\ \ |$ indicates that after you determine $|-7|$, you then need to make it negative.*

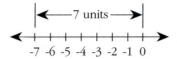

Additive Inverse

The concept of absolute value leads logically to another principle that is fundamental to algebra and is especially important in solving equations. This is the concept of **additive inverse.** The additive inverse of a number is the same number with the opposite sign. The sum of the two numbers is 0. Examples of additive inverses are

$$5 \text{ and } -5 \qquad 12 \text{ and } -12 \qquad 127 \text{ and } -127$$

Additive inverses are also called *opposites*.

Figures 7.3 and 7.4 illustrated that both $+4$ and -4 have absolute values of 4. If you add these two numbers together, the sum is 0. Thus, they are additive inverses, or opposites. The additive inverse of a number is what you must add to it to get a sum of 0.

Figure 7.5 demonstrates this concept. To move from 0 to 4 requires a move to the right, or in a positive direction. To move from 4 back to 0 requires a move to the left, or in a negative direction. Thus, $+4$ (the move to the right) $+ (-4)$ (the move to the left) $= 0$ (the starting point).

FIGURE 7.5

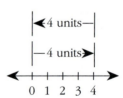

EXAMPLE 3   Determine the additive inverse for each number.

a. -7 **b.** 16 **c.** -0.6

SOLUTIONS

a. $+7$ *Both $+7$ and -7 are 7 units from 0*
b. -16 *Both 16 and -16 are 16 units from 0*
c. $+0.6$ *Both -0.6 and 0.6 are 0.6 unit from 0*

Work the Comprehension Checkpoint problems to assess your understanding of the concepts you studied in this section. If you have difficulty with these problems, review the material in this section before going on to the Practice Sets.

COMPREHENSION CHECKPOINT

Use either < or > to complete a true statement.

1. 3 _____ −5
2. 0 _____ −8
3. −13 _____ 13

Write the absolute value of the following numbers.

4. |−7| _____
5. −|10| _____
6. |−3.8| _____

Give the additive inverse for each number.

7. −5.9 _____
8. $\frac{2}{5}$ _____
9. $-\frac{3}{7}$ _____

7.1A PRACTICE SET

In problems 1–10, insert either < or > to show the proper relationship between the two numbers.

1. −2 _____ 5
2. −3.75 _____ −4.25
3. −6 _____ 3
4. −0.375 _____ 0.625
5. 5 _____ 4
6. 3.8 _____ 0
7. 16 _____ 15
8. 0.9 _____ −0.9
9. 18 _____ −37
10. −2.5 _____ 8.63

In problems 11–18, give the absolute value as indicated.

11. $|7| = $ _____
12. $|-4| = $ _____
13. $-|2.35| = $ _____
14. $|8| = $ _____
15. $\left|-\frac{3}{7}\right| = $ _____
16. $-\left|3\frac{9}{10}\right| = $ _____
17. $-|-18.5| = $ _____
18. $|0| = $ _____

In problems 19–25, write the additive inverse of the given value.

19. 5 _____
20. 1.75 _____
21. −7 _____
22. 12 _____
23. $-\frac{3}{4}$ _____
24. $\frac{4}{9}$ _____
25. −0.35 _____

26. What point is +5 units from −3 on the number line?

_____

27. What point is −6 units from 1 on the number line?

_____

28. Locate on the following number line two points with an absolute value of 3.

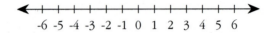

7.1B PRACTICE SET

In problems 1–10, insert either < or > to show the proper relationship between the two numbers.

1. 8 _____ −7
2. 8.45 _____ 6.25
3. −5 _____ −6
4. −14.75 _____ −14.65
5. −1.5 _____ 0
6. 15.6 _____ −2
7. −10 _____ −14
8. 0.003 _____ 0.3
9. 21 _____ −8
10. −0.58 _____ 0.58

In problems 11–18, give the absolute value as indicated.

11. $|6| =$ _____
12. $|-1| =$ _____
13. $|-2.06| =$ _____
14. $-|3| =$ _____
15. $\left|\dfrac{5}{8}\right| =$ _____
16. $-\left|-2\dfrac{6}{17}\right| =$ _____
17. $-|-25.4| =$ _____
18. $|-9| =$ _____

In problems 19–25, write the additive inverse of the given value.

19. 12 _____
20. 12.35 _____
21. −9 _____
22. −87 _____
23. $-\dfrac{2}{5}$ _____
24. $\dfrac{3}{8}$ _____
25. −2.08 _____

26. What point is +3 units from −2 on the number line?

   _____

27. What point is −5 units from −1 on the number line?

   _____

28. Locate two points with an absolute value of 2 on the following number line.

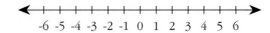

7.2 ADDING SIGNED NUMBERS

Addition of signed numbers requires that the added numbers first be identified as either positive or negative. The steps for adding signed numbers can be shown using three different types of examples.

The first example involves adding numbers with the same sign.

> **TO ADD NUMBERS WITH THE SAME SIGN**
> 1. Add the absolute values of the numbers.
> 2. Attach the common sign to the sum.

The following examples will illustrate this procedure.

EXAMPLE 1   Add: (4) + (6)

SOLUTION

|4| = 4   *Determine the absolute value of each addend.*
|6| = 6

4 + 6 = 10   *Add the absolute values.*

(4) + (6) = 10 *Attach the common sign. Since both addends are positive, the common sign is +.*

This can be shown on the number line:

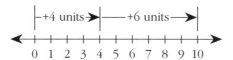

Adding two positive numbers requires beginning at 0 and moving to the right the number of units indicated by the first addend. Since the second addend is also positive, move in a positive direction again. The result is a positive number with an absolute value equal to the sum of the absolute values of the two original addends.

EXAMPLE 2   Add: (−2) + (−5)

SOLUTION

|−2| = 2   *Determine the absolute value of each addend.*
|−5| = 5

2 + 5 = 7   *Add the absolute values.*

(−2) + (−5) = −7   *Attach the common sign. Since both addends are negative, the common sign is −.*

On the number line:

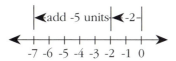

To add two negative numbers, begin at 0 and move to the left (negative direction) until you arrive at the point on the line representing the first addend. Then continue in the same direction the number of spaces represented by the second addend. The result is a negative number with an absolute value equal to the sum of the absolute values of the two original addends.

A second process is required to add two numbers with different signs.

> **TO ADD TWO NUMBERS WITH DIFFERENT SIGNS**
> 1. Find the difference between the absolute values of the numbers.
> 2. Attach the sign of the number with the larger absolute value.

Study the following examples carefully.

EXAMPLE 3   Add: $(4) + (-6)$

SOLUTION

$\|4\| = 4$	*Determine the absolute value of each addend.*
$\|-6\| = 6$	
$6 - 4 = 2$	*Determine the difference between the two absolute values.*
$(4) + (-6) = -2$	*The sign will be negative, since the addend, -6, that produced the larger absolute value is negative.*

On the number line:

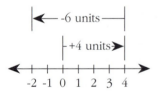

Beginning at 0, move 4 units in a positive direction (to the right). This puts you at 4 on the line. Since the other addend is -6, move from $+4$ in a negative direction (to the left) 6 units. The point where you end is -2.

The next example demonstrates that the sum of two numbers is the same regardless of the order of the addends.

EXAMPLE 4   Add: $(-6) + (+4)$

SOLUTION

$\|-6\| = 6$	*Determine the absolute value of each addend.*
$\|+4\| = 4$	
$6 - 4 = 2$	*Determine the difference between the absolute values.*

Adding Signed Numbers

$(-6) + (4) = -2$ *The sign will be negative, since the addend, -6, that produced the larger absolute value is negative.*

On the number line:

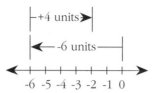

Beginning at 0, move 6 units in a negative direction (to the left) and arrive at -6. From this point, adding $+4$ requires a move in a positive direction (to the right) 4 units. The endpoint is -2.

Examples 3 and 4, using the same addends in a different order, show that adding two numbers will produce the same sum no matter the order of the addends. This point is extremely important for success in working with signed numbers.

It is often necessary to add more than two numbers. The third procedure, for adding more than two signed numbers, is a combination of the two procedures discussed previously for adding two signed numbers.

> **TO ADD MORE THAN TWO SIGNED NUMBERS**
> 1. Add all the positive numbers.
> 2. Add all the negative numbers.
> 3. Using these two sums, follow the procedure for adding two signed numbers with different signs.

EXAMPLE 5   Add: $(6) + (-5) + (-3) + (4)$

SOLUTION

$6 + 4 = 10$ *Add the positive addends.*

$(-5) + (-3) = -8$ *Add the negative addends.*

$10 + (-8) = ?$ *Add the results of the first two steps using the procedures for adding two numbers with different signs.*
$|10| = 10$
$|-8| = 8$
$10 - 8 = 2$
$10 + (-8) = 2$
$(6) + (-5) + (-3) + (4) = 2$

Any number of addends can be added using this procedure, regardless of the signs.

Work the Comprehension Checkpoint problems to practice your ability to add signed numbers. If you have difficulty with these problems, review the material in this section before going on to the Practice Sets.

> **a + b = x**
> # ALGEBRA CONNECTION
>
> Mastering the use of signed numbers is generally considered to be the first step in mastering the concepts of algebra. Throughout the text, problems have been presented in equation form. In this chapter, you have learned to use symbols to compare numbers and to represent absolute value. Being able to use symbols rather than words to express problems is a key to the mastery of algebra. Any of the problems solved in this chapter could have been set up as equations. Study the following examples and notice that they are the same as the problems worked in this chapter.
>
> Remember that a letter can be used to represent the unknown quantity in a problem. Consider the example that you just worked:
>
> $$(-6)+(+4)=\underline{}$$
>
> The result of adding $+4$ to -6 is the unknown quantity in this problem. The problem could also be written as
>
> $$(-6) + (+4) = x$$
>
> The same relationship would be expressed. Observe that the only difference in the two problems is the use of x in place of the blank space. Either way, without regard to how it is written, the result of the calculation will be the same. The value that goes in the blank and the number that x represents is -2.
>
> Any of the other problems in this chapter could be set up using a letter to represent the unknown. The move from basic math to algebra is a matter of understanding the use of symbols or letters to represent numbers.

COMPREHENSION CHECKPOINT

Find the sums in the following problems.

1. $8 + 6 =$ _____

2. $(-6) + (-7) =$ _____

3. $7 + (-12) =$ _____

4. $(-18) + 10 =$ _____

5. $9 + (-10) + 3 + (-5) =$ _____

6. $(-12) + (-9) + 4 + 6 + 15 =$ _____

7.2A PRACTICE SET

Find the sum in each problem.

1. $(-8) + 3 =$ _____

2. $5 + (-9) =$ _____

3. $\left(-\frac{1}{2}\right) + \left(-\frac{3}{4}\right) =$ _____

4. $(-5.7) + 6.25 =$ _____

5. $(-12) + 5.6 =$ _____

6. $(-8.4) + (-2.8) =$ _____

7. $\frac{5}{8} + \left(-\frac{2}{3}\right) =$ _____

8. $112 + (-31) =$ _____

9. $(-8) + 6 + (-3.7) =$ _____

10. $(-9) + (-3) + (-16) =$ _____

11. $12.4 + 3.9 + (-8.2) =$ _____

12. $\frac{2}{3} + \frac{3}{4} + \left(-\frac{5}{6}\right) =$ _____

13. $\left(-\frac{3}{5}\right) + \left(-\frac{2}{7}\right) + \frac{7}{10} =$ _____

14. $4.7 + (-5.3) + (-6.9) =$ _____

15. $42 + (-27) + (-12) =$ _____

16. $(-7.5) + 3.33 + (-5.25) =$ _____

17. $18 + (-7) + 16 + (-5) =$ _____

18. $(-3) + (-8) + (-9) + 12 =$ _____

19. $(-0.5) + 0.67 + (-0.75) + 0.83 =$ _____

20. $(-5.8) + (-6.75) + 15.2 + (-9.7) + 1.6 =$ _____

21. $(-18) + 2.4 + (-9.6) + 6 =$ _____

22. $\left(-\dfrac{5}{9}\right) + \dfrac{3}{7} + \left(-\dfrac{5}{7}\right) + \dfrac{2}{9} =$ _____

23. $\left(-2\dfrac{1}{2}\right) + 5\dfrac{3}{4} + \left(-4\dfrac{5}{12}\right) + 3\dfrac{2}{3} =$ _____

24. $(-12) + 16 + (-14) + (-10) + 20 =$ _____

25. $6.42 + 7.95 + (-8.43) + 10.58 =$ _____

Applications

26. The temperature was 34° at 5:00 AM. It dropped by 5° in the next hour. What was the temperature at 6:00 AM?

   _____

27. A running back gained 106 yards during one game. The quarterback was tackled for 22 yards in losses. What was the net yardage for the two players?

   _____

28. A motorist with a full tank of 19 gallons of gasoline used 7.5 gallons on a trip. How much gas remained in the tank after the trip?

   _____

7.2B PRACTICE SET

Find the sum in each problem.

1. $5 + (-4) =$ _____

2. $7 + (-7) =$ _____

3. $\left(-\frac{2}{3}\right) + \left(-\frac{5}{6}\right) =$ _____

4. $8.4 + 3.6 =$ _____

5. $(-6) + 4.525 =$ _____

6. $(-7.3) + (-10.9) =$ _____

7. $\frac{3}{4} + \left(-\frac{7}{8}\right) =$ _____

8. $(-85) + 19 =$ _____

9. $(-18) + 2.6 + 6 =$ _____

10. $(-8) + (-3) + (-11) =$ _____

11. $12.3 + (-18.4) + (-13.9) =$ _____

12. $\frac{1}{2} + \left(-\frac{3}{5}\right) + \left(-\frac{7}{10}\right) =$ _____

13. $\left(-\frac{5}{6}\right) + \left(-\frac{2}{3}\right) + \frac{5}{9} =$ _____

14. $(-9.2) + 8.7 + 16.5 =$ _____

15. $(-16) + 8 + (-9) =$ _____

16. $4.5 + (-5.33) + 6.25 =$ _____

17. $(-15) + 7 + (-18) + 26 =$ _____

18. $10 + (-3) + (-9) + (-12) =$ _____

19. $0.53 + 0.84 + (-0.92) + (-0.72) =$ _____

20. 12.2 + 15.5 + (−21.6) + (−18.7) = _____

21. (−1) + 14 + 11.4 + (−7.5) = _____

22. $\left(-\frac{1}{4}\right) + \frac{3}{8} + \left(-\frac{1}{2}\right) + \frac{9}{16} =$ _____

23. $\left(-5\frac{2}{5}\right) + \left(-6\frac{7}{8}\right) + \left(-3\frac{7}{10}\right) + \left(-4\frac{11}{20}\right) =$ _____

24. 8 + (−6) + 7 + 3 + (−15) + 19 = _____

25. (−4.95) + (−9.64) + 3.42 + (−18.67) = _____

Applications

26. The temperature began at 5°F and dropped 6° over a two-hour period. What was the new temperature?

_____

27. There were 17 passengers on a bus. At the first stop, four people got off the bus and three got on. At the second stop, another five people got on. How many passengers were on the bus after the second stop?

_____

28. The punter on a football team kicked the ball 47 yards, but it was returned 53 yards. What was the net yardage on the punt?

_____

7.3 SUBTRACTING SIGNED NUMBERS

Recall from Section 1.4 that subtraction is the process of finding the difference between two numbers. The key to subtracting signed numbers is to remember that some of the signs are going to change. In the problem

$$(+8) - (+2) = ?$$

the negative sign indicates subtraction. It is the operation sign in this problem. The object is to subtract $+2$ from $+8$. The answer is $+6$. This is the type of problem that you studied in Chapter 1.

Compare this with a problem like

$$(+8) - (-2) = ?$$

In this problem, -2 is being subtracted from $+8$. The symbol $-$ between the different sets of parentheses indicates a subtraction operation. The minus sign inside the parentheses with the 2 indicates that the 2 is negative. The object of this problem is to subtract a negative number from a positive number. This leads to the procedure for subtracting signed numbers.

> **TO SUBTRACT SIGNED NUMBERS**
> 1. Change the operation sign to addition and change the sign of the number being subtracted to its opposite.
> 2. Follow the rules for adding signed numbers.

A careful reading of the steps for subtraction indicates that subtracting signed numbers actually becomes an addition process. Following this procedure to solve the previous problem gives

$$(+8) - (-2) = ?$$

change the operation sign to +   *change the sign of the number being subtracted to its opposite*

$$(+8) + (+2) = +10$$

Study the next four examples carefully. Note that the procedure is the same in each problem.

EXAMPLE 1   Subtract: $6 - (+3) = ?$

SOLUTION

$6 + (-3) = ?$ *Change the operation sign to positive. The $(+3)$ is being subtracted. Change its sign to negative, the opposite of positive.*

$6 + (-3) = 3$ *Follow the rules for addition.*

$6 - (+3) = 3$

EXAMPLE 2   Subtract: $5 - (-2) = ?$

SOLUTION

$5 + (+2) = ?$ *Change the operation sign to positive. The (-2) is being subtracted. Change its sign to positive, the opposite of negative.*

CHAPTER 7 Signed Numbers

$5 + (+2) = +7$ *Follow the rules for addition.*
$5 - (-2) = +7$

EXAMPLE 3   Subtract: $(-8) - (+4) = ?$

SOLUTION

$(-8) + (-4) = ?$ *Change the operation sign to positive. The $(+4)$ is being subtracted. Change its sign to negative, the opposite of positive.*

$(-8) + (-4) = -12$ *Follow the rules for addition.*
$(-8) - (+4) = -12$

EXAMPLE 4   Subtract: $(-9) - (-7) = ?$

SOLUTION

$(-9) + (+7) = ?$ *Change the operation sign to positive. The (-7) is being subtracted. Change its sign to positive, the opposite of negative.*

$(-9) + (+7) = -2$ *Follow the rules for addition.*
$(-9) - (-7) = -2$

Being able to work with signed numbers is an important skill in mathematics. A more thorough explanation of the rules for subtracting signed numbers is beyond the scope of this text. For now, remember the rules that you learned in this section. Remember that a number without a sign is a positive number.

Work the Comprehension Checkpoint problems to measure your understanding of the procedure for subtracting signed numbers. If you have difficulty with these problems, review the material in this section before going on to the Practice Sets.

COMPREHENSION CHECKPOINT

1. $(-5) - (-4) = $ _____

2. $(-12.6) - (3.9) = $ _____

3. $8\frac{2}{3} - \left(6\frac{1}{2}\right) = $ _____

4. $23 - (-8.75) = $ _____

5. $(-1.04) - (-3.5) = $ _____

7.3A PRACTICE SET

Determine the difference in the following subtraction problems.

1. $9 - (-6) = $ _____
2. $(-7) - (-3) = $ _____
3. $5 - (7) = $ _____
4. $10 - (-10) = $ _____
5. $(-6.4) - (-8.2) = $ _____
6. $\left(-\frac{1}{2}\right) - \left(-\frac{1}{4}\right) = $ _____
7. $\frac{3}{4} - \left(\frac{2}{3}\right) = $ _____
8. $(-24) - (6) = $ _____
9. $18.2 - (-9.6) = $ _____
10. $\left(-12\frac{1}{2}\right) - \left(8\frac{2}{3}\right) = $ _____
11. $10.85 - (-3.76) = $ _____
12. $(-7) - (6.3) = $ _____
13. $(-3.62) - (-12.8) = $ _____
14. $\left(-\frac{5}{8}\right) - \left(\frac{3}{4}\right) = $ _____
15. $(-28) - (17) = $ _____
16. $15\frac{1}{3} - \left(-4\frac{2}{3}\right) = $ _____
17. $8.76 - (3.5) = $ _____
18. $16 - (-5.4) = $ _____

19. $1\frac{7}{8} - \left(-3\frac{5}{12}\right) = $ _____

20. $\left(-6\frac{4}{5}\right) - \left(-8\frac{2}{3}\right) = $ _____

21. $(-18.09) - (10.22) = $ _____

22. $0.01 - (-0.001) = $ _____

23. $(-3.09) - (4.76) = $ _____

24. $\left(-14\frac{2}{7}\right) - \left(-6\frac{1}{2}\right) = $ _____

25. $\left(-31\frac{3}{5}\right) - \left(9\frac{3}{4}\right) = $ _____

Applications

26. What is the change in temperature from 18° to −4°?

27. David jumped out of an airplane at an altitude of 10,000 feet. He pulled the cord at 4,500 feet above the ground. How far was he in free fall?

28. A student scored 148 points on a 200-point test. How many points did he miss?

7.3B PRACTICE SET

Determine the difference in the following subtraction problems.

1. $8 - (-4) =$ _____

2. $(-2) - (-5) =$ _____

3. $3 - (10) =$ _____

4. $5 - (-6) =$ _____

5. $(-7.3) - (-4.7) =$ _____

6. $\left(-\dfrac{3}{8}\right) - \left(-\dfrac{2}{3}\right) =$ _____

7. $\dfrac{5}{9} - \left(\dfrac{5}{6}\right) =$ _____

8. $(-18) - (12) =$ _____

9. $24.4 - (-15.6) =$ _____

10. $\left(-3\dfrac{3}{4}\right) - \left(2\dfrac{1}{3}\right) =$ _____

11. $6.43 - (-8.95) =$ _____

12. $(-12) - (9.2) =$ _____

13. $(-24.65) - (-13.82) =$ _____

14. $\left(-\dfrac{4}{5}\right) - \left(\dfrac{9}{10}\right) =$ _____

15. $(-37) - (23) =$ _____

16. $22\dfrac{1}{2} - \left(-5\dfrac{5}{6}\right) =$ _____

17. $3.49 - (-4.7) =$ _____

18. $12 - (-3.8) =$ _____

19. $2\frac{7}{8} - \left(3\frac{3}{4}\right) = $ _____

20. $\left(-8\frac{7}{10}\right) - \left(-6\frac{1}{4}\right) = $ _____

21. $(-12.6) - (-3.4) = $ _____

22. $0.035 - (-0.002) = $ _____

23. $(-4.75) - (5.82) = $ _____

24. $\left(-18\frac{7}{15}\right) - \left(-12\frac{2}{3}\right) = $ _____

25. $\left(-42\frac{1}{2}\right) - \left(6\frac{3}{5}\right) = $ _____

Applications

26. What is the difference between 52°F and −16°F?

_____

27. What is the difference in elevation between a point 12,375 feet above sea level and another point 1,382 feet below sea level?

_____

28. Regina had a credit card balance of $125. She made additional purchases during the month of $35 and $68. She also made a payment of $200. What is her current balance?

_____

7.4 MULTIPLYING SIGNED NUMBERS

You learned that multiplication is a faster method of repeated addition of the same number. The problem

$$3 \times 4 = 12$$

could be written

$$4 + 4 + 4 = 12$$

or

$$3 + 3 + 3 + 3 = 12$$

These illustrations demonstrate not only that multiplication is a form of addition but also that the order of the factors makes no difference. Multiplying 3 by 4 gives the same product as multiplying 4 by 3.

Multiplying signed numbers uses the same procedure as multiplying any numbers. The question is how to determine the sign of the product. The procedure for multiplying signed numbers with any number of factors is relatively simple.

TO MULTIPLY SIGNED NUMBERS

1. Multiply the absolute values of the factors.
2. Count the negative factors. An odd number of negative factors means that the product is negative. An even number of negative factors means that the product is positive.

You may recall from previous math courses that "a positive times a positive is a positive" or "a negative times a negative is a positive." Both of these statements are true. The method used in this text requires only that you count the negative factors. The sign of the product is determined by the number of negative factors. If there are no negative factors, the product is always positive.

Several examples will serve to illustrate the procedure. (Remember, when groups of parentheses are written side by side, multiplication is indicated.)

EXAMPLE 1   Multiply: (5)(6)

SOLUTION

$|5| = 5$ *Determine the absolute value of each factor.*
$|6| = 6$

$5 \times 6 = 30$ *Find the product of the absolute values.*

There are no negative factors. The product is positive.

$5 \times 6 = 30$

EXAMPLE 2   Multiply: $(7)(-4)$

SOLUTION

$|7| = 7$ *Determine the absolute value of each factor.*
$|-4| = 4$

$7 \times 4 = 28$ *Find the product of the absolute values.*

Count the negative factors. There is one negative factor, -4. Since 1 is an odd number, the product is negative.

$(7)(-4) = -28$

EXAMPLE 3   Multiply: $(3)(-7)(-6)$

SOLUTION

$\left.\begin{array}{r}|3| = 3 \\ |-7| = 7 \\ |-6| = 6\end{array}\right\}$ *Determine the absolute value of each factor.*

$3 \times 7 \times 6 = 126$ *Find the product of the absolute values.*

Count the negative factors. There are two negative factors, -7 and -6. Since 2 is an even number, the product is positive.

$(3)(-7)(-6) = 126$

EXAMPLE 4   Multiply: $(-4)(5)(-3)(2)(-6)$

SOLUTION

$\left.\begin{array}{r}|-4| = 4 \\ |5| = 5 \\ |-3| = 3 \\ |2| = 2 \\ |-6| = 6\end{array}\right\}$ *Determine the absolute value of each factor.*

$4 \times 5 \times 3 \times 2 \times 6 = 720$ *Find the product of the absolute values.*

Count the negative factors. There are three negative factors: -4, -3, and -6. Since 3 is an odd number, the product is negative.

$(-4)(5)(-3)(2)(-6) = -720$

EXAMPLE 5   Multiply: $(-4)(-5)(-3)(2)(-6)$

SOLUTION

The factors have the same absolute values as in Example 4; you know that the product of the absolute values is 720. However, the product here will be positive because there are an even number of negative factors.

$(-4)(-5)(-3)(2)(-6) = 720$

Work the Comprehension Checkpoint problems to measure your understanding of the procedure for multiplying signed numbers. If you have difficulty with these problems, review the material in this section before going on to the Practice Sets.

COMPREHENSION CHECKPOINT

1. $(-3)(7) =$ _____

2. $(-9)(-4) =$ _____

3. $(6.8)(-2)(-0.75) =$ _____

4. $(-8)(-3)(-0.5) =$ _____

5. $(-10)(-7)(5)(4)(-8) =$ _____

7.4A PRACTICE SET

Find the products in the following problems.

1. $(6)(4) =$ _____

2. $(-9)(7) =$ _____

3. $(-6)(-8) =$ _____

4. $(12)(-13) =$ _____

5. $(3.7)(-8.4) =$ _____

6. $(-21)(-15) =$ _____

7. $(-6.9)(-18.7) =$ _____

8. $\left(\frac{2}{3}\right)\left(-\frac{5}{9}\right) =$ _____

9. $\left(-\frac{3}{8}\right)\left(-\frac{1}{3}\right) =$ _____

10. $\left(3\frac{8}{9}\right)\left(-4\frac{2}{7}\right) =$ _____

11. $(-0.75)(-2.6) =$ _____

12. $(-8)(-9)(-3) =$ _____

13. $(-6)(-12)(15) =$ _____

14. $(2)(-5)(18) =$ _____

15. $(3)(14)(5) =$ _____

16. $(1.75)(-2.6)(-8) =$ _____

17. $\left(3\frac{1}{2}\right)\left(-2\frac{2}{3}\right)\left(-4\frac{1}{5}\right) =$ _____

18. $\left(-\frac{4}{5}\right)\left(-\frac{7}{8}\right)\left(-\frac{15}{16}\right) =$ _____

19. $\left(5\tfrac{1}{3}\right)\left(8\tfrac{2}{3}\right)\left(1\tfrac{3}{13}\right) =$ _____

20. $(-0.065)(-9.4)(-3.72) =$ _____

21. $(-6)(3)(-7)(22) =$ _____

22. $(-8)(9)(-11)(-2) =$ _____

23. $(-15)(-5)(-3)(-25) =$ _____

24. $(6.5)(-7.3)(8.2)(4) =$ _____

25. $\left(\tfrac{2}{3}\right)\left(\tfrac{3}{5}\right)\left(\tfrac{1}{2}\right)\left(\tfrac{20}{21}\right) =$ _____

26. What is negative 4 times negative 5?

_____

27. What is the absolute value of -3 times the absolute value of 9?

_____

28. Would the product of 17 negative factors be positive or negative?

_____

7.4B PRACTICE SET

Find the products in the following problems.

1. $(15)(16) = $ _____

2. $(-8)(23) = $ _____

3. $(-19)(-7) = $ _____

4. $(-3)(8) = $ _____

5. $(-4.2)(6.9) = $ _____

6. $(-12)(-18) = $ _____

7. $(-4.003)(-0.56) = $ _____

8. $\left(-\frac{5}{7}\right)\left(-\frac{7}{15}\right) = $ _____

9. $\left(-\frac{2}{7}\right)\left(-\frac{3}{14}\right) = $ _____

10. $\left(-3\frac{1}{8}\right)\left(2\frac{3}{5}\right) = $ _____

11. $(-3.6)(-0.03) = $ _____

12. $(-18)(-12)(-15) = $ _____

13. $(6)(-3)(-22) = $ _____

14. $(1)(3)(-9) = $ _____

15. $(21)(15)(8) = $ _____

16. $(-6)(3.5)(-4.02) = $ _____

17. $\left(-4\frac{4}{5}\right)\left(3\frac{1}{8}\right)\left(-7\frac{1}{2}\right) = $ _____

18. $\left(-\frac{4}{7}\right)\left(-\frac{9}{14}\right)\left(-\frac{7}{3}\right) = $ _____

19. $\left(3\frac{3}{4}\right)\left(4\frac{2}{5}\right)\left(\frac{6}{11}\right) =$ _____

20. $(-1.09)(-0.002)(-6.4) =$ _____

21. $(1)(8)(-5)(-3) =$ _____

22. $(-6)(-12)(-15)(2) =$ _____

23. $(-16)(-22)(-31)(-9) =$ _____

24. $(-5.6)(4.2)(3.8)(2.5) =$ _____

25. $\left(\frac{7}{8}\right)\left(2\frac{1}{2}\right)\left(3\frac{3}{5}\right)\left(4\frac{1}{9}\right) =$ _____

26. What is negative 5 times negative 12?

_____

27. What is the absolute value of -4 times the absolute value of -18?

_____

28. Would the product of 16 negative factors be negative or positive?

_____

7.5 DIVIDING SIGNED NUMBERS

Division is the opposite of multiplication. If you understand the procedures for multiplying two signed numbers, you will find it easy to understand the procedure for dividing. Consider the equation from the previous section

$$3 \times 4 = 12$$

If the product of 3 and 4 is 12, then 12 divided by either of the factors gives the other factor as a quotient.

$$12 \div 3 = 4 \quad \text{or} \quad 12 \div 4 = 3$$

Remember from the previous section on multiplying signed numbers that multiplying two numbers with the same sign gave a positive product. Multiplying two numbers with different signs produced a negative product. When you divide, the quotient is going to be the factor that can be multiplied by the divisor to give the dividend as a product. Study the problem

$$\underset{dividend}{8} \underset{\div}{\div} \underset{divisor}{-2} \underset{=}{=} \underset{quotient}{-4}$$

If the divisor, -2, is multiplied by the quotient, -4, the result is the dividend, 8. Conversely,

$$\underset{dividend}{8} \underset{\div}{\div} \underset{divisor}{-4} \underset{=}{=} \underset{quotient}{-2}$$

Again, multiplying the divisor, -4, by the quotient, -2, gives a result of 8, the dividend.

This leads to the procedure for dividing signed numbers.

TO DIVIDE SIGNED NUMBERS

1. Divide the absolute values.
2. Count the negative signs; the quotient is:
   a. positive if there are an even number of negative signs.
   b. negative if there are an odd number of negative signs.

As with multiplication, if all of the signs are positive, the quotient will be positive.

All division problems may be checked by multiplying the quotient by the divisor. The result will be the dividend if the quotient is correct.

The examples that follow should help you visualize the result of dividing signed numbers.

EXAMPLE 1 Divide: $(10) \div (5)$

SOLUTION

$|10| = 10$ *Determine the absolute value of each term.*

$|5| = 5$

$10 \div 5 = 2$ *Divide the absolute values.*

There are no negative terms. The quotient is positive.

$(10) \div (5) = 2$

Check: $(5)(2) = 10$

EXAMPLE 2 Divide: $(-3.6) \div (-1.2)$

SOLUTION

$|-3.6| = 3.6$ *Determine the absolute value of each term.*
$|-1.2| = 1.2$

$3.6 \div 1.2 = 3$ *Divide the absolute values.*

Count the negative terms: two terms are negative. Since 2 is an even number, the quotient is positive.

$(-3.6) \div (-1.2) = 3$

Check: $(-1.2)(+3) = -3.6$

EXAMPLE 3 Divide: $(-18) \div (3)$

SOLUTION

$|-18| = 18$ *Determine the absolute value of each term.*
$|3| = 3$
$18 \div 3 = 6$ *Divide the absolute values.*

Count the negative terms: there is one negative term. Since 1 is an odd number, the quotient is negative.

$(-18) \div (3) = -6$

Check: $(3)(-6) = -18$

EXAMPLE 4 Divide: $48 \div (-16)$

SOLUTION

$|48| = 48$ *Determine the absolute value of each term.*
$|-16| = 16$

$48 \div 16 = 3$ *Divide the absolute values.*

Count the negative terms: there is one negative term. Since 1 is an odd number, the quotient is negative.

$48 \div (-16) = -3$

Check: $(-16)(-3) = +48$

Work the Comprehension Checkpoint problems to make sure that you know the procedures for dividing signed numbers. If you have difficulty with these problems, review the material in this section before going on to the Practice Sets.

COMPREHENSION CHECKPOINT

1. $(27) \div (3) =$ _____

2. $(-6) \div \left(-\dfrac{1}{3}\right) =$ _____

3. $(-12) \div (2.5) =$ _____

4. $(37.2) \div (-3.1) =$ _____

7.5A PRACTICE SET

Determine the quotient in each problem.

1. $\dfrac{-12}{2} =$ _____

2. $\dfrac{15}{-3} =$ _____

3. $\dfrac{28}{7} =$ _____

4. $\dfrac{-36}{-9} =$ _____

5. $(27) \div (-3) =$ _____

6. $(-120) \div (15) =$ _____

7. $(-96) \div (-16) =$ _____

8. $(75) \div (15) =$ _____

9. $(-64) \div (16) =$ _____

10. $(42) \div (-7) =$ _____

11. $(-4.8) \div (-1.6) =$ _____

12. $(0.55) \div (0.11) =$ _____

13. $(-3.87) \div (1.29) =$ _____

14. $(3.75) \div (-0.03) =$ _____

15. $(-65.4) \div (-5.45) =$ _____

16. $(607.5) \div (24.3) =$ _____

17. $(7.02) \div (-0.27) =$ _____

18. $(0.75) \div (-0.5) =$ _____

19. $\left(-\dfrac{5}{8}\right) \div \left(\dfrac{1}{2}\right) = $ _____

20. $\left(\dfrac{3}{4}\right) \div \left(\dfrac{7}{8}\right) = $ _____

21. $\left(\dfrac{15}{16}\right) \div \left(-\dfrac{1}{4}\right) = $ _____

22. $\left(9\dfrac{3}{5}\right) \div \left(-1\dfrac{1}{5}\right) = $ _____

23. $\left(-6\dfrac{2}{3}\right) \div \left(-1\dfrac{2}{3}\right) = $ _____

24. $\left(-15\dfrac{1}{2}\right) \div \left(\dfrac{5}{8}\right) = $ _____

25. $\left(4\dfrac{4}{5}\right) \div \left(-1\dfrac{3}{5}\right) = $ _____

26. What sign will the quotient of a positive number divided by a negative number have?

_____

27. Will the quotient of 0 divided by a negative number have a positive or a negative sign?

_____

28. What sign does the quotient of a negative number divided by 0 have?

_____

7.5B PRACTICE SET

Determine the quotient in each problem.

1. $\dfrac{-45}{15} = $ _____

2. $\dfrac{-72}{-18} = $ _____

3. $\dfrac{-84}{14} = $ _____

4. $\dfrac{132}{11} = $ _____

5. $(-105) \div (7) = $ _____

6. $(-280) \div (-35) = $ _____

7. $(-320) \div (-20) = $ _____

8. $(-575) \div (25) = $ _____

9. $(676) \div (-26) = $ _____

10. $(-450) \div (-30) = $ _____

11. $(-0.25) \div (-0.05) = $ _____

12. $(-1.89) \div (-0.63) = $ _____

13. $(-1.95) \div (0.15) = $ _____

14. $(-14.4) \div (0.18) = $ _____

15. $(15.5) \div (7.75) = $ _____

16. $(139) \div (-2.5) = $ _____

17. $(-121.5) \div (-48.6) = $ _____

18. $(-0.49) \div (0.07) = $ _____

19. $\left(\dfrac{3}{4}\right) \div \left(\dfrac{1}{3}\right) =$ _____

20. $(-18) \div \left(-\dfrac{2}{3}\right) =$ _____

21. $\left(\dfrac{4}{9}\right) \div (-4) =$ _____

22. $15\dfrac{6}{7} \div \left(-5\dfrac{2}{7}\right) =$ _____

23. $\left(-37\dfrac{1}{2}\right) \div \left(-2\dfrac{1}{2}\right) =$ _____

24. $(5) \div \left(-\dfrac{1}{5}\right) =$ _____

25. $\left(-24\dfrac{1}{3}\right) \div \left(8\dfrac{1}{9}\right) =$ _____

26. What sign will a negative number divided by a negative number have?

   _____

27. Will the quotient of 0 divided by a positive number have a positive or a negative sign?

   _____

28. What sign will the quotient of a positive number divided by 0 have, positive or negative?

   _____

7.6 ORDER OF OPERATIONS

Problems involving signed numbers requiring more than one operation should be solved using the same rules for the order of operations as were used with whole numbers, fractions, and decimals. These rules are repeated here as a review.

RULES FOR THE ORDER OF OPERATIONS
1. Do all multiplication and division in order first, working from left to right.
2. After completing all multiplication and division, do addition and subtraction in order, from left to right.
3. If grouping symbols such as ( ) or [ ] are included, operations within these symbols should be performed first, using the same steps. After simplifying inside the grouping symbols, begin again, using the first two steps for the operations that remain.

EXAMPLE 1 Simplify: $(-7) \times 4 \div (-14)$

SOLUTION

$(-7) \times 4 \div (-14)$
  $\underbrace{}$ *multiply*

$\quad -28 \quad \div \quad -14$
  $\quad\underbrace{}$ *divide*

$\quad\quad\quad 2$

$(-7) \times 4 \div (-14) = 2$

Notice that the rules for operating with signed numbers are always applied.

EXAMPLE 2 Simplify: $(-3) + 8 \div 2$

SOLUTION

$(-3) + 8 \div 2$
  $\quad\quad\underbrace{}$ *divide*

$(-3) + \quad 4$
  $\underbrace{}$ *add*

$\quad\quad 1$

$(-3) + 8 \div 2 = 1$

Notice that the rules for operating with signed numbers are always applied.

CHAPTER 7 Signed Numbers

EXAMPLE 3 Simplify: $[(-4) \times (10) \div (-8)] - [(2) \times (-3)]$

SOLUTION

$[(-4) \times (10) \div (-8)] - [(2) \times (-3)]$
   └─────┬─────┘ multiply   └──────┬──────┘ multiply

$[(-40) \div (-8)] - [-6]$
   └──────┬──────┘ divide

$[5] - [-6]$
└────┬────┘ subtract

$[5] + [+6]$
└────┬────┘ add

11

$[(-4) \times (10) \div (-8)] - [(2) \times (-3)] = 11$

Notice that the rules for operating with signed numbers are always applied.

Work the Comprehension Checkpoint problems to strengthen your skills on problems requiring more than one operation. If you have difficulty with these problems, review the material in this section before going on to the Practice Sets.

COMPREHENSION CHECKPOINT

Simplify:

1. $(3) \times (6) - (-2) = $ _____

2. $(25) \div (-5) \times (-3) + (4) = $ _____

3. $[(-16) \div (-4) - (3)] + (-2) = $ _____

4. $(24) \div [(-3) \times (-4) \div (6)] = $ _____

7.6A PRACTICE SET

Simplify each problem.

1. $(-6) \times (4) \div (-12) =$ _____

2. $(5) - (-8) \div (2) =$ _____

3. $(-3) + (-2) \times (4) =$ _____

4. $[(-3) + (-2)] \times (4) =$ _____

5. $(-8) \times (-4) + (-16) \times (2) =$ _____

6. $(-12) + (-3) - (-4) \times (-5) =$ _____

7. $[(8) - (4)] \div (-2) =$ _____

8. $(-20) \div (-4) + (-3) \times (-6) =$ _____

9. $(16) - [(18) \div (-6) - (-2)] =$ _____

10. $(-20) + [(15) \times (-3) \div (5)] =$ _____

11. $(-9) \times [(-4) \div (2) - (-8)] =$ _____

12. $(7) + (-3) \times (6) \div (-2) =$ _____

13. $[(-12) \div (4)] + [(6) \times (2)] =$ _____

14. $(9) \times [(-24) \div (8) \times (-11)] =$ _____

15. $[(-28) \times (-3)] \div [(14) \times (-6)] =$ _____

16. $(6) \times (-5) \div (2) - (-15) \div (-3) =$ _____

7.6B PRACTICE SET

Simplify each problem.

1. $(5) \times (-6) \div (-10) = $ _____

2. $(18) \div (-6) \times (2) = $ _____

3. $(24) \times (-7) \div (12) = $ _____

4. $(-10) \div [(4) \div (-2)] = $ _____

5. $(9) \times (-5) + (-6) \div (-3) = $ _____

6. $(-20) + (6) - (-12) \times (-1) = $ _____

7. $[(-18) - (24)] \times (3) = $ _____

8. $[(9) \times (-5) + (-3)] \div (-16) = $ _____

9. $(22) - [(-12) + (-18) \div (-9)] = $ _____

10. $(-6) + [(44) \div (-2) - (-11)] = $ _____

11. $(15) \times [(-3) \times (-12) \times (0)] = $ _____

12. $[(-12) - (15)] \div (3) - (4) = $ _____

13. $(25) \times [(-16) \div (8) \times (-3)] = $ _____

14. $[(-24) \times (-3)] \div [(36) \div (-12)] = $ _____

15. $(-6) + (-3) \times (36) \div (-9) + (-5) = $ _____

16. $[(156) + (-12)] \div [(24) \div (-4) \times (3)] = $ _____

SUMMARY OF KEY CONCEPTS

KEY TERMS

signed numbers (7.1): Negative as well as positive numbers.

inequality symbols (7.1): Symbols such as < and > used to compare numbers.

absolute value (7.1): The distance of a number from 0 on the number line.

additive inverse (7.1): The opposite of a number. The sum of a number and its opposite is 0.

KEY RULES

To add signed numbers (7.2):

A. Two numbers with the same sign:
  1. Add the absolute values.
  2. Attach the common sign.
B. Two numbers with different signs:
  1. Find the difference between the absolute values.
  2. Attach the sign of the number with the larger absolute value.
C. More than two addends:
  1. Add all positive numbers.
  2. Add all negative numbers.
  3. Use the sums from steps 1 and 2 and follow the procedure for adding two numbers with different signs.

To subtract signed numbers (7.3):

A. Change the operation sign to + and the sign of the subtrahend to its opposite.
B. Follow the rules for addition.

To multiply signed numbers (7.4):

A. Multiply the absolute values.
B. Count the negative factors. The product is:
  1. positive if the number of negative factors is even.
  2. negative if the number of negative factors is odd.

To divide signed numbers (7.5):

A. Divide the absolute values.
B. Count the negative signs. The quotient is:
  1. positive if there are an even number of negative signs.
  2. negative if there are an odd number of negative signs.

PRACTICE TEST 7A

Insert the appropriate inequality symbol, < or >.

1. -6 _____ 3
2. 0.002 _____ -0.01
3. $\dfrac{7}{8}$ _____ $\dfrac{2}{3}$
4. -2.8 _____ -2.7

Give the indicated absolute value.

5. $|-0.25| =$ _____
6. $-|7.3| =$ _____

Determine the additive inverse of the given value.

7. 8.4 _____
8. $-2\dfrac{2}{3}$ _____

Add:

9. $7 + (-5) =$ _____

10. $(-24) + (-8) =$ _____

11. $6 + (-9) + 14 + (-15) =$ _____

Subtract:

12. $(-7.8) - (-9.35) =$ _____

13. $8 - (-12) =$ _____

14. $12 - (3) =$ _____

Multiply:

15. $(-8)(-15) =$ _____

16. $(-5)(13) =$ _____

17. $(-8)(-3)(25) =$ _____

18. $(-16)(-3)(-5) =$ _____

Divide:

19. $(-15) \div (-3) =$ _____

20. $(72) \div (-6) =$ _____

21. $(-128) \div (32) =$ _____

22. $(245) \div (35) =$ _____

Simplify:

23. $(2) \times (-3) - (-8) =$ _____

24. $[(-14) - (-6)] \times (-5) =$ _____

25. $[(96) \div (-3)] \div [(-40) \div (5)] =$ _____

26. How many units is -8 from $+5$ on the number line?

_____

27. What is the difference between $24°F$ and $-18°F$?

_____

28. Will the product of 43 negative factors be positive or negative?

_____

29. What sign will the quotient of two negative numbers have?

_____

30. What is the absolute value of $[(-24) \times (3)] \div (-4) \times [(-8) - (-6)]$?

_____

PRACTICE TEST 7B

CHAPTER 7

Insert the appropriate inequality symbol, < or >.

1. $\dfrac{3}{5}$ _____ $-\dfrac{7}{8}$

2. 3 _____ 4

3. -0.001 _____ -0.01

4. -3 _____ -4

Give the indicated absolute value.

5. $-|-6.2| =$ _____

6. $|24| =$ _____

Determine the additive inverse of the given value.

7. $-9.85 =$ _____

8. $3\dfrac{1}{2} =$ _____

Add:

9. $6 + 7 =$ _____

10. $(-3) + (-15) =$ _____

11. $(-4) + (-9) + 6 + (-7) + 3 =$ _____

Subtract:

12. $4.09 - (-3.42) =$ _____

13. $(-8) - (-3) =$ _____

14. $36 - (45) =$ _____

Multiply:

15. $(-12)(-6) =$ _____

16. $(18)(-3) =$ _____

433

17. $(-12)(13)(-6) = $ _____

18. $(-15)(-18)(-3) = $ _____

Divide:

19. $(-48) \div (3) = $ _____

20. $(-54) \div (-18) = $ _____

21. $(64) \div (4) = $ _____

22. $(51) \div (-17) = $ _____

Simplify:

23. $(-6) \times (4) - (-10) = $ _____

24. $(18) \times (-3) \div (27) - (2) = $ _____

25. $[(32) + (-19)] \times [(-8) \div (-2)] = $ _____

26. How many units from -19 is $+12$ on the number line?

   _____

27. What is the difference between 15°F and -7°F?

   _____

28. What sign will the product of 26 negative numbers have?

   _____

29. What sign will the quotient of a positive number and a negative number have?

   _____

30. What is the absolute value of $[(96) \div (-16) - (7)] \times (-1)$?

   _____

SKILLS REVIEW

CHAPTERS 1-7

1. Write 23,076 in words.

   _____

2. Write four hundred six thousand, three hundred ninety-four in numbers.

   _____

3. Round 3,496 to the nearest hundred. _____

4. Round 28,423 to the nearest thousand. _____

5. Round 47.826 to the nearest hundredth. _____

Perform the indicated operation. Express answers in lowest terms.

6. $\dfrac{2}{3} + \dfrac{3}{4} + \dfrac{5}{6} =$ _____

7. $\dfrac{8}{15} - \dfrac{3}{10} =$ _____

8. $\dfrac{3}{4} \cdot \dfrac{10}{21} =$ _____

9. $\dfrac{5}{8} \div \dfrac{1}{2} =$ _____

10. $(8.3)(2.06) =$ _____

11. $18.90 \div 1.05 =$ _____

12. Convert $\dfrac{5}{8}$ to a decimal. _____

13. Convert $\dfrac{3}{16}$ to a percent (round to the nearest tenth). _____

14. Convert 45% to a decimal. _____

15. Convert 16% to a fraction in lowest terms. _____

CHAPTER 7 Signed Numbers

16. What is 24% of 250? _____

17. 40% of what number is 32? _____

18. 25 is what percent of 300? _____

19. What is $33\frac{1}{3}$% of 783? _____

20. What is $16\frac{2}{3}$% of 108? _____

21. Write 0.345 in words.

    _____

22. Write 12.73 in words.

    _____

23. Is the following a true proportion?

    $\dfrac{15}{16} \overset{?}{=} \dfrac{42}{48}$ _____

24. Find the missing element in the proportion.

    $\dfrac{12}{17} = \dfrac{x}{85}$ _____

25. Insert the appropriate inequality symbol, < or >.

    a. −2 _____ 2   a. _____
    b. 4.5 _____ −5.4   b. _____
    c. 0.001 _____ 0.01   c. _____

26. Indicate the absolute value.

    a. |−12| a. _____
    b. −|3.89| b. _____

27. Give the additive inverse.

    a. 6   a. _____
    b. 0   b. _____

Perform the indicated operation.

28. 7 + (−9) = _____

Skills Review

29. $(-8) - (-6) =$ _____

30. $(-15)(-4)(2) =$ _____

31. $(-128) \div (-16) =$ _____

32. $[(18) \div (-6) - (3)] \div [(48) \div (-8)] =$ _____

CHAPTER 7 SOLUTIONS

Skills Preview

1. <
2. <
3. >
4. >
5. 9
6. −6.8
7. −15
8. $3\frac{1}{2}$
9. 3
10. −26
11. −5
12. 4
13. −14
14. −29
15. 10
16. −72
17. 52
18. 100
19. −840
20. 2
21. −7
22. 4
23. −3
24. −35
25. −38

Section 7.1A Comprehension Checkpoint

1. >
2. >
3. <
4. 7
5. −10
6. 3.8
7. 5.9
8. $-\frac{2}{5}$
9. $\frac{3}{7}$

7.1A Practice Set

1. <
2. >
3. <
4. <
5. >
6. >
7. >
8. >
9. >
10. <
11. 7
12. 4
13. −2.35
14. 8
15. $\frac{3}{7}$
16. $-3\frac{9}{10}$
17. −18.5
18. 0
19. −5
20. −1.75
21. 7
22. −12
23. $\frac{3}{4}$
24. $-\frac{4}{9}$
25. 0.35

26. 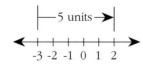 +2 is +5 units from −3.

27. 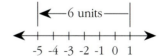 −5 is −6 units from 1.

28. Both +3 and −3 are 3 units from 0.

Section 7.2 Comprehension Checkpoint

1. 14
2. −13
3. −5
4. −8
5. −3
6. 4

7.2A Practice Set

1. $(-8) + (3) = -5.$

2. $(5) + (-9) = -4.$

3. $\left(-\frac{1}{2}\right) + \left(-\frac{3}{4}\right) = \left(-\frac{2}{4}\right) + \left(-\frac{3}{4}\right) = -\frac{5}{4} = -1\frac{1}{4}$

4. $(-5.7) + (6.25) = 0.55$

5. $(-12) + (5.6) = -6.4$

6. $(-8.4) + (-2.8) = -11.2$

7. $\left(\frac{5}{8}\right) + \left(-\frac{2}{3}\right) = \left(\frac{15}{24}\right) - \left(-\frac{16}{24}\right) = -\frac{1}{24}$

8. $(112) + (-31) = 81$

9. $(-8) + (6) + (-3.7)$
   $(-8) + (-3.7) = -11.7 =$ sum of the $-$ addends
   $(-11.7) + (6) = -5.7$

10. $(-9) + (-3) + (-16) = -28$

11. $(12.4) + (3.9) + (-8.2)$
    $(12.4) + (3.9) = 16.3 =$ the sum of the $+$ addends
    $16.3 + (-8.2) = 8.1$

12. $\left(\frac{2}{3}\right) + \left(\frac{3}{4}\right) + \left(-\frac{5}{6}\right)$
    $\left(\frac{8}{12}\right) + \left(\frac{9}{12}\right) = \frac{17}{12} =$ the sum of the $+$ addends
    $\left(\frac{17}{12}\right) + \left(-\frac{10}{12}\right) = \frac{7}{12}$

13. $\left(-\frac{3}{5}\right) + \left(-\frac{2}{7}\right) + \left(\frac{7}{10}\right)$
    $\left(-\frac{21}{35}\right) + \left(-\frac{10}{35}\right) = -\frac{31}{35} =$ the sum of the $-$ addends
    $\left(-\frac{31}{35}\right) + \left(\frac{7}{10}\right) = \left(-\frac{62}{70}\right) + \left(\frac{49}{70}\right) = -\frac{13}{70}$

14. $(4.7) + (-5.3) + (-6.9)$
    $(-5.3) + (-6.9) = -12.2 =$ the sum of the $-$ addends
    $-12.2 + 4.7 = -7.5$

15. $42 - 27 - 12 = 3$

16. $(-7.5) + (3.33) + (-5.25)$
    $(-7.5) + (-5.25) = -12.75 =$ the sum of the $-$ addends
    $(-12.75) + (3.33) = -9.42$

17. $(18) + (-7) + (16) + (-5)$
    $18 + 16 = 34 =$ sum of the $+$ addends
    $(-7) + (-5) = -12 =$ sum of the $-$ addends
    $34 + (-12) = 22$

18. $(-3) + (-8) + (-9) + (+12)$
    $(-3) + (-8) + (-9) = -20 =$ sum of $-$ addends
    $(-20) + 12 = -8$

Solutions 441

19. $(-0.5) + (0.67) + (-0.75) + (0.83)$
    $(-0.5) + (-0.75) = -1.25 =$ sum of $-$ addends
    $0.67 + 0.83 = 1.5 =$ sum of $+$ addends
    $1.5 + (-1.25) = 0.25$

20. $(-5.8) + (-6.75) + (15.2) + (-9.7) + (1.6)$
    $(-5.8) + (-6.75) + (-9.7) = -22.25 =$ sum of $-$ addends
    $15.2 + 1.6 = 16.8 =$ sum of $+$ addends
    $(-22.25) + (16.8) = -5.45$

21. $(-18) + (2.4) + (-9.6) + (6)$
    $(-18) + (-9.6) = -27.6 =$ sum of $-$ addends
    $2.4 + 6 = 8.4 =$ sum of $+$ addends
    $(-27.6) + (8.4) = -19.2$

22. $\left(-\frac{5}{9}\right) + \left(+\frac{3}{7}\right) + \left(-\frac{5}{7}\right) + \left(+\frac{2}{9}\right)$
    $\left(-\frac{5}{9}\right) + \left(-\frac{5}{7}\right) = \left(-\frac{35}{63}\right) + \left(-\frac{45}{63}\right) = -\frac{80}{63} =$ sum of $-$ addends
    $\frac{3}{7} + \frac{2}{9} = \frac{27}{63} + \frac{14}{63} = \frac{41}{63} =$ sum of $+$ addends
    $\left(-\frac{80}{63}\right) + \left(\frac{41}{63}\right) = -\frac{39}{63} = -\frac{13}{21}$

23. $\left(-2\frac{1}{2}\right) + \left(+5\frac{3}{4}\right) + \left(-4\frac{5}{12}\right) + \left(+3\frac{2}{3}\right)$
    $\left(-2\frac{1}{2}\right) + \left(-4\frac{5}{12}\right) = -6\frac{11}{12} =$ sum of $-$ addends
    $5\frac{3}{4} + 3\frac{2}{3} = 8\frac{17}{12} = 9\frac{5}{12} =$ sum of $+$ addends
    $\left(-6\frac{11}{12}\right) + 9\frac{5}{12} = 2\frac{1}{2}$

24. $-12 + 16 - 14 - 10 + 20$
    $(-12) + (-14) + (-10) = -36 =$ sum of $-$ addends
    $16 + 20 = 36 =$ sum of $+$ addends
    $36 + (-36) = 0$

25. $(6.42) + (7.95) + (-8.43) + (10.58)$
    $6.42 + 7.95 + 10.58 = 24.95 =$ sum of $+$ addends
    $24.95 + (-8.43) = 16.52$

26. $34 + (-5) = 34 - 5 = 29 = 29°$

27. $106 + (-22) = 106 - 22 = 84 = 84$ yards

28. $(+19) + (-7.5) = 19 - 7.5 = 11.5 = 11.5$ gallons

Section 7.3 Comprehension Checkpoint

1. -1 2. -16.5 3. $2\frac{1}{6}$ 4. 31.75 5. 2.46

7.3A Practice Set

1. $(9) - (-6) = (9) + (6) = 15$

2. $(-7) - (-3) = (-7) + (3) = -4$

3. $(5) - (+7) = (5) + (-7) = -2$
4. $(10) - (-10) = (10) + (+10) = 20$
5. $(-6.4) - (-8.2) = (-6.4) + (8.2) = 1.8$
6. $\left(-\frac{1}{2}\right) - \left(-\frac{1}{4}\right) = \left(-\frac{2}{4}\right) + \left(+\frac{1}{4}\right) = -\frac{1}{4}$
7. $\left(+\frac{3}{4}\right) - \left(+\frac{2}{3}\right) = \left(+\frac{9}{12}\right) + \left(-\frac{8}{12}\right) = \frac{1}{12}$
8. $(-24) - (+6) = (-24) + (-6) = -30$
9. $(18.2) - (-9.6) = (18.2) + (9.6) = 27.8$
10. $\left(-12\frac{1}{2}\right) - \left(+8\frac{2}{3}\right) = \left(-12\frac{3}{6}\right) + \left(-8\frac{4}{6}\right) = -20\frac{7}{6} = -21\frac{1}{6}$
11. $(10.85) - (-3.76) = (10.85) + (+3.76) = 14.61$
12. $(-7) - (+6.3) = (-7) + (-6.3) = -13.3$
13. $(-3.62) - (-12.8) = (-3.62) + (+12.8) = 9.18$
14. $\left(-\frac{5}{8}\right) - \left(+\frac{3}{4}\right) = \left(-\frac{5}{8}\right) + \left(-\frac{6}{8}\right) = -\frac{11}{8} = -1\frac{3}{8}$
15. $(-28) - (+17) = (-28) + (-17) = -45$
16. $\left(15\frac{1}{3}\right) - \left(-4\frac{2}{3}\right) = \left(15\frac{1}{3}\right) + \left(+4\frac{2}{3}\right) = 19\frac{3}{3} = 20$
17. $(8.76) - (3.5) = (8.76) + (-3.5) = 5.26$
18. $(16) - (-5.4) = (16) + (+5.4) = 21.4$
19. $\left(1\frac{7}{8}\right) - \left(-3\frac{5}{12}\right) = \left(1\frac{21}{24}\right) + \left(+3\frac{10}{24}\right) = 4\frac{31}{24} = 5\frac{7}{24}$
20. $\left(-6\frac{4}{5}\right) - \left(-8\frac{2}{3}\right) = \left(-6\frac{12}{15}\right) + \left(+8\frac{10}{15}\right) = 1\frac{13}{15}$
21. $(-18.09) - (10.22) = (-18.09) + (-10.22) = -28.31$
22. $(+0.01) - (-0.001) = (+0.01) + (+0.001) = 0.011$
23. $(-3.09) - (+4.76) = (-3.09) + (-4.76) = -7.85$
24. $\left(-14\frac{2}{7}\right) - \left(-6\frac{1}{2}\right) = \left(-14\frac{4}{14}\right) + \left(+6\frac{7}{14}\right) = -7\frac{11}{14}$
25. $\left(-31\frac{3}{5}\right) - \left(+9\frac{3}{4}\right) = \left(-31\frac{12}{20}\right) + \left(-9\frac{15}{20}\right) = -40\frac{27}{20} = -41\frac{7}{20}$
26. $18° - (-4°) = 18° + (+4°) = 22°$
27. $10{,}000 - (4{,}500) = 5{,}500$
28. $200 - (48) = 52$

Solutions

Section 7.4 Comprehension Checkpoint

1. -21 2. 36 3. 10.2 4. -12 5. $-11{,}200$

7.4A Practice Set

1. $(+6)(+4) = 24$ (no negative factors)
2. $(-9)(+7) = -63$ (odd number of negative factors)
3. $(-6)(-8) = 48$ (even number of negative factors)
4. $(+12)(-13) = -156$ (odd number of negative factors)
5. $(3.7)(-8.4) = -31.08$ (odd number of negative factors)
6. $(-21)(-15) = 315$ (even number of negative factors)
7. $(-6.9)(-18.7) = 129.03$ (even number of negative factors)
8. $\left(+\dfrac{2}{3}\right)\left(-\dfrac{5}{9}\right) = -\dfrac{10}{27}$ (odd number of negative factors)
9. $\left(-\dfrac{3}{8}\right)\left(-\dfrac{1}{3}\right) = \dfrac{1}{8}$ (even number of negative factors)
10. $\left(+3\dfrac{8}{9}\right)\left(-4\dfrac{2}{7}\right) = \left(\dfrac{35}{9}\right)\left(-\dfrac{30}{7}\right) = -\dfrac{1{,}050}{63} = -16\dfrac{2}{3}$ (odd number of negative factors)
11. $(-0.75)(-2.6) = 1.95$ (even number of negative factors)
12. $(-8)(-9)(-3) = -216$ (odd number of negative factors)
13. $(-6)(-12)(15) = 1{,}080$ (even number of negative factors)
14. $(2)(-5)(18) = -180$ (odd number of negative factors)
15. $(3)(14)(5) = 210$ (no negative factors)
16. $(1.75)(-2.6)(-8) = 36.4$ (even number of negative factors)
17. $\left(3\dfrac{1}{2}\right)\left(-2\dfrac{2}{3}\right)\left(-4\dfrac{1}{5}\right) = \left(\dfrac{7}{2}\right)\left(-\dfrac{8}{3}\right)\left(-\dfrac{21}{5}\right) = \dfrac{1{,}176}{30} = 39\dfrac{1}{5}$ (even number of negative factors)
18. $\left(-\dfrac{4}{5}\right)\left(-\dfrac{7}{8}\right)\left(-\dfrac{15}{16}\right) = -\dfrac{420}{640} = -\dfrac{21}{32}$ (odd number of negative factors)
19. $\left(5\dfrac{1}{3}\right)\left(8\dfrac{2}{3}\right)\left(1\dfrac{3}{13}\right) = \left(\dfrac{16}{3}\right)\left(\dfrac{26}{3}\right)\left(\dfrac{16}{13}\right) = \dfrac{6{,}656}{117} = 56\dfrac{8}{9}$ (no negative factors)
20. $(-0.065)(-9.4)(-3.72) = -2.27292$ (odd number of negative factors)
21. $(-6)(3)(-7)(22) = 2{,}772$ (even number of negative factors)
22. $(-8)(9)(-11)(-2) = -1{,}584$ (odd number of negative factors)
23. $(-15)(-5)(-3)(-25) = 5{,}625$ (even number of negative factors)
24. $(6.5)(-7.3)(8.2)(4) = -1{,}556.36$ (odd number of negative factors)
25. $\left(\dfrac{2}{3}\right)\left(\dfrac{3}{5}\right)\left(\dfrac{1}{2}\right)\left(\dfrac{20}{21}\right) = \dfrac{120}{630} = \dfrac{4}{21}$ (no negative factors)

26. $(-4)(-5) = 20$

27. $|-3| = 3 \quad |9| = 9 \quad 3 \times 9 = 27$

28. negative—there is an odd number of negative factors

Section 7.5 Comprehension Checkpoint

1. 9   2. 18   3. -4.8 4. -12

7.5A Practice Set

1. $\dfrac{-12}{2} = -6$ (odd number of negative terms)

2. $\dfrac{15}{-3} = -5$ (odd number of negative terms)

3. $\dfrac{28}{7} = 4$ (no negative terms)

4. $\dfrac{-36}{-9} = 4$ (even number of negative terms)

5. $(27) \div (-3) = -9$ (odd number of negative terms)

6. $(-120) \div (15) = -8$ (odd number of negative terms)

7. $(-96) \div (-16) = 6$ (even number of negative terms)

8. $(75) \div (15) = 5$ (no negative terms)

9. $(-64) \div (16) = -4$ (odd number of negative terms)

10. $(42) \div (-7) = -6$ (odd number of negative terms)

11. $(-4.8) \div (-1.6) = 3$ (even number of negative terms)

12. $(0.55) \div (0.11) = 5$ (no negative terms)

13. $(-3.87) \div (1.29) = -3$ (odd number of negative terms)

14. $(3.75) \div (-0.03) = -125$ (odd number of negative terms)

15. $(-65.4) \div (-5.45) = 12$ (even number of negative terms)

16. $(607.5) \div (24.3) = 25$ (no negative terms)

17. $(7.02) \div (-0.27) = -26$ (odd number of negative terms)

18. $(0.75) \div (-0.5) = -1.5$ (odd number of negative terms)

19. $\left(-\dfrac{5}{8}\right) \div \left(\dfrac{1}{2}\right) = \left(-\dfrac{5}{8}\right) \cdot \dfrac{2}{1} = -\dfrac{5}{4} = -1\dfrac{1}{4}$ (odd number of negative terms)

20. $\left(\dfrac{3}{4}\right) \div \left(\dfrac{7}{8}\right) = \left(\dfrac{3}{4}\right) \cdot \left(\dfrac{8}{7}\right) = \left(\dfrac{6}{7}\right)$ (no negative terms)

21. $\left(\dfrac{15}{16}\right) \div \left(-\dfrac{1}{4}\right) = \left(\dfrac{15}{16}\right) \cdot \left(-\dfrac{4}{1}\right) = -\left(\dfrac{15}{4}\right) = -3\dfrac{3}{4}$ (odd number of negative terms)

Solutions

22. $\left(9\frac{3}{5}\right) \div \left(-1\frac{1}{5}\right) = \left(\frac{48}{5}\right) \cdot \left(-\frac{5}{6}\right) = -\frac{8}{1} = -8$ (odd number of negative terms)

23. $\left(-6\frac{2}{3}\right) \div \left(-1\frac{2}{3}\right) = \left(-\frac{20}{3}\right) \cdot \left(-\frac{3}{5}\right) = \frac{4}{1} = 4$ (even number of negative terms)

24. $\left(-15\frac{1}{2}\right) \div \left(\frac{5}{8}\right) = \left(-\frac{31}{2}\right) \cdot \left(\frac{8}{5}\right) = \left(-\frac{124}{5}\right) = -24\frac{4}{5}$ (odd number of negative terms)

25. $\left(4\frac{4}{5}\right) \div \left(-1\frac{3}{5}\right) = \left(\frac{24}{5}\right) \cdot \left(-\frac{5}{8}\right) = \left(-\frac{3}{1}\right) = -3$ (odd number of negative terms)

26. Negative—there is only one negative term.

27. No sign—the quotient is 0, and 0 is neither + nor −.

28. Division by 0 is undefined (it cannot be done).

Section 7.6 Comprehension Checkpoint

1. 20   2. 19   3. −1   4. 12

7.6A Practice Set

1. $(-6) \times (4) \div (-12)$
   multiply
   $-24 \div (-12)$
   divide
   2

2. $(5) - (-8) \div (2)$
   divide
   $(5) - (-4)$
   subtract
   $(5) + (+4)$
   add
   9

3. $(-3) + (-2) \times (4)$
   multiply
   $(-3) + (-8)$
   add
   -11

4. $[(-3) + (-2)] \times (4)$
   add
   $(-5) \times (4)$
   multiply
   -20

5. $(-8) \times (-4) + (-16) \times (2)$
   multiply
   $32 + (-32)$
   add
   0

6. $(-12) + (-3) - (-4) \times (-5)$
   multiply
   $(-12) + (-3) - (20)$
   add
   $(-15) - (+20)$
   subtract
   $(-15) + (-20)$
   add
   -35

7. $[(8) - (4)] \div (-2)$ *subtract*
$[(8) + (-4)] \div (-2)$ *add*
$(+4) \div (-2)$ *divide*
-2

8. $(-20) \div (-4) + (-3) \times (-6)$ *divide*, *multiply*
$(5) + (18)$ *add*
23

9. $(16) - [(18) \div (-6) - (-2)]$ *divide*
$(16) - [(-3) - (-2)]$ *subtract*
$(16) - [(-3) + (2)]$
$(16) - (-1)$ *subtract*
$(16) + (+1)$
$+17$

10. $(-20) + [(15) \times (-3) \div (5)]$ *multiply*
$(-20) + [(-45) \div (5)]$ *divide*
$(-20) + (-9)$ *add*
-29

11. $(-9) \times [(-4) \div (2) - (-8)]$ *divide*
$(-9) \times [(-2) - (-8)]$ *subtract*
$(-9) \times [(-2) + (+8)]$ *add*
$(-9) \times (6)$ *multiply*
-54

12. $(7) + (-3) \times (6) \div (-2)$ *multiply*
$(7) + (-18) \div (-2)$ *divide*
$(7) + (9)$ *add*
16

13. $[(-12) \div (4)] + [(6) \times (2)]$ *divide*, *multiply*
$(-3) + (12)$ *add*
9

14. $(9) \times [(-24) \div (8) \times (-11)]$ *divide*
$(9) \times [(-3) \times (-11)]$ *multiply*
$(9) \times (33)$ *multiply*
297

15. $[(-28) \times (-3)] \div [(14) \times (-6)]$ *multiply*, *multiply*
$84 \div (-84)$ *divide*
-1

16. $(6) \times (-5) \div (2) - (-15) \div (-3)$ *multiply*, *divide*
$(-30) \div (2) - (5)$ *divide*
$(-15) - (5)$ *subtract*
$(-15) + (-5)$ *add*
-20

Practice Test 7A

1. $<$
2. $>$
3. $>$
4. $<$
5. 0.25
6. -7.3
7. -8.4
8. $2\frac{2}{3}$
9. 2
10. -32
11. -4
12. 1.55
13. 20
14. 9
15. 120
16. -65
17. 600
18. -240
19. 5
20. -12
21. -4
22. 7
23. 2
24. 40
25. 4
26. 13
27. $42°$
28. negative
29. positive
30. 36

Skills Review   Chapters 1–7

1. twenty-three thousand, seventy-six
2. 406,394
3. 3,500
4. 28,000
5. 47.83
6. $2\frac{1}{4}$
7. $\frac{7}{30}$
8. $\frac{5}{14}$
9. $1\frac{1}{4}$
10. 17.098
11. 18
12. 0.625
13. 18.8%
14. 0.45
15. $\frac{4}{25}$
16. 60
17. 80
18. 8.33%
19. 261
20. 18
21. three hundred forty-five thousandths
22. twelve and seventy-three hundredths
23. no
24. 60
25. a. $<$ b. $>$ c. $<$
26. a. 12 b. -3.89
27. a. -6 b. 0
28. -2
29. -2
30. 120
31. 8
32. 1

8 INTRODUCTION TO ALGEBRA

OUTLINE

8.1 Terminology

8.2 Solving Equations by Addition

8.3 Solving Equations by Division

8.4 Solving Equations by Multiplication

8.5 The Distributive Property

8.6 Solving Multistep Equations

The goal in each of the previous chapters has been to show that algebra is an extension of basic math. The common element among all math problems is the search for a solution to a particular problem. The answer is unknown when the problem is originally set up. The challenge, whether in basic math or in algebra, is to find the solution to the problem. This chapter follows up the concept of using a letter to represent a number as well as the steps required to determine the number that the letter represents.

SKILLS PREVIEW

1. How many terms are there in the following expression?

    $-7x + 3 + 4a - 8y$ _____

2. What is the coefficient of each term?

    a. $\frac{2}{5}x$ b. $6y$ c. $7.4a$

    a. _____   b. _____   c. _____

3. Simplify the following expression by combining like terms:

    $4y - 9 + 6y - 18 - 7y$ _____

In problems 4–9, solve each equation by finding the number that the variable represents.

4. $x - 9 = 6$ _____

5. $3 + x = 8$ _____

6. $9x = 45$ _____

7. $-36 = -4x$ _____

8. $\frac{x}{5} = 6$ _____

9. $\frac{3x}{4} = 54$ _____

10. Use the distributive property to simplify the following expression:

    $-8(-4 + a)$ _____

11. Solve the equation $3(x + 7) = x - 15$. _____

12. Solve the equation $\frac{3x}{5} + 12 = 63$. _____

> **a + b = x**
> ## ALGEBRA CONNECTION
>
> In each of the previous chapters, you have been introduced to an important concept to help you make the transition to algebra. The concepts have included the use of symbols to represent numbers, the importance of understanding fractions, and the use of signed numbers. The objective in this chapter is to tie all of these concepts together to present a workable method for solving problems using equations. The chapters that follow will allow you many opportunities to solve problems by setting up equations and solving them. Algebra is a bridge between solving problems using simple basic math and solving problems that require the use of more involved formulas. As you will see, formulas that you will use in the chapter on geometry (Chapter 10) are merely applications of solving equations. The material in this chapter is essential to solving problems in higher mathematics as well as many other fields.

8.1 TERMINOLOGY

As in any field, you must understand the words that will be used in algebra before you are able to solve problems using algebra. This is true not only for algebra but also for any other math you may eventually study.

Possibly the most important new word to learn is **variable.** A variable is a letter used to represent a number. It is called a variable because its value may vary (or change). Variables are used in equations and formulas. An **equation** is a statement that two quantities have the same value. An example of an equation is

$$x + 3 = 8$$

A **formula** is a rule written in algebraic form to solve particular problems. An example of a formula is

$$A = lw$$

where all three values are represented by variables.

A fundamental skill to be learned in algebra is how to solve an equation. The **solution** (also called the *root*) of an equation is the number that the variable represents. Replacing the variable with the number it represents will complete a true statement. If in the equation

$$x + 3 = 8$$

you replaced the variable, x, with the value 5, getting

$$5 + 3 = 8$$

then you would have a true statement. The various steps for solving equations will be discussed in the sections that follow in this chapter.

A **numerical coefficient** is the number multiplied by a variable, as in the following examples:

$$\begin{aligned} 5x &\quad \text{coefficient is } 5 \\ -3y &\quad \text{coefficient is } -3 \\ b &\quad \text{coefficient is } 1 \\ -a &\quad \text{coefficient is } -1 \end{aligned}$$

Remember that writing a coefficient and a variable together (without a sign between them) indicates that they are factors and are to be multiplied.

A **term** is a combination of a numerical coefficient and one or more variables. Examples of terms are

$$4x \qquad 9ab \qquad -7r$$

It is important to note that a plus or a minus sign is also part of a term. The terms $4x$ and $9ab$ are positive, while $-7r$ is negative.

EXAMPLE 1 Name the coefficient of each term.

a. $8a$ *b.* $-6t$ *c.* $\frac{2}{3}d$

SOLUTIONS

a. 8 *b.* -6 *c.* $\frac{2}{3}$

Terms may be **similar** (or **like**) **terms.** Similar terms have the exact same variable factors (including exponents, if any):

$$3m \text{ and } -4m \text{ are like terms}$$
$$-5xy^2 \text{ and } 8xy^2 \text{ are like terms}$$

Unlike terms do not have the same variable factors:

$$7a \text{ and } 8q \text{ are unlike terms}$$
$$-9bd \text{ and } 10bc \text{ are unlike terms}$$

A **constant term** is a number without a variable, such as

$$6 \qquad -12$$

It is called a constant because its value does not change: 6 always represents the same amount; -12 always indicates the same amount. All constant terms are considered like terms.

An **algebraic expression** is a group of terms including the plus or minus signs. Examples of algebraic expressions are

$$5x + 3y - 12$$
$$8n - 12m + 16pq + 9$$

Algebraic expressions are in simplest form when all like terms have been combined. **Simplifying an expression** is the process of combining like terms in an expression.

Combining Like Terms

Terms can be combined only if they are similar or alike. Once you have determined that terms are alike, combine them by adding the coefficients. When the coefficients are combined, the variable factors are not changed.

EXAMPLE 2 Identify the like terms: $4x \qquad 5y \qquad -6x \qquad -3y^2$

SOLUTION

$4x, -6x$ These are like terms because the variable is the same.

$5y$ and $-3y^2$ are not alike.

Terminology

EXAMPLE 3 Combine: $6t + 5t$

SOLUTION

$6t + 5t$ *Combine like terms by adding the coefficients. The variable*
$6 + 5 = 11$ *part of the term does not change.*
$6t + 5t = 11t$

EXAMPLE 4 Simplify the following algebraic expression by combining like terms: $8x - 10 - 3x + 12$

SOLUTION

$8x, -3x$ *These are like terms.*
$-10, +12$ *These are like terms.*

$8 + (-3) = 5$ *Combine the coefficients.*
$8x + (-3x) = 5x$

$-10 + 12 = +2$ *Combine the constant terms.*

$5x + 2$ *After combining like terms, write the simplified algebraic expression.*

$8x - 10 - 3x + 12$ simplifies to $5x + 2$.

When you simplify an algebraic expression, the normal order is for the variable terms to come first, followed by the constant term. If there are more than two variable terms, it does not matter which one is listed first.

EXAMPLE 5 Simplify the following algebraic expression by combining like terms: $-3x + 7 - 16y + 5x - 4y - 24$

SOLUTION

$-3x, 5x$ *Before you begin work, identify all groups of like terms.*
$-16y, -4y$
$7, -24$

$(-3) + 5 = 2$ *Combine each group of like terms.*
$-3x + 5x = 2x$
$-16 + (-4) = -20$
$-16y + (-4y) = -20y$
$7 + (-24) = -17$

$2x - 20y - 17$ *Write the simplified expression.*

Notice that the constant term is never combined with a variable term.

Work the Comprehension Checkpoint problems to measure your understanding of the terminology discussed in this section. If you have difficulty with these problems, review the material in this section before going on to the Practice Sets.

COMPREHENSION CHECKPOINT

1. Name the coefficient of each term.

   a. $-3y$ b. $7.6m$ c. $\frac{3}{4}x$

   a. _____   b. _____   c. _____

2. Identify the like terms.
   $4x \quad 5xy \quad -6y \quad +7y^2 \quad -12x$ _____

3. Name the constant terms.
   $7z \quad 8y \quad -12 \quad 2a \quad +15$ _____

4. Simplify the algebraic expression by combining like terms.
   $18x - 29x$ _____

5. Simplify the algebraic expression by combining like terms.
   $5y - 7 - 6y - 15x + 18y + 12 + 4x$ _____

8.1A PRACTICE SET

Name the coefficient in each of the following terms.

1. $8.4x$ _____
2. $-9w$ _____
3. $3.5y$ _____
4. $-6z$ _____
5. $-12b$ _____
6. $10t$ _____
7. $-\frac{3}{4}a$ _____
8. $\frac{1}{2}x$ _____

In problems 9–14, identify the like terms.

9. $3x$ $5y$ $-9x$ _____
10. $-4y$ $12y$ -8 $18y$ _____
11. $10x$ $-7a$ $12b$ $-15a$ $8x^2$ _____
12. $3b$ $-6a^2$ $8d$ $9c$ $-4a^2$ _____
13. $16x$ -3 $12xz$ $-5y$ 8 _____
14. $12xyz$ $-7axy$ $9xyz$ $-10axz$ $16xyz$ _____

In problems 15–30, simplify each expression by combining like terms.

15. $8x - 3x$ _____
16. $12y - 12y$ _____
17. $3x + 4x - 8$ _____
18. $-6y + 6 + 18y$ _____
19. $5z + 6z - 12 + 3z$ _____
20. $-10 + 9a + 6 - 3a$ _____
21. $12b + 6a - 7b + 9a$ _____

22. $15x - 7y + 12x + 14y$ _____

23. $18xy + 8 - 3xy + 17xy - 15$ _____

24. $27 - 6c - 9 + 12c - 13$ _____

25. $6x - 7 + x - 15 - 4x$ _____

26. $-y - 2 + y + 2 + 3y - 1$ _____

27. $-4x^2 - 7 + 9x^2 + 18$ _____

28. $18xy^2 - 10 - 12xy^2 + 3xy^2 - 4$ _____

29. $9y + 3x + 8 - 7y + 12 + 15y$ _____

30. $4a + 35 - 9z - 19 + 21a + 6z + 6a - 3z$ _____

8.1B PRACTICE SET

Name the coefficient in each of the following terms.

1. $-5.6a$ _____
2. $18x$ _____
3. $7.2y$ _____
4. $3d$ _____
5. $-12x$ _____
6. $-6b$ _____
7. $-\dfrac{5}{8}t$ _____
8. $\dfrac{2}{3}b$ _____

In problems 9–14, identify the like terms.

9. $5y \quad 7 \quad 9y$ _____

10. $8x \quad -6z \quad -7x \quad 9$ _____

11. $15 \quad -8a \quad 12 \quad -3b \quad -10a^2$ _____

12. $6 \quad 3m \quad -4x^2 \quad -3a \quad 5x^2$ _____

13. $a \quad 6b \quad 15ab \quad -5d \quad a$ _____

14. $20abc \quad -6bc \quad 12ab \quad -16abc \quad 8abc$ _____

In problems 15–30, simplify each expression by combining like terms.

15. $6x + 3x$ _____

16. $5m - 3m$ _____

17. $8t - 5t + 3$ _____

18. $4y + 16 - 8y$ _____

19. $5a + 3 - 4a + 9a$ _____

20. $-6 + 7b + 10 + 15b$ _____

21. $7x + 14y - 2y - 7x$ _____

22. $12a - 6b + 4a - 3b$ _____

23. $4bc - 12bc + 3 - 2bc + 7$ _____

24. $-15 + 6m - 7 + 8m - 9m$ _____

25. $-4t + 3t - 6 - 12 - 8t$ _____

26. $x - 3 + x - 4 - 2x + 7$ _____

27. $3y^2 - 8 + 8y^2 + 8$ _____

28. $5 - 3x^2 - 16 + x^2 - 24$ _____

29. $10 - 7h^2 + 18 - 3h + 19h^2$ _____

30. $16a^2b + 6b - 4a + 9b - 10a^2b + 12$ _____

8.2 SOLVING EQUATIONS BY ADDITION

Knowing how to determine the solution to an equation is a skill that will benefit you in many courses. As you have learned, any math problem may be set up as an equation with a variable representing the unknown answer.

Solving an equation requires finding an equivalent equation. **Equivalent equations** are equations that have the same solution. When solving an equation, your goal is to create an equivalent equation that will be in the form

$$x = a$$

where x is the variable and a is the number it represents in that equation.

The first method for solving equations deals with the addition property of equality.

> **THE ADDITION PROPERTY OF EQUALITY**
> If the same number or term is added to each side of an equation, the resulting equation will be equivalent to the original equation.

When you are working with an equation like

$$x + 3 = 8$$

your goal is to find the number that x represents. You need to isolate the x on one side of the equation. Isolating the x means to have it on one side of the equal sign by itself. Using the addition property of equality is one way to accomplish this goal. Adding -3 to each side of the equation gives

$$\begin{array}{rl} x + 3 = & 8 \quad \text{original equation} \\ + (-3) = & + (-3) \quad \text{add the same term to each side} \\ \hline x = & 5 \quad \text{equivalent equation} \end{array}$$

Notice that the x is now isolated on one side of the equal sign. Remember that isolating the variable is the objective in solving an equation. The solution to the equation appears to be 5.

This is referred to as an **apparent solution,** because it has not yet been checked in the original equation. Every time you solve an equation, the value determined to be the solution should be checked in the original equation. A basic principle of equality is that if two quantities are equal, they may be substituted for each other. Go back to the original equation,

$$x + 3 = 8$$

and replace the x with 5:

$$5 + 3 = 8$$

and determine whether this statement is true. Combining like terms gives

$$8 = 8$$

If your calculations are correct, the result will be a true statement. If your check provides something like $6 = 8$, which is obviously not true, then you know that the apparent solution is incorrect and you will need to solve the equation again to find the correct solution.

When using the addition property of equality, always add the opposite (or additive inverse) of the number you are trying to eliminate from one side of the equation. Study the following examples carefully to see how the addition property is used to solve equations.

CHAPTER 8   Introduction to Algebra

EXAMPLE 1   Solve the equation $x + 5 = 12$

SOLUTION

$$
\begin{aligned}
x + 5 &= 12 &&\text{Write the equation.} \\
+(-5) &= +(-5) &&\text{Add the same term to each side.} \\
\hline
x &= 7 &&\text{The variable is isolated.}
\end{aligned}
$$

The apparent solution is 7.

Check: $x + 5 = 12$
            $7 + 5 = 12$
            $12 = 12$

The solution checks in the original equation, so it is the correct solution.

EXAMPLE 2   Solve the equation $m - 6 = 8$

SOLUTION

$$
\begin{aligned}
m - 6 &= 8 &&\text{Write the equation.} \\
+ 6 &= + 6 &&\text{Add the same term to each side.} \\
\hline
m &= 14 &&\text{The variable is isolated.}
\end{aligned}
$$

The apparent solution is 14.

Check: $m - 6 = 8$
            $14 - 6 = 8$
            $8 = 8$

The solution checks in the original equation, so it is the correct solution.

The key to using the addition property of equality is to remember that the number (or term) to be added to each side of the equation is the additive inverse of the number you are trying to eliminate. Checking each problem that you work should become as much a part of the problem-solving process as finding the solution itself.

Work the Comprehension Checkpoint problems to measure your understanding of the addition property of equality. If you have difficulty with these problems, review the material in this section before going on to the Practice Sets.

COMPREHENSION CHECKPOINT

Solve the following equations. Check your solutions.

1. $x - 7 = 15$ _____

2. $6 + y = 9$ _____

3. $12 + a = 10$ _____

4. $t - 6 = -24$ _____

8.2A PRACTICE SET

Use the addition property of equality to solve each equation. Check each solution.

1. $y - 6 = 8$ _____

2. $12 + m = 9$ _____

3. $t + 7 = -22$ _____

4. $x - 15 = 6$ _____

5. $a + 9 = -3$ _____

6. $c - 18 = 2$ _____

7. $-8 + r = 17$ _____

8. $16 + t = -15$ _____

9. $x + 6 = 6$ _____

10. $-9 + y = -20$ _____

11. $5 = y + 3$ _____

12. $-14 = x - 7$ _____

13. $20 = a + 13$ _____

14. $x - 35 = 70$ _____

15. $y + 42 = -30$ _____

16. $-28 = b - 16$ _____

17. $-32 = -18 + x$ _____

18. $48 = 15 + x$ _____

19. $40 = y - 25$ _____

20. $t - 40 = 100$ _____

8.2B PRACTICE SET

Use the addition property of equality to solve each equation. Check each solution.

1. $r + 5 = 9$ _____
2. $t - 3 = 7$ _____
3. $6 + m = 10$ _____
4. $-8 + n = 15$ _____
5. $x - 18 = 16$ _____
6. $a + 10 = -12$ _____
7. $-8 + r = -8$ _____
8. $9 + t = -8$ _____
9. $n + 24 = -16$ _____
10. $p - 25 = 27$ _____
11. $-6 + r = -3$ _____
12. $17 = x + 16$ _____
13. $24 = -8 + m$ _____
14. $30 = 22 + y$ _____
15. $-50 = -18 + x$ _____
16. $-35 = y - 19$ _____
17. $25 = a + 17$ _____
18. $-34 = q - 16$ _____
19. $45 = -28 + n$ _____
20. $58 = a - 42$ _____

8.3 SOLVING EQUATIONS BY DIVISION

All of the equations in the previous section contained variables with coefficients of 1. This section will consider variables with coefficients that are not 1. The objective for solving equations will not change. Your objective is still to create an equivalent equation of the form $x = a$. The only change is the method that will be used.

> **THE DIVISION PROPERTY OF EQUALITY**
>
> Dividing each side of an equation by the same number (or term) gives a new equation that is equivalent to the original equation.

It should be noted that the number used as a divisor cannot be 0. You have learned that division by 0 cannot be done.

If the variable term in an equation has a coefficient that is not 1, as in the equation

$$3x = 18$$

the division property of equality should be used to isolate the variable. Divide each side of the equation by the coefficient of the variable:

$$\frac{3x}{3} = \frac{18}{3} \quad \text{\textit{original equation}}$$
$$\text{\textit{divide each side by the same term}}$$

The resulting equivalent equation,

$$x = 6$$

is in the form $x = a$, so the apparent solution is 6. Always check the apparent solution in the original equation.

Check: $\quad 3x = 18$

$$(3)(6) = 18$$
$$18 = 18$$

The apparent solution makes a true statement when it is checked in the original equation, so 6 is the solution of the equation.

Study the following examples. Notice that the number used as a divisor is always the coefficient of the variable term.

EXAMPLE 1 Solve the equation $4x = 12$

SOLUTION

$\frac{4x}{4} = \frac{12}{4}$ *Write the equation.*
Divide each side by the coefficient of the variable.

$x = 3$ The apparent solution is 3.

Check: $\quad 4x = 12$

$$(4)(3) = 12$$
$$12 = 12$$

The apparent solution makes a true statement, so 3 is the solution to the equation.

EXAMPLE 2 Solve the equation $-5m = 20$

SOLUTION

$\dfrac{-5m}{-5} = \dfrac{20}{-5}$ *Write the equation.*
Divide each side by the coefficient of the variable.

$m = -4$ The apparent solution is -4.

Check:
$$-5m = 20$$
$$(-5)(-4) = 20$$
$$20 = 20$$

The apparent solution, -4, makes a true statement, so it is the solution to the equation.

Solutions to equations will not necessarily be whole numbers.

EXAMPLE 3 Solve the equation $-3a = 8$

SOLUTION

$\dfrac{-3a}{-3} = \dfrac{8}{-3}$ *Write the equation.*
Divide each side by the coefficient of the variable.

$a = -\dfrac{8}{3}$ The apparent solution is $-\dfrac{8}{3}$.

Check:
$$-3a = 8$$
$$(-3)\left(-\dfrac{8}{3}\right) = 8$$
$$8 = 8$$

The apparent solution, $-\dfrac{8}{3}$, makes a true statement, so it is the solution to the equation.

Example 3 demonstrates that the solution to an equation is not always a whole number. A fraction is a legitimate solution. The only concern is whether or not it solves the equation. The check in Example 3 proves that it does. Do not be hesitant to accept the answer you compute if you have followed the steps properly.

The Comprehension Checkpoint problems provide an opportunity to check your understanding of the division property of equality. If you have difficulty with these problems, review the material in this section before going on to the Practice Sets.

COMPREHENSION CHECKPOINT

Solve the following equations using the division property of equality. Check your solutions.

1. $6x = 24$ _____

2. $-5x = -15$ _____

3. $36 = -9y$ _____

4. $14 = 5b$ _____

8.3A PRACTICE SET

Use the division property of equality to solve the following equations. Check your answers.

1. $3x = 15$ _____

2. $-4r = 28$ _____

3. $6t = 36$ _____

4. $7y = -21$ _____

5. $-8t = -32$ _____

6. $5z = 40$ _____

7. $-9q = 45$ _____

8. $-70 = -14a$ _____

9. $-27 = 3b$ _____

10. $48 = 12x$ _____

11. $15b = 105$ _____

12. $-48 = -16r$ _____

13. $24 = -3c$ _____

14. $-84 = -28n$ _____

15. $54 = -13.5p$ _____

16. $-63 = -12.6m$ _____

17. $8.3t = 49.8$ _____

18. $6y = 27$ _____

19. $8x = -35$ _____

20. $43 = -10z$ _____

8.3B PRACTICE SET

Use the division property of equality to solve the following equations. Check your answers.

1. $8r = 24$ _____

2. $-3p = -9$ _____

3. $5x = 35$ _____

4. $6y = 42$ _____

5. $-56 = 7r$ _____

6. $48 = 4x$ _____

7. $-36 = 12y$ _____

8. $-18z = -54$ _____

9. $99 = -11s$ _____

10. $64 = 16t$ _____

11. $15x = 90$ _____

12. $-22y = 132$ _____

13. $-84 = -28z$ _____

14. $-150 = 25t$ _____

15. $-9.4r = 47$ _____

16. $50.4 = 6.3t$ _____

17. $-33 = -5.5x$ _____

18. $7y = 30$ _____

19. $5p = -49$ _____

20. $-80 = -15x$ _____

8.4 SOLVING EQUATIONS BY MULTIPLICATION

If the coefficient of the variable is a fraction rather than a whole number, the multiplication property of equality can be used.

> **THE MULTIPLICATION PROPERTY OF EQUALITY**
> If each side of an equation is multiplied by the same number (or term), the new equation will be equivalent to the original equation.

Remember that the result of multiplying a whole number, x, by a fraction, $\frac{a}{b}$, is $\frac{ax}{b}$. Thus, in the equation

$$\frac{x}{3} = 6$$

the coefficient of x is $\frac{1}{3}$. Equations of this type can be solved by multiplying by the reciprocal of the coefficient. The **reciprocal** of a number is that number inverted. The reciprocal of $\frac{1}{3}$ is $\frac{3}{1}$, or simply 3. Any fraction can be inverted to determine the reciprocal. The reciprocal of $\frac{4}{5}$ is $\frac{5}{4}$, while the reciprocal of $\frac{6}{7}$ is $\frac{7}{6}$.

EXAMPLE 1 Name the reciprocal of each term.

a. $\frac{2}{3}$ *b.* $\frac{5}{8}$ *c.* 5

SOLUTIONS

a. $\frac{3}{2}$ *b.* $\frac{8}{5}$ *c.* $\frac{1}{5}$

Remember that a whole number can be written as a fraction with a denominator of 1. When you invert a whole number, as in part *c*, the numerator will be 1.

Solving an equation like

$$\frac{x}{3} = 6$$

requires multiplying by the reciprocal of the coefficient:

$$\frac{\cancel{3}}{1} \cdot \frac{x}{\cancel{3}} = \frac{6}{1} \cdot \frac{3}{1}$$

$$x = 18$$

Note the use of the multiplication property of equality. Both sides of the equation were multiplied by the same factor.

As a check,

$$\frac{x}{3} = 6$$

$$\frac{18}{3} = 6$$

$$6 = 6$$

The solution is 6.

The product of reciprocals is always 1:

$$\frac{\cancel{3}^1}{\cancel{4}_1} \cdot \frac{\cancel{4}^1}{\cancel{3}_1} = 1 \qquad \frac{\cancel{6}^1}{1} \cdot \frac{1}{\cancel{6}_1} = 1$$

When you multiply a term with a fractional coefficient by the reciprocal, as in the previous illustration, you get a coefficient of 1 on the variable. That is what you need to solve an equation.

Observe in the following examples that the reciprocal of the coefficient is always used as the factor. Multiplying by this reciprocal has the effect of eliminating the fraction while at the same time creating a coefficient of 1 on the variable term.

EXAMPLE 2 Solve the equation $\frac{x}{5} = 3$

SOLUTION

$\frac{x}{5} = 3$ *The coefficient of the variable is $\frac{1}{5}$.*

$\frac{\cancel{5}^1}{1} \cdot \frac{x}{\cancel{5}_1} = \frac{3}{1} \cdot \frac{5}{1}$ *Multiply both sides of the equation by the reciprocal of the coefficient.*

$x = 15$ The apparent solution is 15.

Check: $\frac{x}{5} = 3$

$\frac{15}{5} = 3$

$3 = 3$

The solution is 15.

EXAMPLE 3 Solve the equation $-\frac{5x}{8} = 10$

SOLUTION

$-\frac{5x}{8} = 10$ *The coefficient of the variable is $-\frac{5}{8}$.*

$\left(-\frac{\cancel{8}^1}{\cancel{5}_1}\right) \cdot \left(-\frac{\cancel{5}^1 x}{\cancel{8}_1}\right) = \frac{\cancel{10}^2}{1} \cdot \left(-\frac{8}{\cancel{5}_1}\right)$ *Multiply both sides of the equation by the reciprocal of the coefficient.*

$x = -16$ The apparent solution is -16.

Solving Equations by Multiplication

Check: $-\dfrac{5x}{8} = 10$

$-\dfrac{(5)(-16)}{8} = 10$

$10 = 10$

The solution is -16.

Remember that the multiplication property of equality is used when the coefficient of the variable is a fraction. *Always multiply by the reciprocal of the coefficient.*

The Comprehension Checkpoint problems offer an opportunity to determine your understanding of the multiplication property of equality. If you have any difficulty with these problems, review the material in this section before going on to the Practice Sets.

COMPREHENSION CHECKPOINT

1. Name the coefficient of each of the following terms:

   a. $\dfrac{5x}{7}$　　　　b. $\dfrac{b}{10}$　　　　c. $-\dfrac{3a}{5}$

   a. _____　　　b. _____　　　c. _____

Solve the following equations. Check your answers.

2. $\dfrac{2x}{3} = 8$ _____

3. $\dfrac{5a}{7} = 15$ _____

4. $12 = \dfrac{3y}{5}$ _____

8.4A PRACTICE SET

1. What is the coefficient of each variable?

   a. $\dfrac{x}{6}$ b. $\dfrac{3x}{4}$ c. $-\dfrac{2}{3}m$

   a. _____   b. _____   c. _____

2. Name the reciprocal of the coefficient of each variable.

   a. $\dfrac{x}{5}$ b. $\dfrac{3p}{7}$ c. $-\dfrac{4x}{5}$

   a. _____   b. _____   c. _____

Solve each equation using the multiplication property of equality. Check each solution.

3. $4 = \dfrac{x}{5}$ _____

4. $\dfrac{2x}{3} = 12$ _____

5. $\dfrac{x}{9} = -16$ _____

6. $\dfrac{3y}{4} = 6$ _____

7. $\dfrac{5a}{8} = 30$ _____

8. $8 = \dfrac{2d}{5}$ _____

9. $-\dfrac{3x}{7} = 9$ _____

10. $\dfrac{t}{4} = -20$ _____

11. $-21 = \dfrac{7a}{15}$ _____

12. $\dfrac{4x}{9} = 12$ _____

13. $-\dfrac{x}{6} = -7$ _____

14. $\dfrac{m}{8} = 0$ _____

15. $60 = \dfrac{5y}{12}$ _____

16. $-\dfrac{3n}{10} = -6$ _____

17. $\dfrac{6x}{11} = 30$ _____

18. $-\dfrac{7n}{10} = 35$ _____

19. $\dfrac{11x}{24} = 33$ _____

20. $\dfrac{3}{4}x = 48$ _____

8.4B PRACTICE SET

1. What is the coefficient of each variable?

   a. $\dfrac{x}{12}$ b. $-\dfrac{y}{8}$ c. $\dfrac{5x}{9}$

   a. _____   b. _____   c. _____

2. Name the reciprocal of the coefficient of each variable.

   a. $-\dfrac{x}{7}$ b. $\dfrac{2b}{3}$ c. $\dfrac{x}{10}$

   a. _____   b. _____   c. _____

Solve each equation using the multiplication property of equality. Check each solution.

3. $\dfrac{x}{6} = 8$ _____

4. $9 = \dfrac{3x}{7}$ _____

5. $-\dfrac{x}{12} = 5$ _____

6. $\dfrac{3m}{8} = 18$ _____

7. $\dfrac{2p}{9} = 8$ _____

8. $\dfrac{6x}{11} = 18$ _____

9. $42 = \dfrac{7t}{12}$ _____

10. $\dfrac{5a}{6} = 40$ _____

11. $\dfrac{4y}{5} = 24$ _____

12. $\dfrac{a}{7} = 13$ _____

13. $-27 = -\dfrac{d}{3}$ _____

14. $\dfrac{7y}{9} = 14$ _____

15. $\dfrac{13r}{24} = 52$ _____

16. $\dfrac{6p}{25} = 30$ _____

17. $21 = \dfrac{7m}{12}$ _____

18. $\dfrac{8x}{9} = 48$ _____

19. $\dfrac{10t}{13} = 30$ _____

20. $-96 = \dfrac{12k}{25}$ _____

8.5 THE DISTRIBUTIVE PROPERTY

Many times, terms will be enclosed in grouping symbols such as ( ) or [ ], that is, parentheses or brackets. When grouping symbols are included in an equation, as in

$$3(x + 7) = 12$$

those symbols must be removed before the solution may be determined. Removing grouping symbols requires an understanding of the distributive property.

DISTRIBUTIVE PROPERTIES OF MULTIPLICATION

If a, b, and c are any numbers, then

$$a(b + c) = ab + ac$$
$$a(b - c) = ab - ac$$

Everything contained within grouping symbols is considered to be a single term. Thus,

$$(a + 6) \quad \text{and} \quad (b - 7)$$

are terms just as

$$4x \quad \text{and} \quad -5y$$

are terms.

Terms inside grouping symbols are like expanded notation of a number. Remember from Chapter 1 that a number can be written in expanded notation. Thus,

$$25 = (20 + 5)$$

Notice that the expanded notation is in the same form as $(a + 6)$; both have two terms inside the parentheses.

Grouping symbols may also have coefficients:

$$3(a + 6) \quad -5(b - 7)$$

It is these coefficients that must be "distributed over" the grouping symbols by multiplying each term inside the symbols by the coefficient. Again, think about multiplication of a multidigit factor by a single-digit factor. Consider the problem 6×25.

Traditional Method:

```
    25
  × 6
  ----
    30 = 6 times 5
   120 = 6 times 20
  ----
   150 = Sum of partial products
```

Using the Distributive Property:

$6(20 + 5)$
$6 \cdot 20 = 120$
$6 \cdot 5 = 30$
Product $= 120 + 30$
Product $= 150$

These two multiplication methods are solutions to the same problem and produce the same result.

To use the distributive property with algebraic terms, use the same procedure as with numbers. Consider the problem $3(a + 6)$. This is a single-digit factor multiplied by a multidigit factor. The a and 6 are added together, as the 20 and 5 were previously, but since they are not like terms, you cannot actually add them. You can, however, multiply them by another factor. After multiplying, you must add the products, just as you did with $120 + 30$ in the example.

3(a + 6) *Multiplication is implied here.*

3 · a = 3a *Multiply 3 by a.*

3 · 6 = 18 *Multiply 3 by 6.*

Total = 3a + 18 *Add the two terms together. They cannot actually be added, because they are not like terms. You can indicate that they are to be added.*

Mastery of the distributive property is an essential skill in solving many equations. Notice in the following examples that the coefficient is multiplied by each term inside the parentheses. Removing grouping symbols is the first step in simplifying an expression.

EXAMPLE 1 Use the distributive property to simplify $5(x - 3)$.

SOLUTION

$5(x - 3)$ *Multiplication is indicated. Multiply each term inside the parentheses by 5. Add the two terms.*

$(5)(x) = 5x$

$(5)(-3) = -15$

Total $= 5x - 15$

$5(x - 3) = 5x - 15$

EXAMPLE 2 Use the distributive property to simplify $-4(y - 9)$.

SOLUTION

$-4(y - 9)$ *Multiplication is indicated. Multiply each term inside the parentheses by -4. Remember that a term includes a sign, either + or $-$.*

$(-4)(y) = -4y$

$(-4)(-9) = 36$

Total $= -4y + 36$

$-4(y - 9) = -4y + 36$

EXAMPLE 3 Use the distributive property to simplify $(x - 7)$.

SOLUTION

$(x - 7)$ *When a number does not appear as a coefficient, the coefficient is 1.*

$x - 7$ *Anytime the coefficient is 1, you can simply drop the parentheses. Nothing else changes.*

EXAMPLE 4 Use the distributive property to simplify $-(m + 3)$.

SOLUTION

$-(m + 3)$ *Again, no coefficient is obvious, but there is a minus sign outside the parentheses. The coefficient is -1.*

$(-1)(m) = -m$ *Multiply both terms inside the parentheses by -1.*

$(-1)(3) = -3$

The Distributive Property

$-m - 3$ *Combine the results.*

$-(m + 3) = -m - 3$

Study the effect the coefficients in Examples 3 and 4 had on the result. When the coefficient was -1 in Example 4, the simplified expression after removing the parentheses was the same as the original expression except that each sign had been changed to its opposite. In Example 3, where the coefficient was $+1$, the parentheses were removed and the terms inside were not changed. Remember the effect of multiplying by $+1$ or -1.

EXAMPLE 5   Use the distributive property to simplify the algebraic expression $7(x + 5) - (x - 3)$.

SOLUTION

$7(x + 5) = 7x + 35$ *Use the distributive property to eliminate the parentheses.*
$-(x - 3) = -x + 3$

$7x + 35 - x + 3$ *Combine like terms.*
$7x + (-x) = 6x$
$35 + 3 = 38$

$6x + 38$ *Create a new algebraic expression with the results of the combined terms.*

$7(x + 5) - (x - 3)$ simplifies to $6x + 38$.

Work the Comprehension Checkpoint problems to measure your understanding of the distributive property of multiplication. If you have difficulty with these problems, review the material in this section before going on to the Practice Sets.

COMPREHENSION CHECKPOINT

Use the distributive property to simplify each term.

1. $-(a + 8)$ _____

2. $(b - 6)$ _____

3. $-8(x - 7)$ _____

4. $10(y + 12)$ _____

5. $9(y - 2) + 4(y - 12)$ _____

8.5A PRACTICE SET

Simplify each expression by applying the distributive property.

1. $(a - 3)$ _____

2. $(b + 7)$ _____

3. $-(x + 9)$ _____

4. $-(-y + 8)$ _____

5. $-2(a - 5)$ _____

6. $6(m - 4)$ _____

7. $7(-n - 8)$ _____

8. $5(q - 10)$ _____

9. $-8(r - 3)$ _____

10. $-\frac{2}{5}(5 + a)$ _____

11. $\frac{3}{7}(a + b)$ _____

12. $\frac{7}{12}(-x - y)$ _____

13. $-5(4m + n)$ _____

14. $-15(3x - 4)$ _____

15. $-18(2d + 7)$ _____

16. $2(a + b + c)$ _____

17. $-3(x - y + z)$ _____

18. $12(a + 13) + 2(a - 6)$ _____

19. $-9(3 - x) + 4(x + 2)$ _____

20. $3(8 + x) - 5(x - 8)$ _____

8.5B PRACTICE SET

Simplify each expression by applying the distributive property.

1. $-(x + 9)$ _____

2. $(a + 2)$ _____

3. $-(-y - 10)$ _____

4. $(4 - z)$ _____

5. $-8(b - 7)$ _____

6. $-4(r + 3)$ _____

7. $6(t + 6)$ _____

8. $7(m - 5)$ _____

9. $-5(-q - 3)$ _____

10. $\frac{2}{3}(-6 - r)$ _____

11. $\frac{5}{8}(b - c)$ _____

12. $-\frac{3}{4}(d + c)$ _____

13. $4(-2t + u)$ _____

14. $-5(-5 - 3k)$ _____

15. $-2(8b - 6)$ _____

16. $3(-y + x + z)$ _____

17. $-(-p + q - r)$ _____

18. $-10(-5 + r) + 8(r - 6)$ _____

19. $8(y - 8) - 4(y - 5)$ _____

20. $-3(m + t) + 2(m + t)$ _____

8.6 SOLVING MULTISTEP EQUATIONS

Each equation that you solved in the previous sections of this chapter required a single step, or one operation, to determine the solution. Frequently, equations will require more than one operation to find the solution.

You need to develop a strategy for solving equations. Remember that solving an equation is the process of finding the number that the variable represents in that situation. The outlined steps are the basics for solving most equations that you will encounter.

> **STEPS FOR SOLVING EQUATIONS**
> 1. Remove any grouping symbols.
> 2. Combine like terms. Move all variable terms to one side of the equation and everything else to the other side.
> 3. Find an equivalent equation in the form $x = a$.

Consider the equation

$$4(a + 6) = 2a - 8$$

There are a number of operations indicated in this equation. The first step is to remove grouping symbols, so distributing the 4 through the parentheses gives

$$4a + 24 = 2a - 8$$

The next step is to combine like terms. All variable terms need to be combined on one side of the equation. All terms not containing a variable need to be combined on the opposite side of the equation. You can combine the variable terms on either side of the equation and everything else on the opposite side using the addition property of equality.

$$\begin{aligned} 4a + 24 &= 2a - 8 \\ +(-2a) &= +(-2a) \\ \hline 2a + 24 &= -8 \end{aligned}$$ *Add the same term to both sides.*

$$\begin{aligned} 2a + 24 &= -8 \\ +(-24) &= +(-24) \\ \hline 2a &= -32 \end{aligned}$$ *Again, add the same term to both sides.*

$$\frac{2a}{2} = \frac{-32}{2}$$ *Now use the division property of equality to isolate the variable.*

$$a = -16$$ The apparent solution is -16.

As always, the solution should be checked in the original equation:

$$4(a + 6) = 2a - 8$$
$$4(-16 + 6) = 2(-16) - 8$$
$$4(-10) = -32 - 8$$
$$-40 = -40$$

The solution is -16.

As more terms or grouping symbols are added to equations, the difficulty level seems to increase. If you learn the steps for solving equations, you may easily solve even the most complex equations. Study the following examples. Notice that the steps for solving equations do not change.

EXAMPLE 1 Solve the equation $3(x - 8) = 9$

SOLUTION

$3(x - 8) = 9$
$3x - 24 = 9$ *First use the distributive property to remove the parentheses.*

$\begin{array}{r} 3x - 24 = 9 \\ +24 +24 \\ \hline 3x = 33 \end{array}$ *Next use the addition property of equality to isolate the variable term.*

$\dfrac{3x}{3} = \dfrac{33}{3}$ *Now use the division property of equality to create a coefficient of 1 on the variable.*

$x = 11$ The apparent solution is 11.

Check: $3(x - 8) = 9$
$3(11 - 8) = 9$
$3(3) = 9$
$9 = 9$

The solution is 11.

EXAMPLE 2 Solve the equation $-5(y - 6) = 3y + 10$

SOLUTION

$-5(y - 6) = 3y + 10$
$-5y + 30 = 3y + 10$ *The first step is to use the distributive property to remove the parentheses.*

$\begin{array}{r} -5y + 30 = 3y + 10 \\ +5y = +5y \\ \hline +30 = 8y + 10 \end{array}$ *Add the same term to each side of the equation.*

$\begin{array}{r} +30 = 8y + 10 \\ +(-10) = + (-10) \\ \hline 20 = 8y \end{array}$ *Add -10 to each side to isolate the variable term.*

$\dfrac{20}{8} = \dfrac{8y}{8}$ *Use the division property of equality to isolate the variable.*

$2.5 = y$ The apparent solution is 2.5.

Check: $-5(y - 6) = 3y + 10$
$-5(2.5 - 6) = 3(2.5) + 10$
$-5(-3.5) = 7.5 + 10$
$17.5 = 17.5$

The solution is 2.5.

EXAMPLE 3 Solve the equation $6(x - 9) = 3(x + 2)$

SOLUTION

$6(x - 9) = 3(x + 2)$
$6x - 54 = 3x + 6$ *Use the distributive property to remove both sets of parentheses at the same time.*

$\begin{array}{r} 6x - 54 = 3x + 6 \\ +(-3x) + 54 = +(-3x) + 54 \\ \hline 3x = 60 \end{array}$ *Use the addition property to combine like terms. You can move all terms that need to be moved at the same time.*

Solving Multistep Equations

$\dfrac{3x}{3} = \dfrac{60}{3}$ Use the division property of equality to isolate the variable.

$x = 20$ The apparent solution is 20.

Check: $6(x - 9) = 3(x + 2)$
$6(20 - 9) = 3(20 + 2)$
$6(11) = 3(22)$
$66 = 66$

The solution is 20.

EXAMPLE 4   Solve the equation $\dfrac{3x}{4} + 12 = 18$

SOLUTION

$\dfrac{3x}{4} + 12 = 18$ *Begin by using the addition property of equality to isolate the variable term.*
$\phantom{\dfrac{3x}{4}} + (-12) = + (-12)$
$\dfrac{3x}{4} = 6$

$\dfrac{\cancel{4}}{\cancel{3}} \cdot \dfrac{\cancel{3}x}{\cancel{4}} = \dfrac{\cancel{6}}{1} \cdot \dfrac{4}{\cancel{3}}$ *Use the multiplication property of equality to isolate the variable.*

$x = 8$ The apparent solution is 8.

Check: $\dfrac{3x}{4} + 12 = 18$
$\dfrac{(3)(8)}{4} + 12 = 18$
$6 + 12 = 18$
$18 = 18$

The solution is 8.

With practice, you can develop the skills necessary to solve any equation. Work the Comprehension Checkpoint problems. Before beginning work on a problem, think about the equivalent equation that you are trying to derive. Try to think your way through the problem to that equivalent equation before you begin working. If you have difficulty with these problems, review the material in this section before going on to the Practice Sets.

COMPREHENSION CHECKPOINT

Solve each equation. Check your answers.

1. $3(x - 5) = 12$ _____

2. $\dfrac{x}{3} + 7 = 14$ _____

3. $-4(a + 3) = a + 18$ _____

4. $2(x + 4) = 3(x - 9)$ _____

8.6A PRACTICE SET

Solve each equation. Check your answers.

1. $3x + 8 = 23$ _____

2. $4y - 9 = 19$ _____

3. $-7a + 6 = 48$ _____

4. $-12 - 5m = 23$ _____

5. $\dfrac{b}{4} - 3 = 6$ _____

6. $\dfrac{x}{8} + 4 = -2$ _____

7. $\dfrac{5z}{8} - 7 = 28$ _____

8. $\dfrac{3c}{10} + 15 = 135$ _____

9. $-(x + 8) = 16$ _____

10. $(9 + a) = 12$ _____

11. $3(a - 4) = 15$ _____

12. $-5(x - 12) = 10$ _____

13. $-(d + 12) - 3 = 18$ _____

14. $(c + 6) + 5 = -11$ _____

15. $8(a - 10) - 3 = 21$ _____

16. $-12(x + 1) - 18 = 42$ _____

17. $4y + 36 = 2y - 8$ _____

18. $5a - 20 = 4a + 3$ _____

19. $6(x - 8) = 4(x - 7)$ _____

20. $7(a - 3) = 5(a - 9)$ _____

8.6B PRACTICE SET

Solve each equation. Check your answers.

1. $5a + 12 = -13$ _____
2. $8x - 3 = 29$ _____
3. $-9 + 3b = 18$ _____
4. $2a - 18 = -16$ _____
5. $\dfrac{x}{3} + 7 = 12$ _____
6. $-\dfrac{y}{9} - 3 = 0$ _____
7. $\dfrac{4m}{7} - 4 = 20$ _____
8. $\dfrac{5t}{6} + 7 = -23$ _____
9. $(3 - 4x) = 13$ _____
10. $-(a + 7) = 25$ _____
11. $4(b - 5) = 24$ _____
12. $-8(y + 3) = 0$ _____
13. $-(c + 8) - 4 = 15$ _____
14. $(a + 12) - 9 = 4$ _____
15. $-6(c + 3) + 12 = 18$ _____
16. $2(-a - 5) + 15 = 29$ _____
17. $5x - 9 = 2x + 18$ _____

18. $6a + 17 = 8a - 25$ _____

19. $5(x - 4) = 3(x + 6)$ _____

20. $-12(t + 6) = 4(t + 8)$ _____

SUMMARY OF KEY CONCEPTS

KEY TERMS

variable (8.1): A letter used to represent a number.

equation (8.1): A statement that two quantities have the same value.

formula (8.1): A rule written with letters and symbols used to solve particular problems.

solution (of an equation) (8.1): The number that the variable represents.

numerical coefficient (8.1): The number multiplied by a variable.

term (8.1): A sign, coefficient, and one or more variables acting together as one unit.

similar (like) terms (8.1): Terms with the exact same variable factor(s).

unlike terms (8.1): Terms with different variable factors.

constant term (8.1): A number without a variable.

algebraic expression (8.1): A group of terms including the plus and/or minus signs.

simplifying an expression (8.1): The process of combining like terms in an algebraic expression.

equivalent equations (8.2): Equations that have the same solutions.

apparent solution (8.2): A proposed solution to an equation. It needs to be checked in the equation to make sure that it is a true solution.

reciprocal (of a number) (8.4): The inverse of a number.

KEY RULES

Addition property of equality (8.2):

Equality is maintained in an equation by adding the same number or term to both sides.

Division property of equality (8.3):

Equality is maintained in an equation by dividing both sides of the equation by the same number or term.

Multiplication property of equality (8.4):

Equality is maintained in an equation by multiplying both sides of the equation by the same number or term.

Distributive property (8.5):

The process of multiplying all terms inside grouping symbols by the coefficient of the grouping symbol.

Steps for solving equations (8.6):

1. Remove any grouping symbols.
2. Combine like terms. Move all variable terms to one side of the equation and all other terms to the other side.
3. Use the addition, division, or multiplication property of equality (or a combination of those properties) to create an equivalent equation in the form $x = a$.

PRACTICE TEST 8A

1. How many terms are there in the following expression?

   $8x + 6y - 9 + 3z - 18w$ _____

2. Name the coefficient of each term.

   a. $4x$ b. $-5y$ c. $\dfrac{5}{9}z$

   a. _____   b. _____   c. _____

3. Simplify the expression by combining like terms.

   $5a - 7a + 6 - 3a - 15$ _____

In problems 4–9, solve each equation.

4. $x - 8 = 12$ _____

5. $4 + y = 11$ _____

6. $15x = 105$ _____

7. $-42 = 3c$ _____

8. $\dfrac{x}{9} = 18$ _____

9. $\dfrac{5y}{8} = 60$ _____

10. Use the distributive property to simplify the expression $+3(q - 6)$.

    _____

11. Solve the equation $4(a - 3) = a - 6$. _____

12. Solve the equation $\dfrac{6d}{7} - 18 = -90$. _____

PRACTICE TEST 8B

1. How many terms are there in the following expression?

   $4y + 8 - 9x - 3z$ _____

2. Name the coefficient of each term.

   a. $-6y$ b. $12b$ c. $\frac{3}{4}c$

   a. _____   b. _____   c. _____

3. Simplify the expression by combining like terms.

   $4t + 3 - 8t + 28 + 5t$ _____

In problems 4–9, solve each equation.

4. $r + 6 = 18$ _____

5. $y - 18 = -2$ _____

6. $24x = 144$ _____

7. $-5a = 65$ _____

8. $12 = \dfrac{t}{3}$ _____

9. $\dfrac{18x}{25} = 72$ _____

10. Use the distributive property to simplify the expression $5(a + 16)$.

    _____

11. Solve the equation $-5(a - 9) = 3a - 11$. _____

12. Solve the equation $\dfrac{4c}{15} + 12 = 48$. _____

CHAPTERS 1-8 SKILLS REVIEW

1. Write 736,394 in words. _____

2. Write three million, two hundred eighty-four thousand, twelve in numbers. _____

3. Round 893,239 to the nearest thousand. _____

4. Round 796,749 to the nearest ten thousand. _____

Perform the indicated operations.

5. $\frac{2}{3} + \frac{4}{5} + \frac{9}{10} =$ _____

6. $\frac{9}{16} - \frac{5}{12} =$ _____

7. $\frac{5}{6} \cdot \frac{12}{25} =$ _____

8. $\frac{7}{8} \div \frac{21}{32} =$ _____

9. $9\frac{1}{3} \cdot 3\frac{3}{5} =$ _____

10. $4\frac{2}{3} \div 2\frac{8}{9} =$ _____

11. $5.65 \times 12.804 =$ _____

12. $6.25 \div 0.025 =$ _____

13. Convert $\frac{5}{16}$ to a decimal. _____

14. Convert $\frac{5}{8}$ to a percent. _____

15. Convert 83% to a decimal. _____

503

16. Convert 28% to a fraction in lowest terms. _____

17. What is 34% of 350? _____

18. 65% of what number is 260? _____

19. What is $8\frac{1}{3}$% of 540? _____

20. Write 0.7632 in words.
    _____

21. Write 28.74 in words.
    _____

22. Is the following a true proportion?
    $\frac{6}{25} \stackrel{?}{=} \frac{30}{150}$ _____

23. Find the missing term in the proportion.
    $\frac{?}{16} = \frac{20}{64}$ _____

24. Insert the appropriate symbol, < or >.
    a. 12 _____ 17
    b. −3.5 _____ −2.6
    c. 1 _____ −1

25. What is the absolute value of 15.8? _____

Perform the indicated operation.

26. $(19) + (-5) =$ _____

27. $(-27) - (-18) =$ _____

28. $(-3)(-6)(+18)(-5) =$ _____

29. $(-30) \div (-6) =$ _____

Solve each equation.

30. $x + 8 = 24$ _____

31. $5x + 18 = 48$ _____

32. $\frac{3}{8}x = 9$ _____

CHAPTER 8 SOLUTIONS

Skills Preview

1. 4
2. a. $\frac{2}{5}$
   b. 6
   c. 7.4
3. $3y - 27$
4. 15
5. 5
6. 5
7. 9
8. 30
9. 72
10. $-8a + 32$
11. -18
12. 85

Section 8.1 Comprehension Checkpoint

1. a. -3
   b. 7.6
   c. $+\frac{3}{4}$
2. $4x$ and $-12x$
3. -12 and $+15$
4. $-11x$
5. $-11x + 17y + 5$

8.1A Practice Set

1. 8.4
2. -9
3. 3.5
4. -6
5. -12
6. 10
7. $-\frac{3}{4}$
8. $\frac{1}{2}$
9. $3x, -9x$
10. $-4y, 12y, 18y$
11. $-7a, -15a$
12. $-6a^2, -4a^2$
13. $-3, 8$
14. $12xyz, 9xyz, 16xyz$
15. $8x - 3x$
    $8 + (-3) = 5$
    $5x$
16. $-12y + 12y$
    $-12 + 12 = 0$
    $0y = 0$
17. $3x + 4x - 8$
    $3 + 4 = 7$
    $7x - 8$
18. $-6y + 6 + 18y$
    $-6 + 18 = 12$
    $12y + 6$
19. $5z + 6z - 12 + 3z$
    $5 + 6 + 3 = 14$
    $14z - 12$
20. $-10 + 9a + 6 - 3a$
    $9 + (-3) = 6$ (variables)
    $-10 + 6 = -4$ (constants)
    $6a - 4$
21. $12b + 6a - 7b + 9a$
    $12 + (-7) = 5$ (b terms)
    $6 + 9 = 15$ (a terms)
    $5b + 15a$
22. $15x - 7y + 12x + 14y$
    $15 + 12 = 27$ (x terms)
    $-7 + 14 = 7$ (y terms)
    $27x + 7y$
23. $18xy + 8 - 3xy + 17xy - 15$
    $18 + (-3) + 17 = 32$ (xy terms)
    $8 - 15 = -7$ (constants)
    $32xy - 7$
24. $27 - 6c - 9 + 12c - 13$
    $27 - 9 - 13 = 5$ (constants)
    $-6 + 12 = 6$ (c terms)
    $6c + 5$
25. $6x - 7 + x - 15 - 4x$
    $6 + 1 - 4 = 3$ (x terms)
    $-7 - 15 = -22$ (constants)
    $3x - 22$
26. $-y - 2 + y + 2 + 3y - 1$
    $-1 + 1 + 3 = 3$ (y terms)
    $-2 + 2 - 1 = -1$ (constants)
    $3y - 1$

507

27. $-4x^2 - 7 + 9x^2 + 18$
    $-4 + 9 = 5$ (x^2 terms)
    $-7 + 18 = 11$
    $5x^2 + 11$

28. $18xy^2 - 10 - 12xy^2 + 3xy^2 - 4$
    $18 - 12 + 3 = 9$ (xy^2 terms)
    $-10 - 4 = -14$ (constants)
    $9xy^2 - 14$

29. $9y + 3x + 8 - 7y + 12 + 15y$
    $9 - 7 + 15 = 17$ (y terms)
    $8 + 12 = 20$ (constants)
    $17y + 3x + 20$

30. $4a + 35 - 9z - 19 + 21a + 6z + 6a - 3z$
    $4 + 21 + 6 = 31$ (a terms)
    $35 - 19 = 16$ (constants)
    $-9 + 6 - 3 = -6$ (z terms)
    $31a - 6z + 16$

Section 8.2   Comprehension Checkpoint

1. 22   2. 3   3. -2 4. -18

8.2A   Practice Set

1. $y - 6 = 8$
   $y - 6 + (+6) = 8 + (+6)$
   $y = 14$

   Check: $y - 6 = 8$
   $14 - 6 = 8$
   $8 = 8$

2. $12 + m = 9$
   $12 + (-12) + m = 9 + (-12)$
   $m = -3$

   Check: $12 + m = 9$
   $12 + (-3) = 9$
   $9 = 9$

3. $t + (7) = -22$
   $t + 7 + (-7) = -22 + (-7)$
   $t = -29$

   Check: $t + 7 = -22$
   $-29 + 7 = -22$
   $-22 = -22$

4. $x - 15 = 6$
   $x - 15 + (+15) = 6 + (+15)$
   $x = 21$

   Check: $x - 15 = 6$
   $21 - 15 = 6$
   $6 = 6$

5. $a + 9 = -3$
   $a + 9 + (-9) = -3 + (-9)$
   $a = -12$

   Check: $a + 9 = -3$
   $-12 + 9 = -3$
   $-3 = -3$

6. $c - 18 = 2$
   $c - 18 + (+18) = 2 + (+18)$
   $c = 20$

   Check: $c - 18 = 2$
   $20 - 18 = 2$
   $2 = 2$

7. $-8 + r = 17$
   $-8 + (+8) + r = 17 + (+8)$
   $r = 25$

   Check: $-8 + r = 17$
   $-8 + 25 = 17$
   $17 = 17$

8. $16 + t = -15$
   $16 + (-16) + t = -15 + (-16)$
   $t = -31$

   Check: $16 + t = -15$
   $16 + (-31) = -15$
   $-15 = -15$

9. $x + 6 = 6$
   $x + 6 + (-6) = 6 + (-6)$
   $x = 0$

   Check: $x + 6 = 6$
   $0 + 6 = 6$
   $6 = 6$

10. $-9 + y = -20$
    $-9 + (+9) + y = -20 + (+9)$
    $y = -11$

    Check: $-9 + y = -20$
    $-9 + (-11) = -20$
    $-20 = -20$

Solutions

11.
$$5 = y + 3$$
$$5 + (-3) = y + 3 + (-3)$$
$$2 = y$$

Check:
$$5 = y + 3$$
$$5 = 2 + 3$$
$$5 = 5$$

12.
$$-14 = x - 7$$
$$-14 + (+7) = x - 7 + (+7)$$
$$-7 = x$$

Check:
$$-14 = x - 7$$
$$-14 = -7 - 7$$
$$-14 = -14$$

13.
$$20 = a + 13$$
$$20 + (-13) = a + 13 + (-13)$$
$$7 = a$$

Check:
$$20 = a + 13$$
$$20 = 7 + 13$$
$$20 = 20$$

14.
$$x - 35 = 70$$
$$x - 35 + (+35) = 70 + (+35)$$
$$x = 105$$

Check:
$$x - 35 = 70$$
$$105 - 35 = 70$$
$$70 = 70$$

15.
$$y + 42 = -30$$
$$y + 42 + (-42) = -30 + (-42)$$
$$y = -72$$

Check:
$$y + 42 = -30$$
$$-72 + 42 = -30$$
$$-30 = -30$$

16.
$$-28 = b - 16$$
$$-28 + (+16) = b - 16 + (+16)$$
$$-12 = b$$

Check:
$$-28 = b - 16$$
$$-28 = -12 - 16$$
$$-28 = -28$$

17.
$$-32 = -18 + x$$
$$-32 + (+18) = -18 + (+18) + x$$
$$-14 = x$$

Check:
$$-32 = -18 + x$$
$$-32 = -18 + (-14)$$
$$-32 = -32$$

18.
$$48 = 15 + x$$
$$48 + (-15) = 15 + (-15) + x$$
$$33 = x$$

Check:
$$48 = 15 + x$$
$$48 = 15 + 33$$
$$48 = 48$$

19.
$$40 = y - 25$$
$$40 + (+25) = y - 25 + (+25)$$
$$65 = y$$

Check:
$$40 = y - 25$$
$$40 = 65 - 25$$
$$40 = 40$$

20.
$$t - 40 = 100$$
$$t - 40 + (+40) = 100 + (+40)$$
$$t = 140$$

Check:
$$t - 40 = 100$$
$$140 - 40 = 100$$
$$100 = 100$$

Section 8.3 Comprehension Checkpoint

1. 4 **2.** 3 **3.** -4 **4.** $\frac{14}{5}$ or $2\frac{4}{5}$

8.3A Practice Set

1.
$$3x = 15$$
$$\frac{3x}{3} = \frac{15}{3}$$
$$x = 5$$

Check:
$$3x = 15$$
$$(3)(5) = 15$$
$$15 = 15$$

2.
$$-4r = 28$$
$$\frac{-4r}{-4} = \frac{28}{-4}$$
$$r = -7$$

Check:
$$-4r = 28$$
$$(-4)(-7) = 28$$
$$28 = 28$$

3. $6t = 36$
$$\frac{6t}{6} = \frac{36}{6}$$
$t = 6$

Check: $6t = 36$
$(6)(6) = 36$
$36 = 36$

4. $7y = -21$
$$\frac{7y}{7} = \frac{-21}{7}$$
$y = -3$

Check: $7y = -21$
$7(-3) = -21$
$-21 = -21$

5. $-8t = -32$
$$\frac{-8t}{-8} = \frac{-32}{-8}$$
$t = 4$

Check: $-8t = -32$
$(-8)(4) = -32$
$-32 = -32$

6. $5z = 40$
$$\frac{5z}{5} = \frac{40}{5}$$
$z = 8$

Check: $5z = 40$
$(5)(8) = 40$
$40 = 40$

7. $-9q = 45$
$$\frac{-9q}{-9} = \frac{45}{-9}$$
$q = -5$

Check: $-9q = 45$
$(-9)(-5) = 45$
$45 = 45$

8. $-70 = -14a$
$$\frac{-70}{-14} = \frac{-14a}{-14}$$
$5 = a$

Check: $-70 = -14a$
$-70 = (-14)(5)$
$-70 = -70$

9. $-27 = 3b$
$$\frac{-27}{3} = \frac{3b}{3}$$
$-9 = b$

Check: $-27 = 3b$
$-27 = (3)(-9)$
$-27 = -27$

10. $48 = 12x$
$$\frac{48}{12} = \frac{12x}{12}$$
$4 = x$

Check: $48 = 12x$
$48 = (12)(4)$
$48 = 48$

11. $15b = 105$
$$\frac{15b}{15} = \frac{105}{15}$$
$b = 7$

Check: $15b = 105$
$(15)(7) = 105$
$105 = 105$

12. $-48 = -16r$
$$\frac{-48}{-16} = \frac{-16r}{-16}$$
$3 = r$

Check: $-48 = -16r$
$-48 = (-16)(3)$
$-48 = -48$

13. $24 = -3c$
$$\frac{24}{-3} = \frac{-3c}{-3}$$
$-8 = c$

Check: $24 = -3c$
$24 = (-3)(-8)$
$24 = 24$

14. $-84 = -28n$
$$\frac{-84}{-28} = \frac{-28n}{-28}$$
$3 = n$

Check: $-84 = -28n$
$-84 = (-28)(3)$
$-84 = -84$

15. $54 = -13.5p$ Check: $54 = -13.5p$
$\dfrac{54}{-13.5} = \dfrac{-13.5p}{-13.5}$ $54 = (-13.5)(-4)$
$-4 = p$ $54 = 54$

16. $-63 = -12.6m$ Check: $-63 = -12.6m$
$\dfrac{-63}{-12.6} = \dfrac{-12.6m}{-12.6}$ $-63 = (-12.6)(5)$
$5 = m$ $-63 = -63$

17. $8.3t = 49.8$ Check: $8.3t = 49.8$
$\dfrac{8.3t}{8.3} = \dfrac{49.8}{8.3}$ $(8.3)(6) = 49.8$
$t = 6$ $49.8 = 49.8$

18. $6y = 27$ Check: $6y = 27$
$\dfrac{6y}{6} = \dfrac{27}{6}$ $(6)\left(4\dfrac{1}{2}\right) = 27$
$y = 4\dfrac{1}{2}$ $27 = 27$

19. $8x = -35$ Check: $8x = -35$
$\dfrac{8x}{8} = \dfrac{-35}{8}$ $(8)\left(-4\dfrac{3}{8}\right) = -35$
$x = -4\dfrac{3}{8}$ $-35 = -35$

20. $43 = -10z$ Check: $43 = -10z$
$\dfrac{43}{-10} = \dfrac{-10z}{-10}$ $43 = (-10)\left(-4\dfrac{3}{10}\right)$
$-4\dfrac{3}{10} = z$ $43 = 43$

Section 8.4 Comprehension Checkpoint

1. a. $\dfrac{5}{7}$ **b.** $\dfrac{1}{10}$ **c.** $-\dfrac{3}{5}$ **2.** 12   **3.** 21   **4.** 20

8.4A Practice Set

1. a. $\dfrac{1}{6}$ **b.** $\dfrac{3}{4}$ **c.** $-\dfrac{2}{3}$ **2. a.** 5   **b.** $\dfrac{7}{3}$ **c.** $-\dfrac{5}{4}$

3. $\dfrac{x}{5} = 4$ Check: $\dfrac{x}{5} = 4$
$\dfrac{\overset{1}{\cancel{5}}}{1} \cdot \dfrac{x}{\cancel{5}} = \dfrac{4}{1} \cdot \dfrac{5}{1}$ $\dfrac{20}{5} = 4$
$x = 20$ $4 = 4$

4. $\dfrac{2x}{3} = 12$ Check: $\dfrac{2x}{3} = 12$
$\dfrac{\overset{1}{\cancel{3}}}{\underset{1}{\cancel{2}}} \cdot \dfrac{\overset{1}{\cancel{2}}x}{\cancel{3}} = \dfrac{\overset{6}{\cancel{12}}}{1} \cdot \dfrac{3}{\cancel{2}}$ $\dfrac{(2)(18)}{3} = 12$
$x = 18$ $12 = 12$

5. $\dfrac{x}{9} = -16$ *Check:* $\dfrac{x}{9} = -16$

$\dfrac{\cancel{9}^1}{1} \cdot \dfrac{x}{\cancel{9}_1} = \dfrac{-16}{1} \cdot \dfrac{9}{1}$ $\dfrac{-144}{9} = -16$

$x = -144$ $-16 = -16$

6. $\dfrac{3y}{4} = 6$ *Check:* $\dfrac{3y}{4} = 6$

$\dfrac{\cancel{4}^1}{\cancel{3}_1} \cdot \dfrac{\cancel{3}^1 y}{\cancel{4}_1} = \dfrac{\cancel{6}^2}{1} \cdot \dfrac{4}{\cancel{3}_1}$ $\dfrac{(3)(8)}{4} = 6$

$y = 8$ $6 = 6$

7. $\dfrac{5a}{8} = 30$ *Check:* $\dfrac{5a}{8} = 30$

$\dfrac{\cancel{8}^1}{\cancel{5}_1} \cdot \dfrac{\cancel{5}^1 a}{\cancel{8}_1} = \dfrac{\cancel{30}^6}{1} \cdot \dfrac{8}{\cancel{5}_1}$ $\dfrac{5(48)}{8} = 30$

$a = 48$ $30 = 30$

8. $\dfrac{2d}{5} = 8$ *Check:* $\dfrac{2d}{5} = 8$

$\dfrac{\cancel{5}^1}{\cancel{2}_1} \cdot \dfrac{\cancel{2}^1 d}{\cancel{5}_1} = \dfrac{\cancel{8}^4}{1} \cdot \dfrac{5}{\cancel{2}_1}$ $\dfrac{2(20)}{5} = 8$

$d = 20$ $8 = 8$

9. $\dfrac{-3x}{7} = 9$ *Check:* $\dfrac{-3x}{7} = 9$

$\dfrac{\cancel{7}^1}{\cancel{-3}_1} \cdot \dfrac{\cancel{-3}^1 x}{\cancel{7}_1} = \dfrac{\cancel{9}^3}{1} \cdot \dfrac{7}{\cancel{-3}_{-1}}$ $\dfrac{(-3)(-21)}{7} = 9$

$x = -21$ $9 = 9$

10. $\dfrac{t}{4} = -20$ *Check:* $\dfrac{t}{4} = -20$

$\dfrac{\cancel{4}^1}{1} \cdot \dfrac{t}{\cancel{4}_1} = \dfrac{-20}{1} \cdot \dfrac{4}{1}$ $\dfrac{-80}{4} = -20$

$t = -80$ $-20 = -20$

11. $\dfrac{7a}{15} = -21$ *Check:* $\dfrac{7a}{15} = -21$

$\dfrac{\cancel{15}^1}{\cancel{7}_1} \cdot \dfrac{\cancel{7}^1 a}{\cancel{15}_1} = \dfrac{\cancel{-21}^3}{1} \cdot \dfrac{15}{\cancel{7}_1}$ $\dfrac{7(-45)}{15} = -21$

$a = -45$ $-21 = -21$

12. $\dfrac{4x}{9} = 12$ Check: $\dfrac{4x}{9} = 12$

$\dfrac{9}{4} \cdot \dfrac{4x}{9} = \dfrac{12}{1} \cdot \dfrac{9}{4}$

$x = 27$

$\dfrac{4(27)}{9} = 12$

$12 = 12$

13. $-\dfrac{x}{6} = -7$ Check: $-\dfrac{x}{6} = -7$

$\dfrac{-6}{1} \cdot \dfrac{x}{-6} = \dfrac{-7}{1} \cdot \dfrac{-6}{1}$

$x = 42$

$-\dfrac{42}{6} = -7$

$-7 = -7$

14. $\dfrac{m}{8} = 0$ Check: $\dfrac{m}{8} = 0$

$\dfrac{8}{1} \cdot \dfrac{m}{8} = \dfrac{0}{1} \cdot \dfrac{8}{1}$

$m = 0$

$\dfrac{0}{8} = 0$

$0 = 0$

15. $\dfrac{5y}{12} = 60$ Check: $\dfrac{5y}{12} = 60$

$\dfrac{12}{5} \cdot \dfrac{5y}{12} = \dfrac{60}{1} \cdot \dfrac{12}{5}$

$y = 144$

$\dfrac{(5)(144)}{12} = 60$

$60 = 60$

16. $\dfrac{-3n}{10} = -6$ Check: $\dfrac{-3n}{10} = -6$

$\dfrac{10}{-3} \cdot \dfrac{-3n}{10} = \dfrac{-6}{1} \cdot \dfrac{10}{-3}$

$n = 20$

$\dfrac{-3(20)}{10} = -6$

$-6 = -6$

17. $\dfrac{6x}{11} = 30$ Check: $\dfrac{6x}{11} = 30$

$\dfrac{11}{6} \cdot \dfrac{6x}{11} = \dfrac{30}{1} \cdot \dfrac{11}{6}$

$x = 55$

$\dfrac{(6)(55)}{11} = 30$

$30 = 30$

18. $\dfrac{7n}{-10} = 35$ Check: $\dfrac{7n}{-10} = 35$

$\dfrac{-10}{7} \cdot \dfrac{7n}{-10} = \dfrac{35}{1} \cdot \dfrac{-10}{7}$

$n = -50$

$\dfrac{(7)(-50)}{-10} = 35$

$35 = 35$

19. $\dfrac{11x}{24} = 33$ Check: $\dfrac{11x}{24} = 33$

$\dfrac{\cancel{24}}{\cancel{11}} \cdot \dfrac{\cancel{11}x}{\cancel{24}} = \dfrac{\cancel{33}}{1} \cdot \dfrac{24}{\cancel{11}}$ $\dfrac{(11)(72)}{24} = 33$

$x = 72$ $33 = 33$

20. $\dfrac{3}{4}x = 48$ Check: $\dfrac{3}{4}x = 48$

$\dfrac{\cancel{4}}{\cancel{3}} \cdot \dfrac{\cancel{3}}{\cancel{4}}x = \dfrac{\cancel{48}}{1} \cdot \dfrac{4}{\cancel{3}}$ $\dfrac{(3)(64)}{4} = 48$

$x = 64$ $48 = 48$

Section 8.5 Comprehension Checkpoint

1. $-a - 8$ 2. $b - 6$ 3. $-8x + 56$ 4. $10y + 120$ 5. $13y - 66$

8.5A Practice Set

1. $+(a - 3) =$
$(+1)(a) + (+1)(-3) =$
$a - 3$

2. $+(b + 7) =$
$(+1)(b) + (+1)(+7) =$
$b + 7$

3. $-(x + 9) =$
$(-1)(x) + (-1)(+9) =$
$-x - 9$

4. $-(-y + 8) =$
$(-1)(-y) + (-1)(+8) =$
$y - 8$

5. $-2(a - 5) =$
$(-2)(a) + (-2)(-5) =$
$-2a + 10$

6. $6(m - 4) =$
$(+6)(m) + (+6)(-4) =$
$6m - 24$

7. $+7(-n - 8) =$
$(+7)(-n) + (+7)(-8) =$
$-7n - 56$

8. $5(q - 10) =$
$(+5)(q) + (+5)(-10) =$
$5q - 50$

9. $-8(r - 3) =$
$(-8)(r) + (-8)(-3) =$
$-8r + 24$

10. $-\dfrac{2}{5}(5 + a) =$
$\left(-\dfrac{2}{5}\right)(+5) + \left(-\dfrac{2}{5}\right)(+a) =$
$-\dfrac{2a}{5} - 2$

11. $\dfrac{3}{7}(a + b) =$
$\left(\dfrac{3}{7}\right)(+a) + \left(\dfrac{3}{7}\right)(+b) =$
$\dfrac{3a}{7} + \dfrac{3b}{7}$

12. $\dfrac{7}{12}(-x - y) =$
$\left(\dfrac{7}{12}\right)(-x) + \left(\dfrac{7}{12}\right)(-y) =$
$-\dfrac{7x}{12} - \dfrac{7y}{12}$

13. $-5(4m + n) =$
$(-5)(4m) + (-5)(+n) =$
$-20m - 5n$

14. $-15(3x - 4) =$
$(-15)(3x) + (-15)(-4) =$
$-45x + 60$

Solutions

15. $-18(2d + 7) =$
$(-18)(2d) + (-18)(+7) =$
$-36d - 126$

16. $2(a + b + c) =$
$(2)(a) + (2)(b) + (2)(c) =$
$2a + 2b + 2c$

17. $-3(x - y + z) =$
$(-3)(x) + (-3)(-y) + (-3)(z) =$
$-3x + 3y - 3z$

18. $12(a + 13) + 2(a - 6) =$
$12a + 156 + 2a - 12 =$
$14a + 144$

19. $-9(3 - x) + 4(x + 2) =$
$-27 + 9x + 4x + 8 =$
$13x - 19$

20. $3(8 + x) - 5(x - 8) =$
$24 + 3x - 5x + 40 =$
$-2x + 64$

Section 8.6 Comprehension Checkpoint

1. 9   **2.** 21   **3.** -6 **4.** 35

8.6A Practice Set

1.
$$3x + 8 = 23$$
$$3x + 8 + (-8) = 23 + (-8)$$
$$3x = 15$$
$$\frac{3x}{3} = \frac{15}{3}$$
$$x = 5$$

Check:
$$3x + 8 = 23$$
$$(3)(5) + 8 = 23$$
$$15 + 8 = 23$$
$$23 = 23$$

2.
$$4y - 9 = 19$$
$$4y - 9 + (+9) = 19 + (+9)$$
$$4y = 28$$
$$\frac{4y}{4} = \frac{28}{4}$$
$$y = 7$$

Check:
$$4y - 9 = 19$$
$$(4)(7) - 9 = 19$$
$$28 - 9 = 19$$
$$19 = 19$$

3.
$$-7a + 6 = 48$$
$$-7a + 6 + (-6) = 48 + (-6)$$
$$-7a = 42$$
$$\frac{-7a}{-7} = \frac{42}{-7}$$
$$a = -6$$

Check:
$$-7a + 6 = 48$$
$$(-7)(-6) + 6 = 48$$
$$42 + 6 = 48$$
$$48 = 48$$

4.
$$-12 - 5m = 23$$
$$-12 + (+12) - 5m = 23 + (+12)$$
$$-5m = 35$$
$$\frac{-5m}{-5} = \frac{35}{-5}$$
$$m = -7$$

Check:
$$-12 - 5m = 23$$
$$-12 - 5(-7) = 23$$
$$-12 + 35 = 23$$
$$23 = 23$$

5.
$$\frac{b}{4} - 3 = 6$$
$$\frac{b}{4} - 3 + (+3) = 6 + (+3)$$
$$\frac{b}{4} = 9$$
$$\frac{\overset{1}{\cancel{4}}}{1} \cdot \frac{b}{\underset{1}{\cancel{4}}} = \frac{9}{1} \cdot \frac{4}{1}$$
$$b = 36$$

Check:
$$\frac{b}{4} - 3 = 6$$
$$\frac{36}{4} - 3 = 6$$
$$9 - 3 = 6$$
$$6 = 6$$

6. $\dfrac{x}{8} + 4 = -2$ Check: $\dfrac{x}{8} + 4 = -2$

$\dfrac{x}{8} + 4 + (-4) = -2 + (-4)$ $\dfrac{-48}{8} + 4 = -2$

$\dfrac{x}{8} = -6$ $-6 + 4 = -2$

$-2 = -2$

$\dfrac{\overset{1}{\cancel{8}}}{1} \cdot \dfrac{x}{\cancel{8}} = \dfrac{-6}{1} \cdot \dfrac{8}{1}$

$x = -48$

7. $\dfrac{5z}{8} - 7 = 28$ Check: $\dfrac{5z}{8} - 7 = 28$

$\dfrac{5z}{8} - 7 + (+7) = 28 + (+7)$ $\dfrac{(5)(56)}{8} - 7 = 28$

$\dfrac{5z}{8} = 35$ $35 - 7 = 28$

$28 = 28$

$\dfrac{\overset{1}{\cancel{8}}}{\underset{1}{\cancel{5}}} \cdot \dfrac{\overset{1}{\cancel{5}} z}{\underset{1}{\cancel{8}}} = \dfrac{\overset{7}{\cancel{35}}}{1} \cdot \dfrac{8}{\cancel{5}}$

$z = 56$

8. $\dfrac{3c}{10} + 15 = 135$ Check: $\dfrac{3c}{10} + 15 = 135$

$\dfrac{3c}{10} + 15 + (-15) = 135 + (-15)$ $\dfrac{(3)(400)}{10} + 15 = 135$

$\dfrac{3c}{10} = 120$ $120 + 15 = 135$

$135 = 135$

$\dfrac{\overset{1}{\cancel{10}}}{\underset{1}{\cancel{3}}} \cdot \dfrac{\overset{1}{\cancel{3}} c}{\underset{1}{\cancel{10}}} = \dfrac{\overset{40}{\cancel{120}}}{1} \cdot \dfrac{10}{\cancel{3}}$

$c = 400$

9. $-(x + 8) = 16$ Check: $-(x + 8) = 16$
   $-x - 8 = 16$ $-(-24 + 8) = 16$
$-x - 8 + (+8) = 16 + (+8)$ $-(-16) = 16$
   $-x = 24$ $16 = 16$
   $\dfrac{-x}{-1} = \dfrac{24}{-1}$
   $x = -24$

10. $(9 + a) = 12$ Check: $(9 + a) = 12$
    $9 + a = 12$ $(9 + 3) = 12$
$9 + a + (-9) = 12 + (-9)$ $12 = 12$
    $a = 3$

11. $3(a - 4) = 15$ Check: $3(a - 4) = 15$
    $3a - 12 = 15$ $3(9 - 4) = 15$
$3a - 12 + 12 = 15 + 12$ $3(5) = 15$
    $3a = 27$ $15 = 15$
    $\dfrac{3a}{3} = \dfrac{27}{3}$
    $a = 9$

Solutions

12.
$$-5(x - 12) = 10$$
$$-5x + 60 = 10$$
$$-5x + 60 + (-60) = 10 + (-60)$$
$$-5x = -50$$
$$\frac{-5x}{-5} = \frac{-50}{-5}$$
$$x = 10$$

Check:
$$-5(x - 12) = 10$$
$$-5(10 - 12) = 10$$
$$-5(-2) = 10$$
$$10 = 10$$

13.
$$-(d + 12) - 3 = 18$$
$$-d - 12 - 3 = 18$$
$$-d - 15 = 18$$
$$-d - 15 + (+15) = 18 + (+15)$$
$$-d = 33$$
$$\frac{-d}{-1} = \frac{33}{-1}$$
$$d = -33$$

Check:
$$-(d + 12) - 3 = 18$$
$$-(-33 + 12) - 3 = 18$$
$$-(-21) - 3 = 18$$
$$21 - 3 = 18$$
$$18 = 18$$

14.
$$(c + 6) + 5 = -11$$
$$c + 6 + 5 = -11$$
$$c + 11 = -11$$
$$c + 11 + (-11) = -11 + (-11)$$
$$c = -22$$

Check:
$$(c + 6) + 5 = -11$$
$$(-22 + 6) + 5 = -11$$
$$-16 + 5 = -11$$
$$-11 = -11$$

15.
$$8(a - 10) - 3 = 21$$
$$8a - 80 - 3 = 21$$
$$8a - 83 + (+83) = 21 + (+83)$$
$$8a = 104$$
$$\frac{8a}{8} = \frac{104}{8}$$
$$a = 13$$

Check:
$$8(a - 10) - 3 = 21$$
$$8(13 - 10) - 3 = 21$$
$$8(3) - 3 = 21$$
$$24 - 3 = 21$$
$$21 = 21$$

16.
$$-12(x + 1) - 18 = 42$$
$$-12x - 12 - 18 = 42$$
$$-12x - 30 + (+30) = 42 + (+30)$$
$$-12x = 72$$
$$\frac{-12x}{-12} = \frac{72}{-12}$$
$$x = -6$$

Check:
$$-12(x + 1) - 18 = 42$$
$$-12(-6 + 1) - 18 = 42$$
$$-12(-5) - 18 = 42$$
$$60 - 18 = 42$$
$$42 = 42$$

17.
$$4y + 36 = 2y - 8$$
$$4y + 36 + (-36) + (-2y) = 2y - 8 + (-2y) + (-36)$$
$$2y = -44$$
$$\frac{2y}{2} = \frac{-44}{2}$$
$$y = -22$$

Check:
$$4y + 36 = 2y - 8$$
$$4(-22) + 36 = 2(-22) - 8$$
$$-88 + 36 = -44 - 8$$
$$-52 = -52$$

18.
$$5a - 20 = 4a + 3$$
$$5a - 20 + (+20) + (-4a) = 4a + 3 + (+20) + (-4a)$$
$$a = 23$$

Check:
$$5a - 20 = 4a + 3$$
$$5(23) - 20 = 4(23) + 3$$
$$115 - 20 = 92 + 3$$
$$95 = 95$$

19.
$$6(x - 8) = 4(x - 7)$$
$$6x - 48 = 4x - 28$$
$$6x - 48 + (+48) + (-4x) = 4x - 28 + (-4x) + (+48)$$
$$2x = 20$$
$$\frac{2x}{2} = \frac{20}{2}$$
$$x = 10$$

Check:
$$6(x - 8) = 4(x - 7)$$
$$6(10 - 8) = 4(10 - 7)$$
$$6(2) = 4(3)$$
$$12 = 12$$

20.
$$7(a - 3) = 5(a - 9)$$
$$7a - 21 = 5a - 45$$
$$7a - 21 + (+21) + (-5a) = 5a - 45 + (+21) + (-5a)$$
$$2a = -24$$
$$\frac{2a}{2} = \frac{-24}{2}$$
$$a = -12$$

Check:
$$7(a - 3) = 5(a - 9)$$
$$7(-12 - 3) = 5(-12 - 9)$$
$$7(-15) = 5(-21)$$
$$-105 = -105$$

Practice Test 8A

1. 5
2. *a.* 4 *b.* −5 *c.* $\frac{5}{9}$
3. $-5a - 9$
4. 20
5. 7
6. 7
7. −14
8. 162
9. 96
10. $3q - 18$
11. 2
12. −84

Skills Review

1. seven hundred thirty-six thousand, three hundred ninety-four
2. 3,284,012
3. 893,000
4. 800,000
5. $2\frac{11}{30}$
6. $\frac{7}{48}$
7. $\frac{2}{5}$
8. $1\frac{1}{3}$
9. $33\frac{3}{5}$
10. $1\frac{8}{13}$
11. 72.3426
12. 250
13. 0.3125
14. 62.5%
15. 0.83
16. $\frac{7}{25}$
17. 119
18. 400
19. 45
20. seven thousand six hundred thirty-two ten-thousandths
21. twenty-eight and seventy-four hundredths
22. no
23. 5
24. *a.* < *b.* < *c.* >
25. 15.8
26. 14
27. −9
28. −1,620
29. 5
30. 16
31. 6
32. 24

9 MEASUREMENT

OUTLINE

9.1 Length Measurement

9.2 Weight Measurement

9.3 Liquid Measurement

9.4 Time and Temperature

9.5 Working with Denominate Numbers

The ability to measure things is a fundamental skill that you will use throughout your life. Whether you want to measure a room for carpeting (length), to weigh a package for shipping (weight), to follow a recipe (liquid), or to convert temperature or time measures, you need good measurement skills.

There are currently two major systems of measurement in use: the United States Customary System (USCS) and the metric system. In this chapter you will learn the basic units of measurement for length, weight, liquid capacity, time, and temperature. You will also learn how to convert from one system to the other.

SKILLS PREVIEW

In problems 1–12, convert each given unit to the other indicated unit of measurement.

1. 8 min = _____ s
2. 400 cm = _____ in
3. 77°F = _____ C
4. 64 fl oz = _____ gal
5. 4,250 lb = _____ tons
6. 650 mm = _____ m
7. 12 pt = _____ L
8. 24 mi = _____ km
9. 50°C = _____ F
10. 49 g = _____ cg
11. 6.5 qt = _____ fl oz
12. 275 L = _____ kL

In problems 13–16, perform the indicated operation. Simplify your answer if possible.

13. 6 ft 8 in
   + 12 ft 6 in

14. 10 h 30 min
   − 2 h 40 min

15. 3 h 50 min 12 s
   ×   5

16. 36 lb 8 oz ÷ 8 = _____

17. The greatest ocean depth is approximately 36,000 ft. How many meters deep is it? _____

18. A packed box weighs 25 kilograms. What is its weight in pounds? _____

9.1 LENGTH MEASUREMENT

United States Customary System (USCS)

There are two measurement systems currently in use in industrialized countries around the world. The USCS is the system more commonly used in the United States today. It is also called the *English system,* because colonists from England brought it to the New World.

There are basically four units of length in the USCS: **inch, foot, yard,** and **mile.** The following chart shows the relationships between these units.

Length Measurement

USCS UNITS OF LENGTH

12 inches (in) = 1 foot (ft)

3 feet = 1 yard (yd)

5,280 feet = 1 mile (mi)

1,760 yards = 1 mile

Converting from one unit to another can be accomplished by setting up a proportion and solving for the unknown.

EXAMPLE 1 How many feet are in 3 miles?

SOLUTION

$\dfrac{5{,}280 \text{ ft}}{1 \text{ mi}} = \dfrac{x}{3 \text{ mi}}$ *The variable in the proportion represents the number you are trying to find. Since you are trying to determine the number of feet in 3 miles, use the ratio of feet to miles.*

$(1)(x) = (5{,}280)(3)$ *Cross-multiply.*
$x = 15{,}840$

There are 15,840 ft in 3 miles.

Any conversion from one unit to another in the USCS can be made using the procedure shown in Example 1. Care must be taken to set the proportion up correctly. The units expressed in each ratio must be in the same order, as in the proportions

$$\dfrac{\text{ft}}{\text{mi}} = \dfrac{\text{ft}}{\text{mi}} \quad \text{or} \quad \dfrac{\text{yd}}{\text{ft}} = \dfrac{\text{yd}}{\text{ft}} \quad \text{or} \quad \dfrac{\text{in}}{\text{ft}} = \dfrac{\text{in}}{\text{ft}}$$

As you study Example 2, pay particular attention to the way the ratios in the proportion are set up.

EXAMPLE 2 Howard Miller is 6 ft tall. How many inches tall is he?

SOLUTION

$\dfrac{1 \text{ ft}}{12 \text{ in}} = \dfrac{6 \text{ ft}}{x}$ *Use a ratio for the number of inches in a foot. The variable x represents the number of inches in 6 ft. Notice the relationship between the measures in the proportion $\dfrac{ft}{in} = \dfrac{ft}{in}$.*

$(12)(6) = (1)(x)$ *Cross-multiply.*
$72 = x$

He is 72 in tall.

Notice that after you set up the proportion, you do not need to show the units when multiplying. Only the numbers are multiplied. You do, however, need to show the units when giving the answer. A number by itself is only a number: 6 is 6, 15 is 15, 194 is 194. It is necessary to put a unit with the result to answer the question properly. In Example 2, Howard is not 72 tall; he is 72 inches tall. A number with some sort of unit following it is called a **denominate number.**

Metric System

All major industrialized countries around the world, except the United States, use the metric system. The metric system is easier to use than the USCS because it is uniform. This will be more readily obvious when you study the next two sections. The metric system is based on units of 10. The standard unit of length in the metric system is the **meter.**

The following chart shows the relationships between length measurements in the metric system.

METRIC SYSTEM LENGTH MEASUREMENTS

1 kilometer = 1,000 meters

1 hectometer = 100 meters

1 dekameter = 10 meters

meter = standard unit

1 decimeter = 0.1 meter

1 centimeter = 0.01 meter

1 millimeter = 0.001 meter

Writing the length units as shown in the following diagram should make converting from one unit to another somewhat easier. Because the metric system is based on units of 10, conversions can be made simply by moving the decimal point.

			standard unit			
kilo-	hecto-	deka-	meter	deci-	centi-	milli-
(km)	(hm)	(dam)	(m)	(dm)	(cm)	(mm)

The units decrease in size as you move from left to right on the chart. (This is the opposite of the number line.) If you are converting from a larger to a smaller unit, move the decimal point the number of places you move from the larger to the smaller value on the chart. For example, if you want to convert 125 hectometers to decimeters, you are moving three places to the right on the chart:

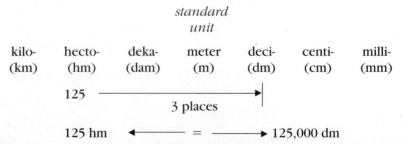

Move the decimal point three places to the right: 125 hectometers is 125,000 decimeters.

Length Measurement

EXAMPLE 3 How many centimeters are in 5 dekameters?

SOLUTION

5 dam = x cm   *You are converting from a larger to a smaller unit.*

5 dam = 5,000 cm   *Move the decimal point three places to the right.*

If you want to convert a smaller unit to a larger unit, move the decimal point to the left the same number of places you move from the smaller unit to the larger unit on the chart. For example, if you convert 318 mm to dam, proceed as follows:

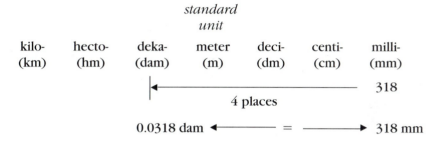

EXAMPLE 4 How many kilometers are in 427 cm?

SOLUTION

427 cm = x km   *You are converting from a smaller to a larger unit.*

427 cm = 0.00427 km   *Move the decimal point five places to the left.*

Converting between USCS and Metric System

The United States is gradually changing to the metric system. Many products are already labeled in both systems of measure (canned goods, soft drinks, sandwich bags, etc.). However, some companies don't use the metric system yet. Thus, until the United States goes to the metric system exclusively, you need to know how to convert between the systems. The following chart shows conversions of length measurements between the two systems. It should be noted that these are approximate equivalent values. The symbol ≈ means "is approximately."

USCS/METRIC LENGTH CONVERSION CHART

1 inch ≈ 2.54 centimeters	1 centimeter ≈ 0.394 inch
1 foot ≈ 0.305 meter	1 meter ≈ 39.37 inches
1 yard ≈ 0.914 meter	1 meter ≈ 3.28 feet
1 mile ≈ 1.61 kilometers	1 meter ≈ 1.09 yards
	1 kilometer ≈ 0.621 mile

Study the following examples to learn how to convert length measurements between the two systems.

EXAMPLE 5 How many miles is 12 km?

SOLUTION

$\dfrac{1 \text{ km}}{0.621 \text{ mi}} = \dfrac{12 \text{ km}}{x \text{ mi}}$ *These problems can be solved using proportions. Since you are converting kilometers to miles, use a ratio of kilometers to miles.*

$(0.621)(12) = x$ *Cross-multiply.*
$7.452 = x$

12 km is approximately 7.452 mi.

EXAMPLE 6 A room is 5 m wide. How many feet wide is it?

SOLUTION

$\dfrac{1 \text{ m}}{3.28 \text{ ft}} = \dfrac{5 \text{ m}}{x \text{ ft}}$ *You want to convert from meters to feet. Use a ratio of meters to feet.*

$(3.28)(5) = x$ *Cross-multiply.*
$16.4 = x$

The room is approximately 16.4 ft wide.

EXAMPLE 7 The distance from Dallas to New Orleans is 525 miles. How many kilometers apart are the cities?

SOLUTION

$\dfrac{1 \text{ mi}}{1.61 \text{ km}} = \dfrac{525 \text{ mi}}{x \text{ km}}$ *You are converting from miles to kilometers. Use a ratio of miles to kilometers.*

$(1.61)(525) = x$ *Cross-multiply.*
$845.25 = x$

The distance between Dallas and New Orleans is about 845.25 km.

EXAMPLE 8 A piece of notebook paper is 11 in long. Express its length in millimeters.

SOLUTION

$\dfrac{1 \text{ in}}{2.54 \text{ cm}} = \dfrac{11 \text{ in}}{x \text{ cm}}$ *You must first convert to centimeters. Then you can convert the centimeters to millimeters. Use a ratio of inches to centimeters.*

$(2.54)(11) = x$ *Cross-multiply.*
$27.94 = x$

27.94 cm = x mm *Convert centimeters to millimeters.*

27.94 cm = 279.4 mm *Move the decimal point one place to the right.*

A sheet of notebook paper is approximately 279.4 mm long.

Example 8 shows that some problems may require more than one conversion before the answer is determined.

Length Measurement

Work the Comprehension Checkpoint problems to make sure you understand length measurement before attempting to work the Practice Sets. If you have difficulty with these problems, review the appropriate material before going on.

COMPREHENSION CHECKPOINT

Convert the USCS units as indicated.

1. 15 yd = _____ ft

2. 4 mi = _____ yd

3. 18 in = _____ ft

Convert the metric units as indicated.

4. 180 m = _____ hm

5. 42 km = _____ cm

6. 24 dm = _____ dam

Convert between the USCS and the metric system as indicated.

7. 24 in = _____ cm

8. 84 m = _____ yd

9. 5,280 yd = _____ km

9.1A PRACTICE SET

Convert the given USCS unit to the indicated unit.

1. 96 in = _____ ft

2. 42 yd = _____ ft

3. 2,640 ft = _____ mi

4. 5 mi = _____ ft

5. 2 mi = _____ in

6. 84 ft = _____ yd

Convert the given metric unit to the indicated unit.

7. 48 m = _____ dm

8. 225 cm = _____ dm

9. 817 mm = _____ hm

10. 19 dam = _____ cm

11. 240 km = _____ m

12. 500 cm = _____ mm

Convert the given unit to the indicated unit.

13. 32 in = _____ cm

14. 18 mi = _____ km

15. 80 ft = _____ m

16. 2,500 m = _____ mi

17. 400 mm = _____ in

18. 27 m = _____ yd

Applications

19. A newspaper is 12 in wide. How many centimeters wide is it?

   _____

20. A room measures 18 ft long by 15 ft wide. Express the measurements of the room in meters.

   _____

21. A bookcase has five shelves. Each shelf is 46 in long. What is the total shelf space in meters?

   _____

22. A bowling lane is 60 ft long. Express the length in meters.

   _____

23. On a blueprint of a building, 1 cm represents 3 m. What is the length of the front of the building, in feet, if the line representing it on the blueprint is 14 cm long?

   _____

24. A marathon race is 26 mi 343 yards. How many kilometers is a marathon?

   _____

25. A road sign shows the distance between city A and city B as 480 km. You know that you get 24 miles per gallon of gas and that you have exactly 12 gallons of gas in your tank. Do you have enough gas to go from city A to city B?

   _____

9.1B PRACTICE SET

Convert the given USCS unit to the indicated unit.

1. 63 yd = _____ ft

2. 16 ft = _____ in

3. 3 mi = _____ yd

4. 13,200 ft = _____ mi

5. 39 yd = _____ in

6. 112 in = _____ ft

Convert the given metric unit to the indicated unit.

7. 370 m = _____ hm

8. 450 cm = _____ m

9. 85 km = _____ dm

10. 24 hm = _____ dam

11. 375 mm = _____ m

12. 2,500 m = _____ mm

Convert the given unit to the indicated unit.

13. 28 cm = _____ in

14. 80 m = _____ ft

15. 75 yd = _____ m

16. 673 mi = _____ km

17. 5 mi = _____ m

18. 59 km = _____ mi

Applications

19. A mirror is 1.5 yd long. How many inches long is the mirror?

20. How many meters long is a football field that is 120 yd long?

21. How many millimeters are in 1 foot?

22. A triathlete swims 6 miles each day in training. How many yards does he swim?

23. The distance between two cities is 250 miles. If your speedometer is metric and you travel 90 km per hour, how long will it take you to complete the drive between the two cities? Round your answer to the nearest half hour.

24. The main span of the Golden Gate Bridge in San Francisco is 4,200 ft long. Express this length in meters.

25. The scale on a map is 1 cm = 50 km. How far apart are two towns, in miles, if the distance on the map is 4 dm?

9.2 WEIGHT MEASUREMENT

United States Customary System (USCS)

Units of weight in the USCS include the **ounce,** the **pound,** and the **ton.** The relationships between these units are given in the following chart.

> **USCS UNITS OF WEIGHT**
>
> 16 ounces (oz) = 1 pound (lb)
>
> 2,000 pounds = 1 ton

Converting from one unit to another can be done with proportions just as the length conversions were done.

EXAMPLE 1 How many ounces are in 4 pounds?

SOLUTION

$$\frac{16 \text{ oz}}{1 \text{ lb}} = \frac{x \text{ oz}}{4 \text{ lb}}$$ *Write a proportion using ounce-to-pound ratios.*

$x = (4)(16)$ *Cross-multiply.*
$x = 64$

There are 64 ounces in 4 pounds.

EXAMPLE 2 A new car weighs $1\frac{1}{2}$ tons. What is its weight in pounds?

SOLUTION

$$\frac{2{,}000 \text{ lb}}{1 \text{ ton}} = \frac{x \text{ lb}}{1\frac{1}{2} \text{ ton}}$$ *Write a proportion using pound-to-ton ratios.*

$x = (2{,}000)\left(1\frac{1}{2}\right)$ *Cross-multiply.*

$x = 3{,}000$

The car weighs 3,000 pounds.

Metric System

> **METRIC SYSTEM WEIGHT MEASUREMENTS**
>
> 1 kilogram = 1,000 grams
>
> 1 hectogram = 100 grams
>
> 1 dekagram = 10 grams
>
> gram = standard unit
>
> 1 decigram = 0.1 gram
>
> 1 centigram = 0.01 gram
>
> 1 milligram = 0.001 gram

The **gram** is the basic unit of weight measurement in the metric system. You learned in Section 9.1 that the metric system is uniform. You can see from the chart on page 531 that the prefixes are the same as those used for length measurements.

To convert from one metric unit to another, use the same procedure that you used with length. If you are converting from a larger to a smaller unit, move the decimal point to the right. If you are converting from a smaller to a larger unit, move the decimal point to the left. Use the following chart to determine how many places to move the decimal point.

			standard unit			
kilo- (kg)	hecto- (hg)	deka- (dag)	gram (g)	deci- (dg)	centi- (cg)	milli- (mg)

EXAMPLE 3   Convert 245 kilograms to dekagrams.

SOLUTION

245 kg = x dag   *You are converting from a larger to a smaller unit.*

245 kg = 24,500 dag   *Move the decimal point two places to the right since dekagram is two places to the right of kilogram on the chart.*

There are 24,500 dekagrams in 245 kilograms.

EXAMPLE 4   Convert 875 milligrams to grams.

SOLUTION

875 mg = x g   *You are converting from a smaller to a larger unit.*

875 mg = 0.875 g   *Move the decimal point three places to the left since gram is three places to the left of milligram.*

875 milligrams is 0.875 gram.

Examples 3 and 4 show that the metric system is consistent. Converting either length or weight measurements is done exactly the same way.

Converting between USCS and Metric System

The chart that follows shows equivalent weights in the two systems. Use these equivalent values to convert between the two systems. As with length measurements, use proportions to make the conversions. These are approximate conversion values.

USCS/METRIC WEIGHT CONVERSION CHART

1 ounce ≈ 28.35 grams   1 gram ≈ 0.035 ounce

1 pound ≈ 454 grams   1 kilogram ≈ 2.2 pounds

1 pound ≈ 0.454 kilogram

Weight Measurement

EXAMPLE 5 A ball weighs 8 oz. What would its weight be in grams?

SOLUTION

$\dfrac{1 \text{ oz}}{28.35 \text{ g}} = \dfrac{8 \text{ oz}}{x}$ *Write a proportion using ounce-to-gram ratios.*

$x = (8)(28.35)$ *Cross-multiply.*
$x = 226.8$

The ball weighs about 226.8 grams.

EXAMPLE 6 A man weighs 78 kg. What is his weight in pounds?

SOLUTION

$\dfrac{1 \text{ kg}}{2.2 \text{ lb}} = \dfrac{78 \text{ kg}}{x}$ *Write a proportion using kilogram-to-pound ratios.*

$x = (2.2)(78)$ *Cross-multiply.*
$x = 171.6$

The man weighs approximately 171.6 pounds.

Work the Comprehension Checkpoint problems to test your understanding of weight measurement. If you have difficulty with these problems, review the appropriate material before going on to the Practice Sets.

COMPREHENSION CHECKPOINT

Convert the USCS units as indicated.

1. 80 oz = _____ lb

2. 5,500 lb = _____ tons

Convert the metric units as indicated.

3. 375 g = _____ kg

4. 4,273 hg = _____ cg

Convert between the USCS and the metric system as indicated.

5. 42 oz = _____ g

6. 220 kg = _____ lb

9.2A PRACTICE SET

Convert the given USCS unit to the indicated unit.

1. 32 oz = _____ lb

2. 400 lb = _____ ton(s)

3. 25 lb = _____ oz

4. 4.5 tons = _____ lb

5. 1 ton = _____ oz

Convert the given metric unit to the indicated unit.

6. 4 g = _____ cg

7. 15 dg = _____ mg

8. 200 kg = _____ g

9. 500 g = _____ kg

10. 250 mg = _____ g

Convert the given unit to the indicated unit.

11. 12 g = _____ oz

12. 38 kg = _____ lb

13. 60 lb = _____ g

14. 100 oz = _____ kg

15. 2 kg = _____ oz

Applications

16. A can of vegetables weighs 15 ounces. Express this weight in grams. Round your answer to the nearest whole gram.

   _____

17. The maximum load for an 18-wheeler is 45,000 lb. Express the maximum load in tons.

   _____

18. A shipping crate has a maximum weight limit of 40 lb. What is the maximum weight in kilograms?

   _____

19. A weightlifter lifted $\frac{1}{4}$ ton on one try. Express the weight in kilograms.

   _____

20. What is the weight in milligrams of a bag of dry beans if the bag weighs 2 lb?

   _____

9.2B PRACTICE SET

Convert the given USCS unit to the indicated unit.

1. 15 lb = _____ oz

2. 0.75 ton = _____ lb

3. 12 oz = _____ lb

4. 112 oz = _____ lb

5. 10,500 lb = _____ ton(s)

Convert the given metric unit to the indicated unit.

6. 400 mg = _____ cg

7. 50 kg = _____ mg

8. 1,000 hg = _____ g

9. 567 g = _____ cg

10. 985 dg = _____ dag

Convert the given unit to the indicated unit.

11. 40 oz = _____ g

12. 25 lb = _____ kg

13. 58 kg = _____ lb

14. 20 lb = _____ cg

15. 8 kg = _____ oz

Applications

16. Fudge costs $0.37 an ounce. What would be the cost of 2 lb of fudge?

17. How many centigrams would be equivalent to 1 lb?

18. How many ounces of flour would be in a 5-lb bag?

19. The weight limit of a small plane is 1 ton. What is the weight limit in kilograms?

20. A standard-size paper clip weighs about 1 g. What would be the weight, in ounces, of a box of 100 paper clips?

9.3 LIQUID MEASUREMENT

United States Customary System (USCS)

Units of liquid measurement in the USCS include the **teaspoon, tablespoon, fluid ounce, cup, pint, quart,** and **gallon.** The relationships between these units are given in the following chart.

USCS UNITS OF LIQUID MEASUREMENT

3 teaspoons (tsp) = 1 tablespoon (tbsp)

2 tablespoons = 1 fluid ounce (fl oz)

8 fluid ounces = 1 cup (c)

2 cups = 1 pint (pt)

2 pints = 1 quart (qt)

4 quarts = 1 gallon (gal)

Use proportions to convert from one unit to another.

EXAMPLE 1 How many quarts are in six gallons?

SOLUTION

$\frac{4 \text{ qt}}{1 \text{ gal}} = \frac{x \text{ qt}}{6 \text{ gal}}$ *Write a proportion using quart-to-gallon ratios.*

$x = (4)(6)$ *Cross-multiply.*
$x = 24$

There are 24 quarts in 6 gallons.

EXAMPLE 2 How many fluid ounces are in one quart?

SOLUTION

$\frac{8 \text{ fl oz}}{1 \text{ c}} = \frac{x}{2 \text{ c}}$ *Begin by writing a proportion using ounce-to-cup ratios.*

$x = (8)(2)$ *Cross-multiply.*
$x = 16$

There are 16 fluid ounces in one pint since 2 cups = 1 pint.

$\frac{16 \text{ fl oz}}{1 \text{ pt}} = \frac{x}{2 \text{ pt}}$ *Write a proportion using ounce-to-pint ratios. Since 2 pt is the same as 1 qt, use 2 pt in the ratio.*

$x = (16)(2)$ *Cross-multiply.*
$x = 32$

There are 32 fluid ounces in 2 pints. Since 2 pints is the same as 1 quart, there must be 32 fluid ounces in 1 quart.

> **a + b = x**
> ## ALGEBRA CONNECTION
>
> Example 2 effectively demonstrates the substitution of equivalent values. For a proportion to be valid, the same units must be used in each ratio. The first step in Example 2 could have been written
>
> $$\frac{8 \text{ fl oz}}{1 \text{ c}} = \frac{x}{1 \text{ pt}}$$
>
> but the first ratio is fluid ounce to cup, while the second ratio is fluid ounce to pint. Written this way, the two ratios do not compare the same units. However, since
>
> $$2 \text{ c} = 1 \text{ pt}$$
>
> these two values are the same. They are, therefore, interchangeable. Substituting 2 c for 1 pt in the original proportion gives
>
> $$\frac{8 \text{ fl oz}}{1 \text{ c}} = \frac{x}{2 \text{ c}}$$
>
> You then have a valid proportion since each ratio compares ounces and cups.

Example 2 demonstrates that converting liquid measurements may require more than one step before the problem is complete.

Metric System

The **liter** is the basic unit of liquid measurement in the metric system. Use the same prefixes that you used for length and weight to find the units for liquid measurement.

METRIC SYSTEM LIQUID MEASUREMENTS

1 kiloliter (kL) = 1,000 liters (L)

1 hectoliter (hL) = 100 liters

1 dekaliter (daL) = 10 liters

liter = standard unit

1 deciliter (dL) = 0.1 liter

1 centiliter (cL) = 0.01 liter

1 milliliter (mL) = 0.001 liter

Convert from one unit to another by moving the decimal point. Move it to the left if you are converting from a smaller to a larger unit. Move it to the right if you are converting from a larger to a smaller unit. Use the following chart to determine how many places to move the decimal point.

			standard unit			
kilo-	hecto-	deka-	liter	deci-	centi-	milli-
(kL)	(hL)	(daL)	L	(dL)	(cL)	(mL)

Liquid Measurement

EXAMPLE 3 Convert 42 deciliters to dekaliters.

SOLUTION

42 dL = x daL *Write an equation showing the desired equality.*

42 dL = 0.42 daL *Move the decimal point two places to the left since daL is two places to the left of dL on the chart.*

There is 0.42 dekaliter in 42 deciliter.

EXAMPLE 4 Convert 75 kiloliters to centiliters.

SOLUTION

75 kL = x cL *Write an equation showing the desired equality.*

75 kL = 7,500,000 cL *Move the decimal point five places to the right since cL is five places to the right of kL on the chart.*

There are 7,500,000 centiliters in 75 kiloliters.

Converting between USCS and Metric System

Use the following chart to convert liquid measures from the USCS to the metric system or from the metric system to the USCS.

USCS/METRIC LIQUID CONVERSION CHART

1 pint ≈ 0.474 liter	1 liter ≈ 2.11 pints
1 quart ≈ 0.946 liter	1 liter ≈ 1.06 quarts
1 gallon ≈ 3.785 liters	1 liter ≈ 0.264 gallon
	1 liter ≈ 33.8 fluid ounces

Use proportions to convert from one system to the other. As with length and weight, these conversions are approximate.

EXAMPLE 5 How many liters are in 5 gallons?

SOLUTION

$\dfrac{1 \text{ gal}}{3.785 \text{ L}} = \dfrac{5 \text{ gal}}{x}$ *Write a proportion using gallon-to-liter ratios.*

$x = (3.785)(5)$ *Cross-multiply.*
$x = 18.925$

There are approximately 18.925 liters in 5 gallons.

EXAMPLE 6 How many quarts are equivalent to 1.2 kiloliters?

SOLUTION

1.2 kL = x L *To use the chart, you must know how many liters you have.*

1.2 kL = 1,200 L *Move the decimal point three places to the right.*

$$\frac{1 \text{ L}}{1.06 \text{ qt}} = \frac{1{,}200 \text{ L}}{x \text{ qt}}$$ Write a proportion using liter-to-quart ratios.

$x = (1.06)(1{,}200)$ Cross-multiply.
$x = 1{,}272$ quarts

There are approximately 1,272 quarts in 1.2 kiloliters.

Work the Comprehension Checkpoint problems to test your understanding of liquid measurements. If you have difficulty with these problems, review the appropriate material before going on to the Practice Sets.

COMPREHENSION CHECKPOINT

Convert the USCS units as indicated.

1. 4 qt = _____ fl oz

2. 1 gal = _____ c

Convert the metric units as indicated.

3. 1,475 mL = _____ daL

4. 85 hL = _____ L

Convert between the USCS and the metric system as indicated.

5. 8 qt = _____ L

6. 950 daL = _____ gal

9.3A PRACTICE SET

Convert the given USCS unit to the indicated unit.

1. 3 qt = _____ pt

2. 28 fl oz = _____ tsp

3. 30 qt = _____ gal

4. 1 gal = _____ c

5. 7 pt = _____ qt

6. 512 fl oz = _____ gal

Convert the given metric unit to the indicated unit.

7. 5 L = _____ kL

8. 456 cL = _____ L

9. 25 hL = _____ dL

10. 100 mL = _____ kL

11. 20 kL = _____ cL

12. 68 daL = _____ dL

Convert the given unit to the indicated unit.

13. 6 pt = _____ L

14. 10 qt = _____ cL

15. 15 L = _____ gal

16. 20 gal = _____ L

17. 5 daL = _____ qt

18. 825 mL = _____ pt

Applications

19. Many soft drinks are packaged in 2-L bottles. How many fluid ounces are in each bottle?

_____

20. How many liters would it take to fill a 12-oz glass?

_____

21. A swimming pool holds 10,000 gal of water. How many liters of water does it hold?

_____

22. How many pints would be contained in a 4-L jug of milk?

_____

9.3B PRACTICE SET

Convert the given USCS unit to the indicated unit.

1. 5 qt = _____ c

2. 8 fl oz = _____ tbsp

3. 5 gal = _____ qt

4. 16 pt = _____ gal

5. 1 pt = _____ fl oz

6. 256 fl oz = _____ gal

Convert the given metric unit to the indicated unit.

7. 12 cL = _____ L

8. 890 mL = _____ dL

9. 84 daL = _____ dL

10. 100 L = _____ hL

11. 15 kL = _____ L

12. 300 mL = _____ daL

Convert the given unit to the indicated unit.

13. 8 qt = _____ L

14. 10 pt = _____ mL

15. 24 L = _____ qt

16. 500 cL = _____ pt

17. 250 kL = _____ gal

18. 384 fl oz = _____ L

Applications

19. Tequila can be purchased in 750-mL bottles. How many fluid ounces of tequila is that? Round your answer to the nearest tenth.

20. How many fluid ounces of juice are in a half-gallon container?

21. A recipe calls for two cups of milk. Express this in metric units.

22. A recipe calls for a cup of water. The only measuring device you have available is a set of spoons. How many tablespoons of water should you use?

9.4 TIME AND TEMPERATURE

Time

Units of time are standard around the world. These units include the **second, minute, hour, day, week, month,** and **year.** The following chart shows the relationships between the standard units of time.

STANDARD UNITS OF TIME

60 seconds (s) = 1 minute (min)

60 minutes = 1 hour (h)

24 hours = 1 day

7 days = 1 week (wk)

52 weeks = 1 year (yr)

365 days = 1 year

12 months (mo) = 1 year

You can use proportions to convert from one unit of time to another. Some conversions may take more than one step.

EXAMPLE 1 How many hours are in one week?

SOLUTION

$\dfrac{24 \text{ h}}{1 \text{ day}} = \dfrac{x}{7 \text{ days}}$ *Write a proportion using hour-to-day ratios.*

$x = (24)(7)$ *Cross-multiply.*
$x = 168 \text{ h}$

There are 168 hours in one week.

EXAMPLE 2 How many seconds are in an hour?

SOLUTION

$\dfrac{60 \text{ s}}{1 \text{ min}} = \dfrac{x}{60 \text{ min}}$ *Write a proportion using second-to-minute ratios.*

$x = (60)(60)$ *Cross-multiply.*
$x = 3{,}600$

There are 3,600 seconds in one hour.

Temperature

There are two scales used to measure temperature: **Fahrenheit** (F) and **Celsius** (C). Celsius is the metric system measure. A comparison of the two scales is shown in the following table.

	F	C
Boiling point of water	212°	100°
Normal body temperature	98.6°	37°
Freezing point of water	32°	0°

Formulas are used to convert from one system to another.

> To convert from Fahrenheit to Celsius, use the formula
> $$C = \frac{5}{9}(F - 32)$$

EXAMPLE 3 Convert 86°F to Celsius.

SOLUTION

$C = \frac{5}{9}(F - 32)$ Write the formula.

$C = \frac{5}{9}(86 - 32)$ Substitute the given Fahrenheit temperature.

$C = \frac{5}{9}(54)$ Use the order of operations to evaluate the formula.

$C = 30°$

A temperature of 86°F is the same as 30°C.

> To convert from Celsius to Fahrenheit, use the formula
> $$F = 1.8C + 32$$

EXAMPLE 4 Convert 25°C to Fahrenheit.

SOLUTION

$F = 1.8C + 32$ Write the formula.

$F = 1.8(25) + 32$ Substitute the given Celsius temperature.

$F = 45 + 32$ Use the order of operations to evaluate the formula.

$F = 77°$

A temperature of 25°C is the same as 77°F.

Work the Comprehension Checkpoint problems to make sure you understand time and temperature measurements. If you have difficulty with these problems, review the appropriate material before going on to the Practice Sets.

COMPREHENSION CHECKPOINT

Convert the units of time as indicated.

1. 96 h = _____ days

2. 84 days = _____ wk

Convert the temperatures as indicated.

3. 40°C = _____ F

4. 95°F = _____ C

9.4A PRACTICE SET

Convert the given units of time as indicated.

1. 6 min = _____ s

2. 30 days = _____ h

3. 3 yr = _____ days

4. 133 days = _____ wk

5. 1 day = _____ min

6. 1 day = _____ s

Convert the given Fahrenheit temperature to Celsius.

7. 50°F = _____ C

8. 122°F = _____ C

9. 203°F = _____ C

Convert the given Celsius temperature to Fahrenheit.

10. 30°C = _____ F

11. 75°C = _____ F

12. 160°C = _____ F

Applications

13. Bob Richards ran a mile in 4 min 25 s. How many seconds did it take him to run the mile?

    _____

14. An overseas flight takes 16 h. Express this time in days.

    _____

15. The all-time record for marathon bowling is 21 days and 20 hours. How many consecutive hours is the record?

_____

16. Express 1 billion seconds in years. Round to the nearest tenth of a year.

_____

17. The average temperature in Chicago in July is 73°F. What is the average temperature on the Celsius scale?

_____

18. The highest recorded temperature in the world occurred at El Azizia, Libya, in 1922. The reading was 58°C. What was the temperature on the Fahrenheit scale?

_____

9.4B PRACTICE SET

Convert the given units of time as indicated.

1. 240 s = _____ min

2. 3 h = _____ min

3. 16 wk = _____ days

4. 1 yr = _____ h

5. 13 wk = _____ yr

6. 18 h = _____ day

Convert the given Fahrenheit temperature to Celsius.

7. 95°F = _____ C

8. 14°F = _____ C

9. 41°F = _____ C

Convert the given Celsius temperature to Fahrenheit.

10. 60°C = _____ F

11. 125°C = _____ F

12. 84°C = _____ F

Applications

13. A 250-mile airplane flight takes about 45 min. Express this time in hours.

   _____

14. The longest career for a professional football player was 26 years. Express this time in months.

   _____

15. Express an 8-hour work shift in terms of a day.

16. The longest dry spell (without rain) ever recorded in Dallas was 41 days. Express this time span in weeks. Round to the nearest whole week.

17. The coldest recorded temperature in the United States occurred at Prospect Creek, Alaska, in 1971. The temperature was −80°F. What does this convert to on the Celsius scale?

18. The highest annual normal temperature in the United States is in Death Valley, California, at 78.2°F. The lowest annual normal temperature in the United States is at Barrow, Alaska, at 9.3°F. What is the difference between these two extremes on the Celsius Scale?

9.5 WORKING WITH DENOMINATE NUMBERS

You learned in Section 9.1 that a denominate number is a number with some unit of measure included. Examples of denominate numbers include 6 inches, 8 years, and 17 pounds. An **abstract number** is a number without a unit of measure included. The sum of 6 and 5 is 11. The 6, 5, and 11 are abstract numbers. When you work with denominate numbers, you should simplify the answer. A number such as 75 minutes could be simplified by writing 1 h 15 min. A measurement of 25 inches could be simplified as 2 ft 1 in.

> To simplify a denominate number, change smaller units to large units.

EXAMPLE 1 Simplify 5 ft

SOLUTION

1 yd = 3 ft   *A yard is 3 ft.*

5 ft = 3 ft + 2 ft   *Rewrite the given units in terms of the next larger unit.*
5 ft = 1 yd 2 ft

5 ft simplifies to 1 yd 2 ft.

EXAMPLE 2 Simplify 125 min

SOLUTION

60 min = 1 h   *One hour is 60 min.*
120 min = 2 h   *Two hours are 120 min.*
125 min = 120 min + 5 min *Rewrite the given units in terms of the next*
125 min = 2 h 5 min   *larger unit.*

125 min simplifies to 2 h 5 min.

EXAMPLE 3 Simplify 124 in

SOLUTION

12 in = 1 ft   *12 in are 1 ft.*
120 in = 10 ft   *120 in are 10 ft.*
124 in = 120 in + 4 in   *Rewrite the given units in terms of the next larger unit.*

124 in = 10 ft + 4 in

10 ft can be simplified.

3 ft = 1 yd
10 ft = 9 ft + 1 ft
10 ft = 3 yd 1 ft   (9 ft = 3 ft × 3)

124 in simplifies to 3 yd 1 ft 4 in.

Study the final answers in Examples 1 through 3. None of the units is large enough to be converted to a larger unit. In Example 3, 4 in is not enough to be converted to a foot; 1 ft is not enough to be converted to a yard. The answer is

Adding Denominate Numbers

> **TO ADD DENOMINATE NUMBERS**
> 1. Add like units.
> 2. Simplify the answer if possible.

EXAMPLE 4 Add 3 ft 6 in and 7 ft 9 in

SOLUTION

$\quad\;\;$ 3 ft $\;\;$ 6 in $\quad$ *Line up like units: ft over ft, in over in, etc. Add like units.*
$+\;$ 7 ft $\;\;$ 9 in
$\quad\;$ 10 ft 15 in
$\qquad\quad\;\overbrace{}$
$\qquad\;\;$ 1 ft 3 in $\quad$ *Simplify where possible.*
$\quad\;$ 11 ft 3 in
$\;\overbrace{}$
3 yd 2 ft $\qquad\quad$ *Simplify the ft.*

3 yd 2 ft 3 in is the answer in simplified form.

EXAMPLE 5 Add 12 yr 50 wk 6 days and 8 yr 10 wk 4 days

SOLUTION

$\quad\;\;$ 12 yr 50 wk $\;\;$ 6 days $\quad$ *Line up like units.*
$+\quad$ 8 yr 10 wk $\;\;$ 4 days $\quad$ *Add like units.*
$\quad\;\;$ 20 yr 60 wk 10 days
$\qquad\qquad\qquad\;\overbrace{}$
$\qquad\qquad\;\;$ 1 wk 3 days $\quad$ *Simplify the days.*
$\quad\;$ 20 yr 61 wk 3 days
$\qquad\;\;\overbrace{}$
$\quad\;$ 1 yr 9 wk $\qquad\qquad\quad$ *Simplify the weeks.*

21 yr 9 wk 3 days is the answer in simplified form.

Subtracting Denominate Numbers

> **TO SUBTRACT DENOMINATE NUMBERS**
> 1. Subtract like units.
> 2. Simplify the answer if possible.

EXAMPLE 6 Subtract 3 h 20 min from 5 h 45 min

SOLUTION

$\quad\;$ 5 h 45 min $\quad$ *Line up like units.*
$-\;$ 3 h 20 min $\quad$ *Subtract like units.*
$\quad\;$ 2 h 25 min

The answer is in simplest form.

Working with Denominate Numbers

EXAMPLE 7 Subtract 4 gal 3 qt from 9 gal

SOLUTION

   9 gal
− 4 gal 3 qt

Align like units for subtraction. Since you can subtract only like units, you must borrow 1 gal from the 9 gal and convert it to quarts.

   8 gal 4 qt
− 4 gal 3 qt

Rewrite the problem to show the result of borrowing.

   8 gal 4 qt
− 4 gal 3 qt
   4 gal 1 qt

Now you can subtract.

The answer is in simplest form.

Multiplying Denominate Numbers

TO MULTIPLY A DENOMINATE NUMBER BY AN ABSTRACT NUMBER
1. Multiply each part of the denominate number by the abstract number.
2. Simplify the answer if possible.

EXAMPLE 8 Multiply 3 ft 6 in by 5

SOLUTION

  3 ft 6 in
×   5
15 ft 30 in

Set the problem up for multiplication. Multiply each unit.

  2 ft 6 in

Simplify the answer.

17 ft 6 in

Simplify 17 ft.

5 yd 2 ft

5 yd 2 ft 6 in is the answer in simplified form.

EXAMPLE 9 A commercial airplane pilot flies a route that takes 4 h 20 min. What would be her total flying time if she made the trip eight times one month?

SOLUTION

  4 h 20 min
×   8
32 h 160 min

Set the problem up for multiplication. Multiply each unit.

  2 h 40 min

Simplify the minutes.

34 h 40 min

1 day 10 h

Simplify the hours.

1 day 10 h 40 min

She would be in the air a total of 1 day 10 h 40 min.

CHAPTER 9   Measurement

Multiplying a denominate number by a denominate number is a different type of problem. It will be covered in Chapter 10, Introduction to Geometry.

Dividing Denominate Numbers

> **TO DIVIDE A DENOMINATE NUMBER BY AN ABSTRACT NUMBER**
> 1. Divide the largest unit of the denominate number by the abstract number. If the division comes out even, go on to the next unit of the denominate number and divide.
> 2. If the division does not come out even, convert the remainder to the next smaller unit, add it to the units already there, and divide.
> 3. Continue until all units of the denominate number have been divided.
> 4. If the smallest denominate number does not divide evenly, express the answer as a fraction.

EXAMPLE 10   Divide 20 ft 8 in by 4

SOLUTION

4)20 ft 8 in   *Set the problem up for division.*

$$\underline{5\text{ ft }2\text{ in}}$$
4)20 ft 8 in   *Divide each unit by the divisor. Each division comes out even.*

20 ft 8 in divided by 4 is 5 ft 2 in.

EXAMPLE 11   Divide 9 h 12 min by 6

SOLUTION

6)9 h 12 min   *Set the problem up for division.*

$$\underline{\phantom{9\text{ h }}1\text{ h}}$$
6)9 h 12 min   *Divide 9 by 6. There is a remainder of 3 h. Convert it to min*
$$\underline{6\text{ h}}$$ *and add it to the 12 min already in the dividend. (3 h =*
$$3\text{ h}$$ *180 min. 180 + 12 = 192)*
$$\phantom{3\text{ h}}192\text{ min}$$

$$\underline{\phantom{9\text{ h }}1\text{ h }32\text{ min}}$$
6)9 h 12 min   *Dividing 192 min by 6 gives 32 min.*
$$\underline{6\text{ h}}$$
$$3\text{ h}$$
$$\phantom{3\text{ h}}192\text{ min}$$
$$\phantom{3\text{ h}}\underline{192\text{ min}}$$

The result of dividing 9 h 12 min by 6 is 1 h 32 min.

Sometimes you may have to make conversions before dividing.

Working with Denominate Numbers

EXAMPLE 12   How many 6-oz glasses could you fill from one gallon of juice?

SOLUTION

1 gal = x fl oz   *Convert one gallon to fluid ounces.*
1 gal = 4 qt
1 qt = 2 pt, so 1 gal = 8 pt (2 pt × 4)
1 pt = 2 c, so 1 gal = 16 c (2 c × 8)
1 c = 8 fl oz, so 1 gal = 128 fl oz (8 fl oz × 16)

There are 128 fl oz in one gallon. Divide 128 fl oz by 6 oz.

6 fl oz$\overline{)128\text{ fl oz}}$ *Set up the division problem.*

$$\begin{array}{r} 21 \\ 6\text{ fl oz}\overline{)128\text{ fl oz}} \\ \underline{12} \\ 8 \\ \underline{6} \\ 2\text{ fl oz left over} \end{array}$$

The answer here is an abstract number.

You can fill twenty-one 6-oz glasses of juice from a gallon of juice.

Work the Comprehension Checkpoint problems to make sure you understand working with denominate numbers. If you have difficulty with these problems, review the appropriate material before going on to the Practice Sets.

COMPREHENSION CHECKPOINT

Simplify each denominate number.

1.   9 ft 20 in = _____

2.   23 h 150 min = _____

Perform the indicated operation.

3.   2 qt 3 pt 18 fl oz
      8 qt 5 pt
   + 5 qt 6 pt 24 fl oz

4.   5 mi 400 yd
 − 3 mi 700 yd

5.   12 ft 8 in
 ×   5

6.   19 h 15 min ÷ 15 = _____

9.5A PRACTICE SET

Perform the indicated operation. Simplify each answer as needed.

1.   5 h 30 min
     + 4 h 15 min

2.   7 days 8 h 20 min
     + 10 days 15 h 40 min

3.   9 gal 3 qt 2 pt
       7 gal 2 qt 1 pt
     + 1 gal 3 qt 3 pt

4.   12 ft 6 in
     − 8 ft 1 in

5.   9 min 20 s
     − 7 min 35 s

6.   4 pt 2 c
     − 3 pt 3 c

7.   6 lb 4 oz
     ×   8

8.   25 min 30 s
     ×   9

9.   8 gal 3 qt 5 pt
     ×   6

10. 8 lb 10 oz ÷ 3 = _____

11. 16 h 45 min ÷ 5 = _____

12. 24 gal 3 qt ÷ 6 = _____

Applications

13. A carpenter uses studs that are 8 ft 3 in long. What would be the total length of 24 studs?

14. Pepper costs $0.245 per ounce. What would be the total cost of 3 pounds of pepper?

15. If the average person weighs 160 lb, how many people would an elevator with a 2,000-lb capacity be able to carry?

16. How many 6-oz hamburger patties could be made from 40 lb of hamburger meat?

_____

17. Each shelf on a bookcase is 3 ft 5 in long. What is the total length of the shelving of a bookcase with six shelves?

_____

18. Marie worked at her first job for 3 yr 2 mo 15 days. She worked at her second job for 2 yr 10 mo 26 days. What was the total amount of time she spent on the two jobs? (Assume 30 days per month.)

_____

19. Julio cut a 3-ft 7-in strip of plasterboard from a sheet that was 7 ft 4 in long. How much of the original piece of plasterboard remained?

_____

20. The 1993 cost of postage in the United States was $0.29 per ounce. If you had an important package that had to be delivered the next day, you could pay $13.95 for 2 lb. How much more would you pay to have the package delivered the next day than you would normally pay for the same 2-lb package?

_____

9.5B PRACTICE SET

Perform the indicated operation. Simplify each answer as needed.

1.   7 lb 4 oz
  + 9 lb 9 oz

2.   12 min 38 s
  + 38 min 25 s

3.   7 gal 5 qt 6 pt
     5 gal 2 qt 8 pt
 + 10 gal   7 pt

4.   12 days 8 h
  − 5 days 10 h

5.   3 gal
 − 1 gal 1 pt 9 fl oz

6.   18 lb 3 oz
 − 12 lb 8 oz

7.   6 h 20 min
  ×   4

8.   15 gal 3 qt 10 fl oz
  ×   8

9.   12 ft 4 in
  ×   15

10. 7 days 2 h ÷ 5 = _____

11. 4 qt 3 pt 4 fl oz ÷ 6 = _____

12. 5 yd 2 ft ÷ 4 = _____

Applications

13. Cans of vegetables that weigh 17 oz each are packed 24 cans to a case. What is the total weight of the case?

14. Boy Scouts want to build a rope ladder. They have pieces of rope that are 22 ft 7 in long, 16 ft 12 in long, and 8 ft 3 in long. What is the total length of rope available for the rope ladder?

15. If the rope ladder in problem 14 requires 50 ft of rope, how much more rope do these Boy Scouts need?

16. Two men on a heavyweight wrestling tag team weigh 225 lb 10 oz and 243 lb 6 oz, respectively. What is the total weight of the team?

17. If a person drinks eight glasses of water each day, and each glass is 12 fl oz, how many gallons of water does the person drink per week?

18. The wall of a room is 18 ft long. Two windows, each from the floor to the ceiling, are both 2 ft 9 in wide. If wallpaper comes in rolls that are 15 in wide, how many strips would be needed to cover the entire wall?

19. How many 110-yard sprints would a football player need to run to go 1 mile?

20. Sugar is packaged in bags that weigh 1.81 kg. How many bags would be required to package 1 ton of sugar?

SUMMARY OF KEY CONCEPTS

KEY TERMS

inch, foot, yard, mile (9.1): The length units of measure in the US Customary System (USCS).

denominate number (9.1): A number followed by some unit (for example, 7 in or 5 gal).

meter (9.1): The standard unit of length measure in the metric system.

ounce, pound, ton (9.2): The weight units of measure in the US Customary System.

gram (9.2): The standard unit of weight measure in the metric system.

teaspoon, tablespoon, fluid ounce, cup, pint, quart, gallon (9.3): The units of liquid measure in the US Customary System.

liter (9.3): The standard unit of liquid measure in the metric system.

second, minute, hour, day, week, month, year (9.4): The standard units of time used around the world.

Fahrenheit scale (9.4): A temperature scale used in some countries.

Celsius scale (9.4): The scale for temperature measurement in the metric system.

abstract number (9.5): A number without a unit attached; for example, the number 5 or the number 12.

KEY RULES

— Units of Measurement

United States Customary System	Metric System
Length Measures (9.1)	
12 inches = 1 foot	1 kilometer = 1,000 meters
3 feet = 1 yard	1 hectometer = 100 meters
5,280 feet = 1 mile	1 dekameter = 10 meters
1,760 yards = 1 mile	meter = standard unit
	1 decimeter = 0.1 meter
	1 centimeter = 0.01 meter
	1 millimeter = 0.001 meter
Weight Measures (9.2)	
16 ounces = 1 pound	1 kilogram = 1,000 grams
2,000 pounds = 1 ton	1 hectogram = 100 grams
	1 dekagram = 10 grams
	gram = standard unit
	1 decigram = 0.1 gram
	1 centigram = 0.01 gram
	1 milligram = 0.001 gram
Liquid Measures (9.3)	
3 teaspoons = 1 tablespoon	1 kiloliter = 1,000 liters
2 tablespoons = 1 fluid ounce	1 hectoliter = 100 liters
8 fluid ounces = 1 cup	1 dekaliter = 10 liters
2 cups = 1 pint	liter = standard unit
2 pints = 1 quart	1 deciliter = 0.1 liter
4 quarts = 1 gallon	1 centiliter = 0.01 liter
	1 milliliter = 0.001 liter

Time (9.4)
60 seconds = 1 minute
60 minutes = 1 hour
24 hours = 1 day
7 days = 1 week
52 weeks = 1 year
365 days = 1 year
12 months = 1 year

Conversion Factors

From USCS to Metric   From Metric to USCS

Length (9.1)

From USCS to Metric	From Metric to USCS
1 inch ≈ 2.54 centimeters	1 centimeter ≈ 0.394 inch
1 foot ≈ 0.305 meter	1 meter ≈ 39.37 inches
1 yard ≈ 0.914 meter	1 meter ≈ 3.28 feet
1 mile ≈ 1.61 kilometers	1 meter ≈ 1.09 yards
	1 kilometer ≈ 0.621 mile

Weight (9.2)

From USCS to Metric	From Metric to USCS
1 ounce ≈ 28.35 grams	1 gram ≈ 0.035 ounce
1 pound ≈ 454 grams	1 kilogram ≈ 2.2 pounds
1 pound ≈ 0.454 kilogram	

Liquid (9.3)

From USCS to Metric	From Metric to USCS
1 pint ≈ 0.474 liter	1 liter ≈ 2.11 pints
1 quart ≈ 0.946 liter	1 liter ≈ 1.06 quarts
1 gallon ≈ 3.785 liters	1 liter ≈ 0.264 gallon
	1 liter ≈ 33.8 ounces

Formula for converting from Fahrenheit to Celsius (9.4):

$$C = \frac{5}{9}(F - 32)$$

Formula for converting from Celsius to Fahrenheit (9.4):

$$F = 1.8C + 32$$

To simplify denominate numbers (9.5):

Change smaller units to large units.

To add denominate numbers (9.5):

1. Add like units.
2. Simplify the answer if possible.

To subtract denominate numbers (9.5):

1. Subtract like units.
2. Simplify the answer if possible.

To multiply a denominate number by an abstract number (9.5):

1. Multiply each part of the denominate number by the abstract number.
2. Simplify the answer if possible.

To divide a denominate number by an abstract number (9.5):

1. Divide the largest unit of the denominate number by the abstract number. If the division comes out even, go on to the next unit of the denominate number and divide.
2. If the division does not come out even, convert the remainder to the next smaller unit, add it to the units already there, and divide.
3. Continue until all units of the denominate number have been divided.
4. If the smallest denominate number does not divide evenly, express the answer as a fraction.

PRACTICE TEST 9A

CHAPTER 9

In problems 1–12, convert each given unit of measurement to the indicated unit of measurement.

1. 6 gal = _____ qt
2. 5 L = _____ cL
3. 12 h = _____ min
4. 60 cm = _____ in
5. 12 lb = _____ oz
6. 500 dg = _____ hg
7. 42 days = _____ wk
8. 95°F = _____ C
9. 24 lb = _____ kg
10. 1.5 m = _____ cm
11. 5 c = _____ fl oz
12. 30 m = _____ ft

In problems 13–16, perform the indicated operation. Simplify your answer if possible.

13.   16 lb 9 oz
    +   8 lb 7 oz

14.   25 gal
    − 17 gal 1 qt

15.   5 pt 12 fl oz
    ×   7

16. 9 qt 6 pt ÷ 2 = _____

17. A recipe calls for 3 cups of water. Your measuring cup is in metric units. How many cL of water should you use?

18. A carpenter needs three boards, each 2 ft 9 in long. He has one board that is 10 ft long. If he cuts the three boards from the 10-ft board, how much of the 10-ft board will be left?

PRACTICE TEST 9B

In problems 1–12, convert each given unit of measurement to the indicated unit of measurement.

1. 30 m = _____ mm
2. 4.5 min = _____ s
3. 20 gal = _____ L
4. 15 m = _____ ft
5. 28°F = _____ C
6. 38 yd = _____ m
7. 24 km = _____ mi
8. 250 g = _____ lb
9. 800 hL = _____ cL
10. 5 pt = _____ fl oz
11. 80 oz = _____ lb
12. 40°C = _____ F

In problems 13–16, perform the indicated operation. Simplify your answer if possible.

13.   2 gal 3 qt 12 fl oz
     + 10 gal 5 qt 9 fl oz

14.   12 h 25 min
    −   8 h 49 min

15.   16 lb 3 oz
    ×   4

16. 18 ft ÷ 4 = _____

17. Melvin, a bricklayer, needs to brick a wall. The wall is 20 ft long. If each brick is 9 in long, how many bricks will he need for each row? (Disregard the mortar between the bricks.)

    _____

18. A driveway is 42 ft long. How many meters long is it?

    _____

SKILLS REVIEW

CHAPTERS 1-9

1. Write 875,206 in words. _____

2. Round 74,551 to the nearest hundred. _____

3. Round 9,950 to the nearest thousand. _____

4. Round 8.909 to the nearest tenth. _____

Perform the indicated operation. Express answers in lowest terms.

5. $\frac{3}{5} + \frac{8}{3} + \frac{4}{9} =$ _____

6. $\frac{6}{7} - \frac{7}{12} =$ _____

7. $\frac{2}{15} \times \frac{5}{8} =$ _____

8. $\frac{12}{25} \div \frac{3}{5} =$ _____

9. $1.85 \times 0.09 =$ _____

10. $22.5 \div 0.15 =$ _____

11. Convert $\frac{17}{40}$ to a decimal. _____

12. Convert $\frac{18}{25}$ to a percent. _____

13. Convert 32% to a decimal. _____

14. Convert 95% to a fraction in lowest terms. _____

15. What is 27% of 900? _____

16. What is $\frac{1}{3}$% of 600? _____

17. 25 is what percent of 400? _____

18. Write 0.89 in words.
    _____

19. Write 7.007 in words.
    _____

20. Find the missing element in the proportion.
    $\dfrac{19}{6} = \dfrac{95}{x}$ _____

21. Insert the appropriate symbol, < or >.

    a. −9 _____ −8   a. _____

    b. 0 _____ 5   b. _____

    c. 3 _____ 2   c. _____

22. Indicate the absolute value.

    a. $|-3.6|$ a. _____

    b. $\left|8\dfrac{4}{5}\right|$ b. _____

Perform the indicated operation.

23. (−3)(−5)(6)(2) = _____

24. (−8) − (−20) = _____

25. (24) ÷ (−8) = _____

26. (18) + (−47) + (−82) + (38) = _____

27. Solve: $8x - 3 = 29$ _____

28. Solve: $9(x - 5) = 3(x + 7)$ _____

29. Solve: $\dfrac{x}{3} - 20 = 45$ _____

30. Convert 12 lb to ounces. _____

31. Convert 678 cL to liters. _____

32. Convert 59°F to Celsius. _____

CHAPTER 9   SOLUTIONS

Skills Preview

1. 480 s
2. 157.6 in
3. 25°C
4. $\frac{1}{2}$ gal
5. $2\frac{1}{8}$ tons
6. 0.65 m
7. 5.688 L
8. 38.64 km
9. 122°F
10. 4,900 cg
11. 208 fl oz
12. 0.275 kL
13. 6 yd 1 ft 2 in
14. 7 h 50 min
15. 19 h 11 min
16. 4 lb 9 oz
17. 10,980 m
18. 55 lb

Section 9.1   Comprehension Checkpoint

1. 45 ft
2. 7,040 yd
3. 1.5 ft
4. 1.8 hm
5. 4,200,000 cm
6. 0.24 dam
7. 60.96 cm
8. 91.56 yd
9. 4.83 km

9.1A   Practice Set

1. $\frac{1 \text{ ft}}{12 \text{ in}} = \frac{x \text{ ft}}{96 \text{ in}}$
$12x = 96$
$x = 8 \text{ ft}$

2. $\frac{1 \text{ yd}}{3 \text{ ft}} = \frac{42 \text{ yd}}{x \text{ ft}}$
$x = (42)(3)$
$x = 126 \text{ ft}$

3. $\frac{5{,}280 \text{ ft}}{1 \text{ mi}} = \frac{2{,}640 \text{ ft}}{x \text{ mi}}$
$5{,}280x = 2{,}640$
$x = 0.5 \text{ mi}$

4. $\frac{1 \text{ mi}}{5{,}280 \text{ ft}} = \frac{5 \text{ mi}}{x \text{ ft}}$
$x = (5)(5{,}280)$
$x = 26{,}400 \text{ ft}$

5. $\frac{1 \text{ ft}}{12 \text{ in}} = \frac{5{,}280 \text{ ft (1 mi)}}{x \text{ in}}$
$x = (12)(5{,}280)$
$x = 63{,}360 \text{ in per mile}$
$\frac{63{,}360 \text{ in}}{1 \text{ mi}} = \frac{x \text{ in}}{2 \text{ mi}}$
$x = (2)(63{,}360)$
$x = 126{,}720 \text{ inches in 2 miles}$

6. $\frac{3 \text{ ft}}{1 \text{ yd}} = \frac{84 \text{ ft}}{x \text{ yd}}$
$3x = 84$
$x = 28 \text{ yd}$

7. 48 m = x dm
48 m = 480 dm
Move the decimal point one place to the right.

8. 225 cm = x dm
225 cm = 22.5 dm
Move the decimal point one place to the left.

9. 817 mm = x hm
817 mm = 0.00817 hm
Move the decimal point five places to the left.

10. 19 dam = x cm
19 dam = 19,000 cm
Move the decimal point three places to the right.

11. 240 km = x m
240 km = 240,000 m
Move the decimal point three places to the right.

12. 500 cm = x mm
500 cm = 5,000 mm
Move the decimal point one place to the right.

13. $\frac{1 \text{ in}}{2.54 \text{ cm}} = \frac{32 \text{ in}}{x \text{ cm}}$
$x = (32)(2.54)$
$x = 81.28 \text{ cm}$

14. $\frac{1 \text{ mi}}{1.61 \text{ km}} = \frac{18 \text{ mi}}{x \text{ km}}$
$x = (18)(1.61)$
$x = 28.98 \text{ km}$

15. $\dfrac{1 \text{ ft}}{0.305 \text{ m}} = \dfrac{80 \text{ ft}}{x \text{ m}}$
$x = (80)(0.305)$
$x = 24.4 \text{ m}$

16. $2{,}500 \text{ m} = x \text{ km}$
$2{,}500 \text{ m} = 2.5 \text{ km}$
$\dfrac{1 \text{ km}}{0.621 \text{ mi}} = \dfrac{2.5 \text{ km}}{x \text{ mi}}$
$x = (2.5)(0.621)$
$x = 1.55 \text{ mi (rounded)}$

17. $400 \text{ mm} = x \text{ cm}$
$400 \text{ mm} = 40 \text{ cm}$
$\dfrac{1 \text{ cm}}{0.394 \text{ in}} = \dfrac{40 \text{ cm}}{x \text{ in}}$
$x = (40)(0.394)$
$x = 15.76 \text{ in}$

18. $\dfrac{1 \text{ m}}{1.09 \text{ yd}} = \dfrac{27 \text{ m}}{x \text{ yd}}$
$x = (27)(1.09)$
$x = 29.43 \text{ yd}$

19. $\dfrac{1 \text{ in}}{2.54 \text{ cm}} = \dfrac{12 \text{ in}}{x \text{ cm}}$
$x = (12)(2.54)$
$x = 30.48 \text{ cm}$

20. $\dfrac{1 \text{ ft}}{0.305 \text{ m}} = \dfrac{18 \text{ ft}}{x \text{ m}}$
$x = (18)(0.305)$
$x = 5.49 \text{ m}$
$\dfrac{1 \text{ ft}}{0.305 \text{ m}} = \dfrac{15 \text{ ft}}{x \text{ m}}$
$x = (15)(0.305)$
$x = 4.575 \text{ m}$
The room is 5.49 m by 4.575 m.

21. $\dfrac{1 \text{ in}}{2.54 \text{ cm}} = \dfrac{46 \text{ in}}{x \text{ cm}}$
$x = (46)(2.54)$
$x = 116.84 \text{ cm}$
$116.84 \text{ cm} = x \text{ m}$
$116.84 \text{ cm} = 1.1684 \text{ m}$
Each shelf is 1.1684 m long.
$(5)(1.1684) = \text{total meters}$
$5.84 = \text{total length in meters}$

22. $\dfrac{1 \text{ ft}}{0.305 \text{ m}} = \dfrac{60 \text{ ft}}{x \text{ m}}$
$x = (60)(0.305)$
$x = 18.3 \text{ m}$

23. $(14)(3) = \text{length of the building in m}$
$42 \text{ m} = \text{length of the building}$
$\dfrac{1 \text{ m}}{3.28 \text{ ft}} = \dfrac{42 \text{ m}}{x \text{ ft}}$
$x = (42)(3.28)$
$x = 137.76 \text{ ft}$

24. $\dfrac{1 \text{ mi}}{1{,}760 \text{ yd}} = \dfrac{26 \text{ mi}}{x \text{ yd}}$
$x = (26)(1{,}760)$
$x = 45{,}760 \text{ yd}$
$45{,}760 \text{ yd} + 343 \text{ yd} = 46{,}103$ total yards in the race
$\dfrac{1 \text{ yd}}{0.914 \text{ m}} = \dfrac{46{,}103 \text{ yd}}{x \text{ m}}$
$x = (46{,}103)(0.914)$
$x = 42{,}138.1 \text{ m (rounded)}$
$42{,}138.1 \text{ m} = x \text{ km}$
$42{,}138.1 \text{ m} = 42.1 \text{ km (rounded)}$

25. $\dfrac{1 \text{ km}}{0.621 \text{ mi}} = \dfrac{480 \text{ km}}{x \text{ mi}}$
$x = (480)(0.621)$
$x = 298$ miles (rounded) between the two cities
$(12)(24) = $ number of miles that can be driven on 12 gal of gas
$288 = $ number of miles that can be driven
You need more gas!!

Solutions

Section 9.2 Comprehension Checkpoint

1. 5 lb **2.** $2\frac{3}{4}$ tons **3.** 0.375 kg **4.** 42,730,000 cg **5.** 1,190.7 g **6.** 484 lb

9.2A Practice Set

1. $\dfrac{16 \text{ oz}}{1 \text{ lb}} = \dfrac{32 \text{ oz}}{x \text{ lb}}$
   $16x = 32$
   $x = 2$ lb

2. $\dfrac{2{,}000 \text{ lb}}{1 \text{ ton}} = \dfrac{400 \text{ lb}}{x \text{ ton}}$
   $2{,}000x = 400$
   $x = \dfrac{400}{2{,}000} = \dfrac{1}{5}$ ton

3. $\dfrac{1 \text{ lb}}{16 \text{ oz}} = \dfrac{25 \text{ lb}}{x \text{ oz}}$
   $x = (16)(25)$
   $x = 400$ oz

4. $\dfrac{1 \text{ ton}}{2{,}000 \text{ lb}} = \dfrac{4.5 \text{ tons}}{x \text{ lb}}$
   $x = (4.5)(2{,}000)$
   $x = 9{,}000$ lb

5. $\dfrac{1 \text{ lb}}{16 \text{ oz}} = \dfrac{2{,}000 \text{ lb (1 ton)}}{x \text{ oz}}$
   $x = (2{,}000)(16)$
   $x = 32{,}000$ oz

6. 4 g = x cg
   4 g = 400 cg
   Move the decimal point two places to the right.

7. 15 dg = x mg
   15 dg = 1,500 mg
   Move the decimal point two places to the right.

8. 200 kg = x g
   200 kg = 200,000 g
   Move the decimal point three places to the right.

9. 500 g = x kg
   500 g = 0.5 kg
   Move the decimal point three places to the left.

10. 250 mg = x g
    250 mg = 0.25 g
    Move the decimal point three places to the left.

11. $\dfrac{1 \text{ g}}{0.035 \text{ oz}} = \dfrac{12 \text{ g}}{x \text{ oz}}$
    $x = (12)(0.035)$
    $x = 0.42$ oz

12. $\dfrac{1 \text{ kg}}{2.2 \text{ lb}} = \dfrac{38 \text{ kg}}{x \text{ lb}}$
    $x = (2.2)(38)$
    $x = 83.6$ lb

13. $\dfrac{1 \text{ lb}}{454 \text{ g}} = \dfrac{60 \text{ lb}}{x \text{ g}}$
    $x = (60)(454)$
    $x = 27{,}240$ g

14. $\dfrac{1 \text{ oz}}{28.35 \text{ g}} = \dfrac{100 \text{ oz}}{x \text{ g}}$
    $x = (100)(28.35)$
    $x = 2{,}835$ g
    $2{,}835$ g = x kg
    $2{,}835$ g = 2.835 kg

15. $\dfrac{1 \text{ kg}}{2.2 \text{ lb}} = \dfrac{2 \text{ kg}}{x \text{ lb}}$
    $x = (2)(2.2)$
    $x = 4.4$ lb
    $\dfrac{1 \text{ lb}}{16 \text{ oz}} = \dfrac{4.4 \text{ lb}}{x \text{ oz}}$
    $x = (16)(4.4)$
    $x = 70.4$ oz

16. $\dfrac{1 \text{ oz}}{28.35 \text{ g}} = \dfrac{15 \text{ oz}}{x \text{ g}}$
    $x = (15)(28.35)$
    $x = 425.25 = 425$ g (rounded)

17. $\dfrac{1 \text{ ton}}{2{,}000 \text{ lb}} = \dfrac{x \text{ tons}}{45{,}000 \text{ lb}}$
    $2{,}000x = 45{,}000$
    $x = 22.5$ tons

18. $\dfrac{1 \text{ lb}}{0.454 \text{ kg}} = \dfrac{40 \text{ lb}}{x \text{ kg}}$
    $x = (40)(0.454)$
    $x = 18.16$ kg

19. $\dfrac{1 \text{ ton}}{2{,}000 \text{ lb}} = \dfrac{\frac{1}{4} \text{ ton}}{x \text{ lb}}$

$x = (2{,}000)\left(\dfrac{1}{4}\right)$
$x = 500 \text{ lb}$
$\dfrac{1 \text{ lb}}{0.454 \text{ kg}} = \dfrac{500 \text{ lb}}{x \text{ kg}}$
$x = (500)(0.454)$
$x = 227 \text{ kg}$

20. $\dfrac{1 \text{ lb}}{0.454 \text{ kg}} = \dfrac{2 \text{ lb}}{x \text{ kg}}$

$x = (2)(0.454)$
$x = 0.908 \text{ kg}$
$0.908 \text{ kg} = x \text{ mg}$
$0.908 \text{ kg} = 908{,}000 \text{ mg}$

Section 9.3   Comprehension Checkpoint

1. 128 fl oz   **2.** 16 c   **3.** 0.1475 daL   **4.** 8,500 L   **5.** 7.568 L   **6.** 2,508 gal

9.3A   Practice Set

1. $\dfrac{1 \text{ qt}}{2 \text{ pt}} = \dfrac{3 \text{ qt}}{x \text{ pt}}$
$x = (3)(2)$
$x = 6 \text{ pt}$

2. $\dfrac{1 \text{ fl oz}}{2 \text{ tbsp}} = \dfrac{28 \text{ fl oz}}{x \text{ tbsp}}$
$x = (2)(28)$
$x = 56 \text{ tbsp}$
$\dfrac{1 \text{ tbsp}}{3 \text{ tsp}} = \dfrac{56 \text{ tbsp}}{x \text{ tsp}}$
$x = (3)(56)$
$x = 168 \text{ tsp}$

3. $\dfrac{1 \text{ gal}}{4 \text{ qt}} = \dfrac{x \text{ gal}}{30 \text{ qt}}$
$4x = 30$
$x = \dfrac{30}{4} = 7\dfrac{1}{2} \text{ gal}$

4. $1 \text{ gal} = 4 \text{ qt}$
$1 \text{ qt} = 2 \text{ pt}$
$1 \text{ gal} = 8 \text{ pt} \ (4 \times 2)$
$1 \text{ pt} = 2 \text{ c}$
$1 \text{ gal} = 16 \text{ c} \ (8 \times 2)$

5. $\dfrac{2 \text{ pt}}{1 \text{ qt}} = \dfrac{7 \text{ pt}}{x \text{ qt}}$
$2x = 7$
$x = \dfrac{7}{2} = 3\dfrac{1}{2} \text{ qt}$

6. $8 \text{ fl oz} = 1 \text{ c}$
$2 \text{ c} = 1 \text{ pt} = 16 \text{ fl oz} \ (8 \times 2)$
$2 \text{ pt} = 1 \text{ qt} = 32 \text{ fl oz} \ (16 \times 2)$
$1 \text{ gal} = 4 \text{ qt} = 128 \text{ fl oz} \ (32 \times 4)$
$\dfrac{1 \text{ gal}}{128 \text{ fl oz}} = \dfrac{x \text{ gal}}{512 \text{ fl oz}}$
$128x = 512$
$x = 4 \text{ gal}$

7. $5 \text{ L} = x \text{ kL}$
$5 \text{ L} = 0.005 \text{ kL}$
Move the decimal point three places to the left.

8. $456 \text{ cL} = x \text{ L}$
$456 \text{ cL} = 4.56 \text{ L}$
Move the decimal point two places to the left.

9. $25 \text{ hL} = x \text{ dL}$
$25 \text{ hL} = 25{,}000 \text{ dL}$
Move the decimal point three places to the right.

10. $100 \text{ mL} = x \text{ kL}$
$100 \text{ mL} = 0.0001 \text{ kL}$
Move the decimal point six places to the left.

11. $20 \text{ kL} = x \text{ cL}$
$20 \text{ kL} = 2{,}000{,}000 \text{ cL}$
Move the decimal point five places to the right.

12. $68 \text{ daL} = x \text{ dL}$
$68 \text{ daL} = 6{,}800 \text{ dL}$
Move the decimal point two places to the right.

13. $\dfrac{1 \text{ pt}}{0.474 \text{ L}} = \dfrac{6 \text{ pt}}{x \text{ L}}$
    $x = (6)(0.474)$
    $x = 2.844 \text{ L}$

14. $\dfrac{1 \text{ qt}}{0.946 \text{ L}} = \dfrac{10 \text{ qt}}{x \text{ L}}$
    $x = (10)(0.946)$
    $x = 9.46 \text{ L}$
    $9.46 \text{ L} = x \text{ cL}$
    $9.46 \text{ L} = 946 \text{ cL}$

15. $\dfrac{1 \text{ L}}{0.264 \text{ gal}} = \dfrac{15 \text{ L}}{x \text{ gal}}$
    $x = (15)(0.264)$
    $x = 3.96 \text{ gal}$

16. $\dfrac{1 \text{ gal}}{3.785 \text{ L}} = \dfrac{20 \text{ gal}}{x \text{ L}}$
    $x = (20)(3.785)$
    $x = 75.7 \text{ L}$

17. $5 \text{ daL} = x \text{ L}$
    $5 \text{ daL} = 50 \text{ L}$
    $\dfrac{1 \text{ L}}{1.06 \text{ qt}} = \dfrac{50 \text{ L}}{x \text{ qt}}$
    $x = (50)(1.06)$
    $x = 53 \text{ qt}$

18. $825 \text{ mL} = x \text{ L}$
    $825 \text{ mL} = 0.825 \text{ L}$
    $\dfrac{1 \text{ L}}{2.11 \text{ pt}} = \dfrac{0.825 \text{ L}}{x \text{ pt}}$
    $x = (2.11)(0.825)$
    $x = 1.74 \text{ pt (rounded)}$

19. $\dfrac{1 \text{ L}}{33.8 \text{ fl oz}} = \dfrac{2 \text{ L}}{x \text{ fl oz}}$
    $x = (2)(33.8)$
    $x = 67.6 \text{ fl oz}$

20. $\dfrac{1 \text{ L}}{33.8 \text{ oz}} = \dfrac{x \text{ L}}{12 \text{ oz}}$
    $x = 0.355 \text{ L}$

21. $\dfrac{1 \text{ gal}}{3.785 \text{ L}} = \dfrac{10{,}000 \text{ gal}}{x \text{ L}}$
    $x = (3.785)(10{,}000)$
    $x = 37{,}850 \text{ L}$

22. $\dfrac{1 \text{ L}}{2.11 \text{ pt}} = \dfrac{4 \text{ L}}{x \text{ pt}}$
    $x = (4)(2.11)$
    $x = 8.44 \text{ pt}$

Section 9.4 Comprehension Checkpoint

1. 4 days
2. 12 weeks
3. 104°F
4. 35°C

9.4A Practice Set

1. $\dfrac{1 \text{ min}}{60 \text{ s}} = \dfrac{6 \text{ min}}{x \text{ s}}$
   $x = (60)(6)$
   $x = 360 \text{ s}$

2. $\dfrac{1 \text{ day}}{24 \text{ h}} = \dfrac{30 \text{ days}}{x \text{ h}}$
   $x = (30)(24)$
   $x = 720 \text{ h}$

3. $\dfrac{1 \text{ yr}}{365 \text{ days}} = \dfrac{3 \text{ yr}}{x \text{ days}}$
   $x = (3)(365)$
   $x = 1{,}095 \text{ days}$

4. $\dfrac{7 \text{ days}}{1 \text{ wk}} = \dfrac{133 \text{ days}}{x \text{ wk}}$
   $7x = 133$
   $x = 19 \text{ wk}$

5. $\dfrac{1 \text{ h}}{60 \text{ min}} = \dfrac{24 \text{ h (1 day)}}{x \text{ min}}$
   $x = (60)(24)$
   $x = 1{,}440 \text{ min}$

6. $\dfrac{1 \text{ min}}{60 \text{ s}} = \dfrac{60 \text{ min (1 h)}}{x \text{ s}}$
   $x = (60)(60)$
   $x = 3{,}600 \text{ s}$
   $\dfrac{1 \text{ h}}{3{,}600 \text{ s}} = \dfrac{24 \text{ h (1 day)}}{x \text{ s}}$
   $x = (24)(3{,}600)$
   $x = 86{,}400 \text{ s}$

7. $C = \frac{5}{9}(F - 32)$
   $C = \frac{5}{9}(50 - 32)$
   $C = \frac{5}{9}(18)$
   $C = 10°$

8. $C = \frac{5}{9}(F - 32)$
   $C = \frac{5}{9}(122 - 32)$
   $C = \frac{5}{9}(90)$
   $C = 50°$

9. $C = \frac{5}{9}(F - 32)$
   $C = \frac{5}{9}(203 - 32)$
   $C = \frac{5}{9}(171)$
   $C = 95°$

10. $F = 1.8C + 32$
    $F = 1.8(30) + 32$
    $F = 54 + 32$
    $F = 86°$

11. $F = 1.8C + 32$
    $F = 1.8(75) + 32$
    $F = 135 + 32$
    $F = 167°$

12. $F = 1.8C + 32$
    $F = 1.8(160) + 32$
    $F = 288 + 32$
    $F = 320°$

13. $\frac{1 \text{ min}}{60 \text{ s}} = \frac{4 \text{ min}}{x \text{ s}}$
    $x = (4)(60)$
    $x = 240$ s
    240 s + 25 s = total seconds
    265 s = total seconds

14. $\frac{1 \text{ day}}{24 \text{ h}} = \frac{x \text{ day}}{16 \text{ h}}$
    $24x = 16$
    $x = \frac{16}{24} = \frac{2}{3}$ day

15. $\frac{1 \text{ day}}{24 \text{ h}} = \frac{21 \text{ days}}{x \text{ h}}$
    $x = (24)(21)$
    $x = 504$ h
    504 h + 20 h = total time in hours
    524 h = total time in hours

16. From problem 6, you know that there are 86,400 seconds per day.
    $\frac{86,400 \text{ s}}{1 \text{ day}} = \frac{x \text{ s}}{365 \text{ days (1 yr)}}$
    $x = (86,400)(365)$
    $x = 31,536,000$ s per year
    $1,000,000,000 \div 31,536,000 =$ number of years
    $31.7 =$ number of years

17. $C = \frac{5}{9}(F - 32)$
    $C = \frac{5}{9}(73 - 32)$
    $C = \frac{5}{9}(41)$
    $C = 22.8°$

18. $F = 1.8C + 32$
    $F = 1.8(58) + 32$
    $F = 104.4 + 32$
    $F = 136.4°$

Section 9.5 Comprehension Checkpoint

1. 3 yd 1 ft 8 in
2. 1 day 1 h 30 min
3. 5 gal 3 qt 10 fl oz
4. 1 mi 1,460 yd
5. 21 yd 4 in
6. 1 h 17 min

9.5A Practice Set

1. 5 h 30 min
    $\underline{+\ 4\ \text{h}\ 15\ \text{min}}$
     9 h 45 min

2. 7 days 8 h 20 min
    $\underline{+\ 10\ \text{days}\ 15\ \text{h}\ 40\ \text{min}}$
    17 days 23 h 60 min
    $\phantom{17 \text{ days } 23 \text{ h}}\overbrace{\phantom{60 \text{ min}}}^{1\ h}$
    17 days 24 h
    $\phantom{17 \text{ days}}\overbrace{\phantom{24 \text{ h}}}^{1\ day}$
    18 days

3. 9 gal 3 qt 2 pt
     7 gal 2 qt 1 pt
    $\underline{+\ 1\ \text{gal}\ 3\ \text{qt}\ 3\ \text{pt}}$
    17 gal 8 qt 6 pt
    $\phantom{17\ \text{gal}\ 8\ \text{qt}}\overbrace{\phantom{6\ \text{pt}}}^{3\ qt}$
    17 gal 11 qt
    $\phantom{17\ \text{gal}}\overbrace{\phantom{11\ \text{qt}}}^{2\ gal\ 3\ qt}$
    19 gal 3 qt

4. 12 ft 6 in
    $\underline{-\ 8\ \text{ft}\ 1\ \text{in}}$
    1 yd 1 ft 5 in

5. $\overset{8}{9}$ min $\overset{80}{2\!\!\!/\!\!0}$ s
    $\underline{-\ 7\ \text{min}\ 35\ \text{s}}$
     1 min 45 s

6. $\overset{3}{4\!\!\!/}$ pt $\overset{4}{2\!\!\!/}$ c
    $\underline{-\ 3\ \text{pt}\ 3\ \text{c}}$
    $\phantom{-3 \text{ pt }}$ 1 c

7. 6 lb 4 oz
    $\underline{\times8}$
    48 lb 32 oz
    $\phantom{48\ \text{lb}}\overbrace{\phantom{32\ \text{oz}}}^{2\ lb}$
    50 lb

8. 25 min 30 s
    $\underline{\times9}$
    225 min 270 s
    $\phantom{225\ \text{min}}\overbrace{\phantom{270\ \text{s}}}^{4\ min\ 30\ s}$
    229 min 30 s
    $\overbrace{\phantom{229\ \text{min}}}^{3\ h\ 49\ min}$
    3 h 49 min 30 s

9. 8 gal 3 qt 5 pt
    $\underline{\times6}$
    48 gal 18 qt 30 pt
    $\phantom{48\ \text{gal}\ 18\ \text{qt}}\overbrace{\phantom{30\ \text{pt}}}^{15\ qt}$
    48 gal 33 qt
    $\phantom{48\ \text{gal}}\overbrace{\phantom{33\ \text{qt}}}^{8\ gal\ 1\ qt}$
    56 gal 1 qt

10. 2 lb 14 oz
    3$\overline{)\ 8\ \text{lb}\ 10\ \text{oz}}$
    $\underline{6}$
    2 lb = 32 oz
    $\overline{42\ \text{oz}}$
    $\underline{42\ \text{oz}}$

11. 3 h 21 min
    5$\overline{)\ 16\ \text{h}\ 45\ \text{min}}$
    $\underline{15}$
    1 h = 60 min
    $\overline{105\ \text{min}}$
    $105\ \text{min}$

12. 4 gal 1 pt
    6$\overline{)\ 24\ \text{gal}\ 3\ \text{qt}}$
    $\underline{24}$
    3 qt = 6 pt
    $\underline{6\ \text{pt}}$

13. 8 ft 3 in
    $\underline{\times24}$
    192 ft 72 in
    $\phantom{192\ \text{ft}}\overbrace{\phantom{72\ \text{in}}}^{6\ ft}$
    198 ft

14. $\dfrac{1\ \text{lb}}{16\ \text{oz}} = \dfrac{3\ \text{lb}}{x\ \text{oz}}$
    $x = (3)(16)$
    $x = 48$ oz
    $(48\ \text{oz})(\$0.245\ \text{per ounce}) =$ total cost
    $\$11.76 =$ total cost

15. 12
    160 lb$\overline{)\ 2{,}000\ \text{lb}}$
    $\phantom{160\ \text{lb})\ }\underline{1\ 60}$
    $\phantom{160\ \text{lb})\ \ \ }$400
    $\phantom{160\ \text{lb})\ \ \ }\underline{320}$
    $\phantom{160\ \text{lb})\ \ \ \ \ }$80
    The elevator can carry only 12 people (average size).

16. $\dfrac{1\ \text{lb}}{16\ \text{oz}} = \dfrac{40\ \text{lb}}{x\ \text{oz}}$
    $x = (16)(40)$
    $x = 640$ oz
    106
    6 oz$\overline{)\ 640\ \text{oz}}$
    $\phantom{6\ \text{oz})\ }\underline{6}$
    $\phantom{6\ \text{oz})\ \ }$40
    $\phantom{6\ \text{oz})\ \ }\underline{36}$
    $\phantom{6\ \text{oz})\ \ \ \ }$4
    106 hamburger patties weighing 6 oz each could be made from 40 lb of hamburger meat.

17. 3 ft 5 in
    $\underline{\times6}$
    18 ft 30 in
    $\phantom{18\ \text{ft}}\overbrace{\phantom{30\ \text{in}}}^{2\ ft\ 6\ in}$
    20 ft 6 in

18. 3 yr 2 mo 15 days
    $\underline{+\ 2\ \text{yr}\ 10\ \text{mo}\ 26\ \text{days}}$
    5 yr 12 mo 41 days
    $\phantom{5\ \text{yr}\ 12\ \text{mo}}\overbrace{\phantom{41\ \text{days}}}^{1\ yr}$
    6 yr 41 days
    $\phantom{6\ \text{yr}\ }\overbrace{\phantom{41\ \text{days}}}^{1\ mo\ 11\ days}$
    6 yr 1 mo 11 days

19.
$$\begin{array}{r} \overset{6}{\cancel{7}} \text{ ft } \overset{16}{\cancel{4}} \text{ in} \\ -\ 3 \text{ ft } 7 \text{ in} \\ \hline 3 \text{ ft } 9 \text{ in} \end{array}$$

20. $\dfrac{1 \text{ lb}}{16 \text{ oz}} = \dfrac{2 \text{ lb}}{x \text{ oz}}$
   $x = (2)(16)$
   $x = 32 \text{ oz}$
   $(32 \text{ oz})(\$0.29 \text{ per ounce}) = \text{regular cost}$
   $\$9.28 = \text{regular cost}$
   $\$13.95 - \$9.28 = \$4.67$
   $\$4.67$ is the increased amount for next-day delivery.

Practice Test 9A

1. 24 qt
2. 500 cL
3. 720 min
4. 23.64 in
5. 192 oz
6. 0.5 hg
7. 6 wk
8. 35°C
9. 10.896 kg
10. 150 cm
11. 40 fl oz
12. 98.4 ft
13. 25 lb
14. 7 gal 3 qt
15. 5 gal 4 fl oz
16. 6 qt
17. 71.1 cL
18. 1 ft 9 in

Skills Review   Chapters 1–9

1. eight hundred seventy-five thousand, two hundred six
2. 74,600
3. 10,000
4. 8.9
5. $3\dfrac{32}{45}$
6. $\dfrac{23}{84}$
7. $\dfrac{1}{12}$
8. $\dfrac{4}{5}$
9. 0.1665
10. 150
11. 0.425
12. 72%
13. 0.32
14. $\dfrac{19}{20}$
15. 243
16. 2
17. $6\dfrac{1}{4}\%$
18. eighty-nine hundredths
19. seven and seven thousandths
20. 30
21. a. <
    b. <
    c. >
22. a. 3.6
    b. $8\dfrac{4}{5}$
23. 180
24. 12
25. −3
26. −73
27. 4
28. 11
29. 195
30. 192 oz
31. 6.78 L
32. 15°C

10 INTRODUCTION TO GEOMETRY

OUTLINE

10.1　Terminology

10.2　Polygons

10.3　Circles

10.4　Area

10.5　Volume

In Chapter 9 you learned the fundamentals of measurement systems currently in use around the world. Measuring objects of different shapes and sizes is a skill that virtually everyone employs at one time or another. Being able to recognize geometric forms and then to use formulas to find the perimeter, area, or volume of these forms helps you to standardize and simplify the measurement process. Geometric formulas will help you answer questions such as, How much fencing do I need to enclose the yard? or How many square yards of carpeting are needed for the living room? or How much will a storage box hold? Each of these types of problems will be discussed in this chapter.

SKILLS PREVIEW

1. Write the proper designation for the angle shown in the drawing.

   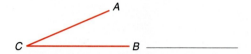 _____

2. Identify each set of lines as either parallel, perpendicular, or neither.

   a.    a. _____

   b.    b. _____

   c. 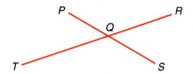   c. _____

3. Identify the two sets of angles that are equal.

   _____

   _____

4. Give the name of each polygon.

   a.    a. _____

   b.   b. _____

   c.    c. _____

   d. 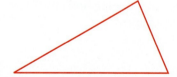   d. _____

   e.   e. _____

Skills Preview

5. Draw and label one diameter and one radius in the circle.

6. Calculate the circumference of a circle if the diameter is 14 in. _____

7. Calculate the radius of a circle if the circumference is 20 ft. _____

8. Find the area of each polygon or circle.

a. 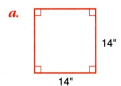 *a.* _____

b. 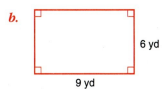 *b.* _____

c. 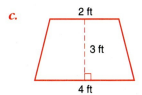 *c.* _____

d. 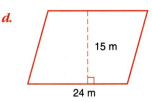 *d.* _____

e. 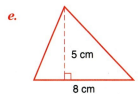 *e.* _____

f.   *f.* _____

9. Calculate the volume of each geometric solid.

a. 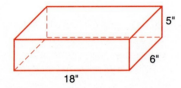 a. _____

b. 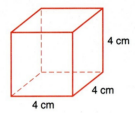 b. _____

c. 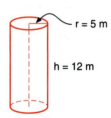 c. _____

d. 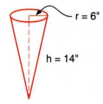 d. _____

e.   e. _____

10.1 TERMINOLOGY

Geometry is the measurement of different shapes. Before studying any formulas, you need to learn some vocabulary terms.

A **point** is represented by a dot and is designated by a capital letter placed next to it. A point has neither length nor width nor height. A **line segment** connects two points; it is the shortest distance between those points. Two points, A and B, are connected by a straight line segment in the following drawing:

The line segment is designated $\overline{AB}$.

Parallel lines are lines that are the same distance apart throughout their length. These lines never intersect. Parallel lines are illustrated in the next drawing:

A **ray** is a straight line extending indefinitely from a point. The next drawing shows a ray beginning at point C and extending through point D. The symbol for a ray is $\overrightarrow{CD}$. Notice the difference between the symbol for a ray and the symbol for a straight line.

An **angle** is formed when two rays have a common endpoint. The common endpoint is called the **vertex**. The drawing shows an angle created by two rays that meet at point C.

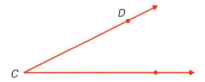

The angle created by the two rays can be designated in either of two ways. The first way is by ∠DCE, which is read "angle DCE." Note that the letter at the vertex is the middle letter. The second way to designate the same angle is as ∠C, meaning the angle formed by two rays meeting at point C. This is read simply "angle C."

When two lines intersect to create angles of 90°, the lines are said to be **perpendicular.** An angle of 90° is called a **right angle.** The symbol used to indicate that two lines are perpendicular is ⊥. The lines in the following drawing are perpendicular.

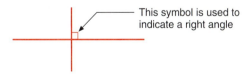

This symbol is used to indicate a right angle

Lines may intersect without being at right angles. When this happens, the opposite angles are equal. In the next drawing, ∠AOB = ∠COD. Also, ∠AOC = ∠BOC.

588   CHAPTER 10   Introduction to Geometry

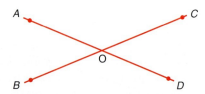

EXAMPLE 1   Name the angle shown in the following drawing and identify the vertex.

SOLUTION

The angle is ∠ ABC or ∠B.   *The letter at the vertex must be the middle letter of three or the one letter used by itself.*

The vertex is at point B.

EXAMPLE 2   Label each set of lines as either parallel, perpendicular, or neither.

a.   *b.*   *c.*

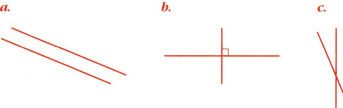

SOLUTIONS

a.   Parallel   *These lines never intersect.*
b.   Perpendicular   *These lines intersect at right angles.*
c.   Neither   *These lines intersect but not at right angles.*

EXAMPLE 3   Name the angles that are equal in the drawing.

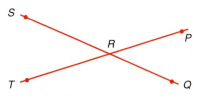

SOLUTION

∠SRT = ∠PRQ   *These angles are on opposite sides of the intersection point.*

∠PRS = ∠QRT

Note that the angles could not be identified by the vertex letter in this case, because all angles have the same vertex.

Work the Comprehension Checkpoint problems to make sure you understand the terminology in this section. If you have difficulty with any of these problems, review the appropriate material before trying the Practice Sets.

Terminology

COMPREHENSION CHECKPOINT

1. Name the following angle using all three letters.

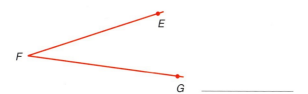

_____

2. Name the following angle using only the vertex letter.

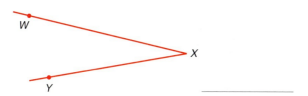

_____

3. Identify each set of lines as either parallel, perpendicular, or neither.

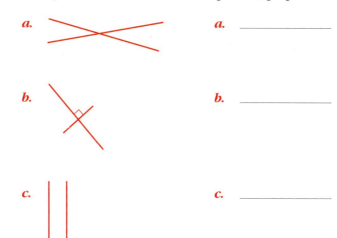

a. _____

b. _____

c. _____

4. Identify the two sets of angles that are equal.

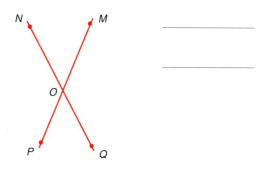

_____

_____

10.1A PRACTICE SET

In problems 1–5, use two methods to identify each angle. First use the method where all three letters are used, and second use the vertex letter method.

1.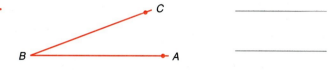

    _____

    _____

2.

    _____

    _____

3.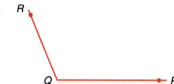

    _____

    _____

4.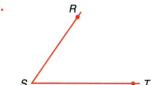

    _____

    _____

5.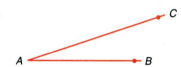

    _____

    _____

In problems 6–10, identify each set of lines as either parallel, perpendicular, or neither.

6.

    _____

7.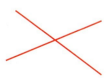

    _____

8.

9.

10.

In problems 11–15, list each set of equal angles.

11.

12.

13.

14.

15.

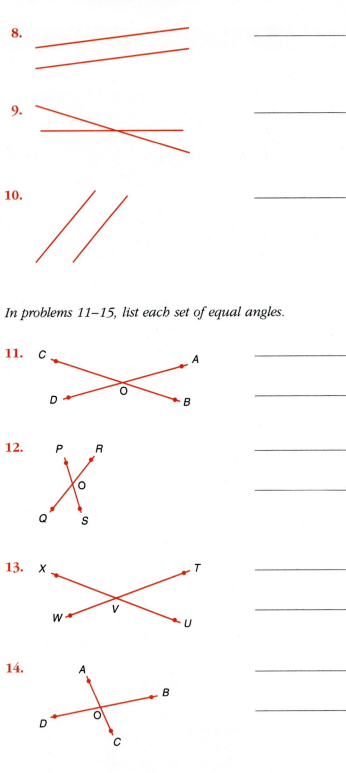

10.1B PRACTICE SET

In problems 1–5, use two methods to identify each angle. First use the method where all three letters are used, and second use the vertex letter method.

1.

2.

3.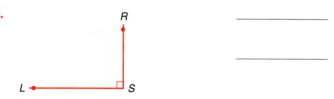

4.

5.

In problems 6–10, identify each set of lines as either parallel, perpendicular, or neither.

6.

7.

8.

9.

10.

In problems 11–15, list each set of equal angles.

11.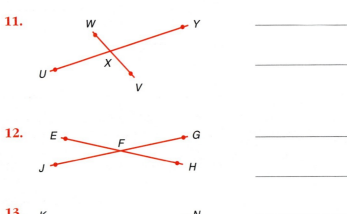

12.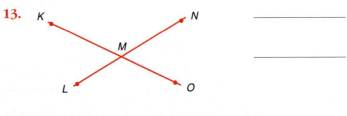

13.

14.

15.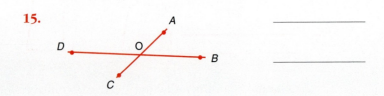

10.2 POLYGONS

A **polygon** is a geometric figure with three or more sides. Each side intersects with two other sides to close the polygon.

A **square** has four sides of equal length. The opposite sides are parallel and each of the angles is a right angle.

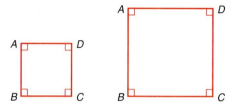

A **rectangle** has four sides. Opposite sides are equal in length and parallel. Each angle is a right angle. In each rectangle pictured in the following drawing, side $\overline{AB}$ is equal in length to side $\overline{DC}$. Also, side $\overline{AD}$ is equal in length to side $\overline{BC}$.

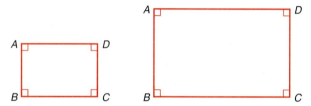

A **parallelogram** is similar to a rectangle. Opposite sides are equal and parallel. The difference is that the angles in a parallelogram are not right angles. In each of the following parallelograms, side $\overline{AB}$ is equal in length and parallel to side $\overline{DC}$. Also, side $\overline{AD}$ is equal in length and parallel to side $\overline{BC}$.

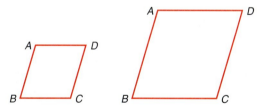

It is important to note that opposite angles in a parallelogram are equal. In the preceding parallelograms, $\angle ABC = \angle ADC$. Also, $\angle BCD = \angle DAB$.

A final type of polygon with four sides is a **trapezoid.** Two of the sides are parallel but not equal in length. The angles are not necessarily equal. Some trapezoids are pictured here.

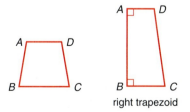
right trapezoid

A **triangle** is a three-sided polygon. If two of the sides are perpendicular, a **right triangle** is formed. The sides of a triangle can all be the same length (an **equilateral triangle**), or all of the sides may be different lengths (a **scalene triangle**). Several types of triangles are illustrated.

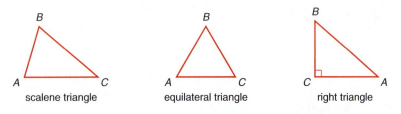

scalene triangle	equilateral triangle	right triangle

EXAMPLE 1 Identify each polygon by name.

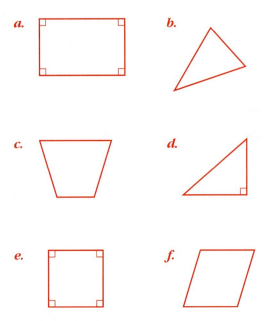

SOLUTIONS

a. Rectangle. Opposite sides are equal and parallel, and the angles are all right angles.
b. Triangle. There are only three sides.
c. Trapezoid. Two sides are parallel.
d. Right triangle. There is a right angle in the triangle.
e. Square. Four sides are the same length. All angles are right angles.
f. Parallelogram. Opposite sides are equal and parallel. None of the angles are right angles.

Perimeter

The **perimeter** is the distance around the edge of a polygon. For any polygon, the perimeter is the sum of the lengths of the sides. A formula could be derived for the perimeter of each type of polygon. However, one formula will work for all polygons:

$$p = s_1 + s_2 + s_3 + \cdots + s_n$$

The p stands for the perimeter of any polygon. The number to the lower right of each s is called a **subscript.** Subscripts are used to distinguish between different objects—in this case, sides of a polygon. The subscript n simply means the total number of sides. The formula simply says to add the lengths of the sides together until you have added all the sides. Then you have the perimeter.

Polygons

EXAMPLE 2   What is the perimeter of the following square?

SOLUTION

$p = s_1 + s_2 + \cdots + s_n$ *Write the formula.*

$p = 5 + 5 + 5 + 5$ *Substitute the given lengths. Because it is a square, all sides have the same length.*

$p = 20$ ft   *Add the lengths.*

The perimeter of the square is 20 ft.

EXAMPLE 3   What is the perimeter of the following rectangle?

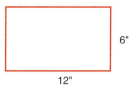

SOLUTION

$p = s_1 + s_2 + \cdots + s_n$ *Write the formula.*

$p = 12 + 12 + 6 + 6$ *Substitute the given lengths.*

$p = 36$ in   *Add the lengths.*

The perimeter is 36 in.

 Note that you are given the measurements of only two sides, but the sides are equal and parallel. This must be a rectangle. The lengths of the unlabeled sides must be the same as the lengths of the labeled sides.

EXAMPLE 4   What is the perimeter of the trapezoid?

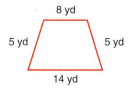

SOLUTION

$p = s_1 + s_2 + \cdots + s_n$ *Write the formula.*

$p = 14 + 5 + 8 + 5$ *Substitute the given lengths.*

$p = 32$ yd   *Add the given lengths.*

The perimeter is 32 yd.

It is often helpful with word problems to draw a sketch of the problem. It does not have to be drawn to scale. The idea here is to help you visualize the problem.

EXAMPLE 5 A rectangular city park measures 80 yards by 140 yards. How many times would a person have to walk around the outside edge of the park to walk one mile?

SOLUTION

The park is rectangular. What is the distance around the outside edge?

$p = s_1 + s_2 + \cdots + s_n$ *Write the formula for perimeter.*

$p = 140 + 80 + 140 + 80$ *Substitute the given distances.*

$p = 440$ yd   *The distance around the park is 440 yards.*

$1{,}760 \div 440 = 4$ *There are 1,760 yards in one mile. You must walk around the park four times to walk one mile.*

EXAMPLE 6 Joanne wants to put a fence around her garden. The dimensions of the garden are shown. If fencing costs $4.95 per yard, what will it cost to put a fence around the garden?

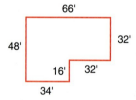

SOLUTION

$p = s_1 + s_2 + \cdots + s_n$ *First find the perimeter of the garden.*

$p = 66 + 32 + 32 + 16 + 34 + 48$

$p = 228$ ft   *She needs 228 feet of fencing.*

$228 \div 3 = 76$ yd   *There are 3 feet per yard. She needs 76 yards of fencing.*

$76 \times \$4.95 = \$376.20$

The cost of the fencing is $376.20.

Work the Comprehension Checkpoint problems to check your understanding of the material in this section. If you have difficulty with these problems, review the appropriate material before going on to the Practice Sets.

Polygons

COMPREHENSION CHECKPOINT

1. Label each polygon with its correct name.

   a.

   b.

   a. _____   b. _____

   c.

   d.

   c. _____   d. _____

   e.

   f.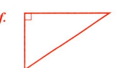

   e. _____   f. _____

2. What is the perimeter of each polygon?

   a.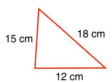

   15 cm, 18 cm, 12 cm

   b.

   6", 6"

   a. _____   b. _____

   c.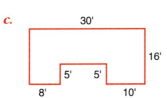

   30', 16', 5', 5', 8', 10'

   d.

   18 m, 24 m

   c. _____   d. _____

   e.

   30 yd, 20 yd, 20 yd, 50 yd

   e. _____

3. What is the perimeter of an equilateral triangle if one side measures 28 meters in length? _____

10.2A PRACTICE SET

In problems 1–5, label each polygon with its proper name.

1.
   _____

2.
   _____

3.
   _____

4.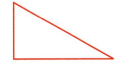
   _____

5.
   _____

Applications

6. What is the perimeter of a square with sides of 8 cm?

7. What is the perimeter of a rectangle with sides of 42 m and 29 m?

8. What is the perimeter of a trapezoid with sides of 19 in, 24 in, 19 in, and 12 in?

9. What is the perimeter of a parallelogram if the length of the longer side is 30 yd and the length of the shorter side is 18 yd?

10. What is the perimeter of a triangle with sides of 50 ft, 42 ft, and 24 ft?

11. What is the perimeter of a computer monitor screen that is 12 in by 9 in?

12. What is the perimeter of a football field if the longer sides are 120 yd long and the shorter sides are 42 yd wide?

13. A square card table has metal edging around the perimeter. If the edging costs $0.38 per foot, how much will it cost to edge a card table if the length of a side is 3.5 ft?

14. A standard-size door for a house is about 3 ft by 6 ft 8 in. What is the perimeter of the door?

15. A classroom is a rectangle 25 ft by 18 ft. There is one door leading into the room that is 3 ft wide. A chalkboard, mounted on the wall 2.5 ft from the floor, is 8 ft long. The principal wants to put molding around the room, 3 ft from the floor, to keep the tables from hitting the walls. How much molding is needed?

16. A triangular field is 150 yd on one side, 125 yd on another side, and 165 yd on the third side. How many times would you have to jog around the field to jog 3 miles?

17. The perimeter of a square is 48 ft. What would be the length of each side of an equilateral triangle if the perimeter of the triangle is to be the same as that of the square?

18. Fence posts are spaced every 4 ft around a field. How many posts are there if the field is rectangular and measures 40 yd by 30 yd?

19. An executive desk measures 6 ft by 3 ft. A small pencil tray is cut around the desk 3 inches in from each short side and 2 inches in from each long side. What is the perimeter of the pencil tray?

20. Mr. Escobedo wants to fence in his backyard. In addition to the measurements shown, he knows that the perimeter of the rectangular house is 160 ft. If fencing costs $8.90 per yard and he can buy it only by the yard, how much will it cost him to buy the fencing for the backyard?

10.2B PRACTICE SET

In problems 1–5, label each polygon with its proper name.

1. _____

2. _____

3. _____

4. _____

5. _____

Applications

6. What is the perimeter of a rectangle with sides of 30 m and 24 m?
   _____

7. What is the perimeter of a square with sides of 75 cm?
   _____

8. What is the perimeter of a parallelogram with sides of 48 yd and 32 yd?
   _____

9. The sides of a triangle are 26 ft, 32 ft, and 29 ft. What is the perimeter of the triangle?
   _____

10. What is the perimeter of a trapezoid that has sides of 40 m, 36 m, 6 m, and 6 m?
    _____

11. How much metal stripping would be required to go around the edge of a blackboard that is 8 ft by 4 ft?
    _____

603

12. How much would it cost to put up a fence around a rectangular field that measured 78 ft by 60 ft if fencing cost $2.58 per foot?

13. How much edging would be needed to go around the base of a mobile home that measures 70 ft by 14 ft?

14. The distance between bases on a professional baseball field is 90 ft. The distance between bases on a Little League field is 60 ft. How much farther must a batter run around the bases in major league baseball than in Little League baseball?

15. How long would a sidewalk need to be if it was to be built around the park pictured in the drawing?

60'
40'
30'
45'
18'

16. Chad jogs around a city block each morning. The block measures 50 yd by 60 yd. How many times would he have to run around the block to run 1 mile?

17. What is the perimeter (in yards) of an airport runway that is 3 mi long and 125 yd wide?

18. A skyscraper sits on a city block that is 120 ft by 140 ft. The skyscraper is 15 ft from the edge of the block on all sides. What is the perimeter of the building?

19. A dresser has nine drawers arranged in three rows of three drawers each. Each drawer measures 20 in by 7.5 in. There is a 1-in space between drawers and 1 in between the drawers and the edges of the dresser. What is the perimeter of the front of the dresser?

20. A ship on a training mission sailed 25 miles due east and then turned due north and traveled 40 miles. If the captain made two more left turns and ended up exactly where the voyage started,

  a. What polygon would best describe the ship's journey?

  _____

  b. How far did the ship travel?

  _____

10.3 CIRCLES

A **circle** is a curved line connecting a series of points, each of which is the same distance from a central point. The length of the curved line is called the **circumference** (instead of *perimeter,* as in a polygon). A line that connects two points on the circumference and passes through the center is called a **diameter.** A **radius** is one-half of a diameter; it is a line that goes from the center of the circle to a point on the circumference. There are an infinite number of radii (plural of radius) in a circle.

EXAMPLE 1 Calculate the diameter of circle *a* and the radius of circle *b*.

a.   b.

SOLUTIONS

a. 5 in × 2 = 10 in   *Since the radius is one-half of the diameter, multiply the radius by 2 to find the length of the diameter.*

The diameter is 10 in.

b. 12 cm ÷ 2 = 6 cm   *Divide the diameter by 2 to find the length of the radius.*

The radius is 6 cm.

Calculating the circumference requires some explanation. Take any circular object such as a glass, a plate, or a can. Measure the circumference. (You can do this with a piece of string. Wrap the string around the object one full time and then measure the length of the string.) Next, measure the diameter. This will at best be an estimate, because of the difficulty of determining the exact center of the object. (There is a method for finding the exact center of a circle, but it is beyond the scope of this text.) Finally, divide the length of the circumference by the length of the diameter. You should get approximately 3.14. Any circle that you use will produce approximately the same result. (These results are approximate because of the inaccuracy of the measurements.)

Remember from Chapter 6 that the decimal fraction 3.1415927 . . . has a special name and symbol. It is called *pi* (from the Greek alphabet), and the symbol is π. Technically, this is not π but a very good approximation. Thus, the following formula can be derived:

$$\frac{C}{d} = \pi$$

In the formula, *C* is the circumference of a circle and *d* is the diameter. From this formula, it follows that

$$\pi d = C \quad \text{and} \quad \frac{C}{\pi} = d$$

These formulas will be used to solve problems involving circles. In working problems, π is usually rounded to 3.14. If you prefer to use a fraction, you can use $\frac{22}{7}$.

EXAMPLE 2 What is the circumference of a circle with a diameter of 15 meters?

SOLUTION

$\pi d = C$ *Write the formula needed to solve the problem.*

$(3.14)(15) = C$ *Substitute the given values.*

$47.1 \text{ m} = C$ *Solve the problem.*

The circumference is 47.1 m.

EXAMPLE 3 What is the diameter of a circle if the circumference is 94.2 cm?

SOLUTION

$d = \dfrac{C}{\pi}$ *Write the formula needed to solve the problem.*

$d = \dfrac{94.2}{3.14}$ *Substitute the given values.*

$d = 30 \text{ cm}$ *Solve the problem.*

The diameter is 30 cm.

When solving a word problem, you must first understand exactly what you are trying to find. Consider the information that you are given. Determine the relationship between what you know and what you are trying to find. Select a formula (or formulas) that will give you the desired answer.

EXAMPLE 4 Phil Anderson wants to put up a circular sign with a 3-ft radius to advertise his restaurant. How much neon lighting must he buy to go around the outside edge of the sign if it can be bought only in 1-ft-long sections?

SOLUTION

$2r\pi = C$ *Because twice the radius is the diameter, you can substitute $2r$ for d in the formula $\pi d = C$.*

$(2)(3)(3.14) = C$ *Substitute the given values.*

$18.84 = C$ *Solve the problem.*

The circumference is 18.84 ft. Since he can buy the edging only in 1-ft increments, he must buy 19 ft of edging to encircle the sign.

Work the Comprehension Checkpoint problems to measure your understanding of the material in this section. If you have difficulty with these problems, review this section before going on to the Practice Sets.

COMPREHENSION CHECKPOINT

1. Calculate the diameter and the circumference of the circle.

   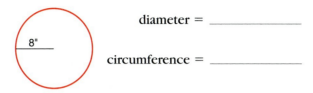

   diameter = _____

   circumference = _____

2. Determine the diameter of a circle if the circumference is 20 in. Round your answer to the nearest tenth.

   _____

3. What is the radius of a circle if the circumference is 24 in? Round your answer to the nearest tenth.

   _____

10.3A PRACTICE SET

In problems 1–4, calculate the circumference of each circle.

1.  _____

2.  _____

3. _____

4.  _____

In problems 5 and 6, calculate the diameter and radius of a circle with the given circumference. Round each answer to the nearest tenth.

5. $C = 28$ in $d =$ _____

     $r =$ _____

6. $C = 42$ ft $d =$ _____

     $r =$ _____

Applications

In each word problem, round the answer to the nearest tenth.

7. What is the diameter of a silver dollar if the circumference is $4\frac{7}{8}$ in?

   _____

8. An oscillating fan has a diameter of 12 in. What is the length of the metal strip around the outside edge of the safety guard?

   _____

9. Emma Blackburn uses a circular magnifying glass when reading. What is the diameter of the reading glass if the circumference is 6.25 in?

   _____

10. The circular mouth of a vase has a radius of 3 in. What is the circumference of the mouth of the vase?

    _____

11. What is the radius of a 16-oz soft drink bottle if the circumference is 9.5 in?

12. What is the diameter of a circular conference table if the circumference is 25.12 ft?

13. The radius of the earth is about 4,000 miles. A communications satellite is in a circular orbit 150 miles above the earth. How many miles does the satellite travel in one full revolution around the earth?

14. A flower garden has a circumference of 78.5 ft. There is a 2-ft-wide cement walking path around the garden. What is the circumference of the outside edge of the walking path?

15. The minute hand on a courthouse clock is 24 in long. What is the length of the arc the minute hand will make in 30 minutes?

16. An airplane is in a circular holding pattern above an airport. The plane travels 18 miles each time it circles the airport. If the airport is the center of the circle, how far is the plane from the airport?

17. One curved piece of track for a model railroad is 8 in long. It takes 20 of the curved pieces to make a circular track. If the track is to be mounted on a piece of fiberboard, what is the minimum width of the board? Round your answer to the nearest tenth of an inch.

18. A car tire has a radius of 14 in. How far will the car travel with one complete revolution of the tire?

19. How many revolutions per minute will the tire in problem 18 make if the car is traveling 60 miles per hour (mph)?

20. A bicycle racecourse is shown. The cyclists begin at point A, travel around the curve to point B, and then go around the second curve to point C. From point C they return on the straight line to point A. The curves between points A and B and points B and C are semicircles (half circles). If the distance between A and C is 250 yd and B is the midpoint between A and C, how many miles will the cyclists ride if the race is 20 laps?

10.3B PRACTICE SET

In problems 1–4, calculate the circumference of each circle.

1. 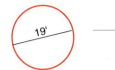 _____

2.  _____

3. _____

4.  _____

In problems 5 and 6, calculate the diameter and radius of a circle with the given circumference. Round each answer to the nearest tenth.

5. $C = 14$ m $d =$ _____

                     $r =$ _____

6. $C = 60$ cm $d =$ _____

                     $r =$ _____

Applications

In each word problem, round the answer to the nearest tenth.

7. Calculate the circumference of a dinner plate with a 10-in diameter.

   _____

8. What is the circumference of a table with a radius of 2 ft?

   _____

9. What is the diameter of a circle with a circumference of 25 ft?

   _____

10. What is the radius of a circle with a circumference of 160 cm?

    _____

11. A model gasoline-powered plane is controlled by a stout line. The maximum length of the line is 40 ft. What is the maximum circumference of the circle that the plane will fly around?

12. How many times would a string that is 42 cm long wrap around a spool that has a radius of 0.5 cm?

13. Jennifer wanted to place circular stones from her house to her garage door, a distance of 20 ft. Each stone has a circumference of 56.52 in. If she uses only whole stones, how many stones will she need?

14. A merry-go-round has a radius of 8 ft. For safety reasons, there can be only one child every 2 ft while the merry-go-round is working. How many children can ride at one time?

15. A satellite makes a circular orbit of 27,004 mi around the earth. If the radius of the earth is 4,000 mi, how high above the earth is the satellite?

16. A computer diskette for one of the early computers had a radius of 12 in. What was the circumference of the diskette?

17. A ball makes 12 revolutions while rolling 612.3 cm. What is the radius of the ball?

18. A circle is inscribed in a square (it is inside the square and touches each side of the square). The circumference of the circle is 94.2 m. What is the perimeter of the square?

19. Framing for a mirror costs $2.50 per foot. What would it cost to frame a mirror that has a radius of 3 ft?

20. A cross section of pipe is 1 in thick. The circumference of the inside edge of the pipe is 47.1 in. What is the circumference of the outside edge of the pipe?

10.4 AREA

Polygons

Area is the amount of space inside a polygon or circle. Area is always measured in square units. Remember that a square is a polygon with four equal sides. One square inch is pictured below.

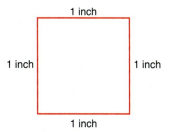

How many square inches, like the one pictured in the drawing, would be in a square with a side 2 inches long?

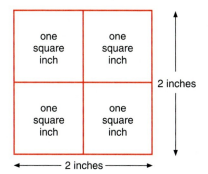

As you can see, there are 4 square inches in a square that measures 2 inches on each side. The area of the square then is 4 square inches, or 4 in^2.

How many square inches would be in a rectangle 4 inches long and 3 inches wide?

There are 12 square inches in a rectangle 4 inches long and 3 inches wide. The area of this rectangle is 12 square inches, or 12 in^2.

These first two examples should not imply that area is always measured in square inches. It is measured in square units. If a polygon is measured in meters, the area will be given in square meters. If a polygon is measured in feet, the area will be given in square feet.

Formulas are used to compute the area of polygons and circles. From the information you have just read, you can see that the formula for the area of a square is

$$A = s^2$$

where A represents area and s represents the length of the side of the square. The formula for the area of a rectangle is

$$A = lw$$

where A represents area, l represents the length of the rectangle, and w represents the width of the rectangle. In each formula for area that you learn, A will represent the area of the polygon or circle.

EXAMPLE 1 What is the area of a window that is a 3-foot square?

SOLUTION

$A = s^2$ *Select the formula needed to solve the problem. The window is a square, so use the formula for area of a square.*

$A = 3^2$ *Substitute the given value.*

$A = (3)(3)$ *Solve the problem.*
$A = 9 \text{ ft}^2$

The area of the window is 9 ft^2.

EXAMPLE 2 A city park is 48 yd long and 32 yd wide. What is the area of the park?

SOLUTION

$A = lw$ *Select the formula needed to solve the problem. The park is a rectangle, so use the formula for area of a rectangle.*

$A = (48)(32)$ *Substitute the given values.*

$A = 1,536 \text{ yd}^2$ *Solve the problem.*

The area of the city park is $1,536 \text{ yd}^2$.

Each polygon has a different formula for determining its area. Different polygons and the area formula for each are discussed below.

Parallelogram

Since a parallelogram can be converted to an equivalent rectangle, as will be demonstrated, the formula for its area is similar to that for the area of a rectangle. If the shaded portion in the following drawing could be moved to the other end of the parallelogram, a rectangle would be formed. It should be noted, though, that the height of a parallelogram is the perpendicular distance between the sides. It should not be confused with the length of the shorter side.

Area

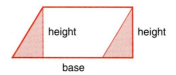

EXAMPLE 3 What is the area of the parallelogram?

SOLUTION

$A = bh$ *Write the formula for area of a parallelogram.*

$A = (18)(5)$ *Substitute the given values.*

$A = 90 \text{ in}^2$ *Solve the problem.*

The area of the parallelogram is 90 in².

Triangle

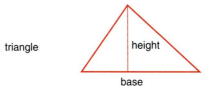

It can be shown that any triangle is one-half of a rectangle. Since this is true, the area of a triangle is one-half the area of a rectangle.

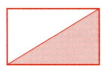 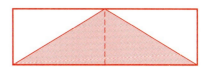

The base of a triangle can be any of the three sides. The height is the perpendicular distance from a vertex to the opposite base.

EXAMPLE 4 What is the area of the triangle?

SOLUTION

$A = \dfrac{1}{2} bh$ *Write the formula for the area of a triangle.*

$A = \dfrac{1}{2}(8)(6)$ *Substitute the given values.*

$A = 24 \text{ cm}^2$ *Solve the problem.*

The area of the triangle is 24 cm².

Trapezoid

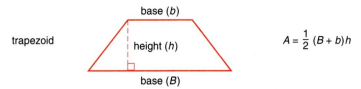

A trapezoid is different from the other polygons in that it is not similar to a rectangle (except that both have four sides). The height of a trapezoid is the perpendicular distance between the parallel sides.

EXAMPLE 5   Find the area of the trapezoid.

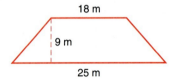

SOLUTION

$A = \frac{1}{2}(B + b)h$ *Write the formula for area of a trapezoid.*

$A = \frac{1}{2}(25 + 18)9$ *Substitute the given values.*

$A = 193.5 \text{ m}^2$ *Solve the problem.*

The area of the trapezoid is 193.5 m^2.

Area of Any Geometric Figure

The total area of any geometric figure is the sum of the areas of its parts.

EXAMPLE 6   What is the area of the end of a barn as shown?

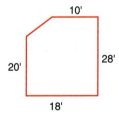

SOLUTION

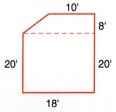

Divide the figure into geometric shapes that you have formulas for. In this case, you can divide it into a rectangle and a trapezoid. Now compute the area of each figure.

$A = lw$
$A = (20)(18)$ *The area of the rectangular bottom portion is 360 square feet.*
$A = 360 \text{ ft}^2$

$A = \frac{1}{2}(B + b)h$

Area

$$A = \frac{1}{2}(18 + 10)8$$

$A = 112 \text{ ft}^2$ *The area of the trapezoid portion is 112 square feet.*

The area of the side of the barn is $360 \text{ ft}^2 + 112 \text{ ft}^2 = 472 \text{ ft}^2$.

Circles

The formula for the area of a circle is

$$A = \pi r^2$$

Like the area of polygons, circle area is given in square units.

EXAMPLE 7   What is the area of a circle with a radius of 9 inches?

SOLUTION

$A = \pi r^2$ *Write the formula for the area of a circle.*

$A = (3.14)(9)^2$ *Substitute the given values.*

$A = (3.14)(9)(9)$ *Solve the problem.*

$A = 254.34 \text{ in}^2$

The area of the circle is 254.34 in^2.

Sometimes you may have to calculate the radius before you can use the formula.

EXAMPLE 8   What is the area of a circle with a circumference of 84 ft?

SOLUTION

$\dfrac{C}{\pi} = d$ *You need to know the radius. Find the diameter and divide it in half.*

$\dfrac{84}{3.14} = d$

$26.75 = d$
$26.75 \div 2 = r$
$13.38 \text{ (rounded)} = r$

$A = \pi r^2$ *Now that you know the radius, you can use the area formula.*

$A = (3.14)(13.38)^2$
$A = (3.14)(13.38)(13.38)$
$A = 562.14 \text{ ft}^2$

The area of the circle is 562.14 ft^2. It should be noted that the area is at best an approximation because of rounding.

Work the Comprehension Checkpoint problems to make sure you understand the concept of area and how to use the formulas in this section. If you have difficulty with any of these problems, review the appropriate material before going on to the Practice Sets.

COMPREHENSION CHECKPOINT

Find the area of each figure.

1.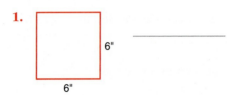

2.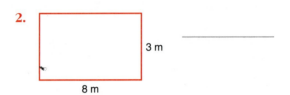

3.

4.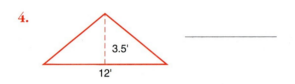

5.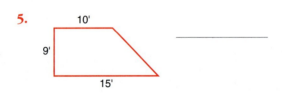

6.

10.4A PRACTICE SET

In problems 1–12, calculate the area of the polygon or circle from the given information.

1.

2.

3.

4.

5.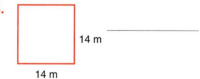

6.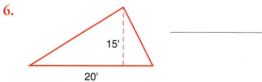

7. Square: $s = 12$ m _____

8. Rectangle: $l = 8$ in, $w = 4.75$ in _____

9. Parallelogram: $b = 14$ cm, $h = 7$ cm _____

10. Triangle: $b = 16$ yd, $h = 17$ yd _____

11. Trapezoid: $B = 10$ ft, $b = 7$ ft, $h = 3$ ft _____

12. Circle: $r = 15$ cm _____

Applications

In each word problem, round the answer to the nearest tenth.

13. What is the area of a basketball court if it is 94 ft long and 42 ft wide?
    _____

14. The base of the awning on an electronics store is 80 ft long. The top of the awning is 70 ft long. The awning is 5 ft tall. What is the area of the awning?
    _____

15. How many 4-in square tiles will be needed to cover a shower wall that is 3 ft 4 in by 4 ft 8 in?
    _____

16. A garden in the shape of a right triangle is 20 ft long by 16 ft wide at the widest part. What is the area of the garden?
    _____

17. A circular walkway around a fountain in a park is 5 ft wide. If the fountain has a radius of 10 ft, what is the area of the walkway?
    _____

18. A round racetrack has an inside circumference of 500 ft. What is the area of the infield inside the track?
    _____

19. Panes of glass can be purchased in 1-square-foot sections. How many panes of glass would be needed for the nine windows in a house if each window is 6 ft by 3 ft?
    _____

20. A sheet metal worker needs circular pieces of metal with a diameter of 3 in. He has some scrap metal that has a trapezoidal shape. The longer side is 36 in. The side parallel to the long side is 28 in. The distance between those two sides is 12 in. What is the maximum number of the circular pieces of metal he can cut from this scrap?
    _____

10.4B PRACTICE SET

In problems 1–12, calculate the area of the polygon or circle from the given information.

1.  _____

2.  _____

3. 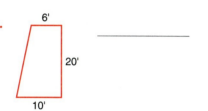 _____

4.  _____

5.  _____

6.  _____

7. Square: $s = 3.5$ cm _____

8. Rectangle: $l = 9$ in, $w = 2.5$ in _____

9. Parallelogram: $b = 24$ yd, $h = 19$ yd _____

10. Triangle: $b = 24$ m, $h = 15$ m _____

11. Trapezoid: $B = 15$ ft, $b = 10$ ft, $h = 12$ ft _____

12. Circle: $d = 25$ in _____

Applications

In each word problem, round the answer to the nearest tenth.

13. A conference table is 10 ft by 4 ft. What is its area?
_____

14. What is the area of a circular window that has a circumference of 24 ft?
_____

15. What is the area inside the bases on a baseball field if the bases are 90 ft apart?
_____

16. What is the surface area of the tabletop shown? Each edge of the octagon is 17 in.

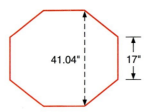

_____

17. Enrico's garage is 6 yd wide and 8 yd long. Shingles to cover the flat roof are 9 in by 5 in and cost $0.67 each. What will it cost him to buy enough shingles to cover his roof?
_____

18. Marianne wants to paint the room that will be used for a nursery. The room is 12 ft long and 10 ft wide. The walls are 7 ft tall. There is one window that is 2.5 ft by 5 ft, and there is one door that is 3 ft by 6.5 ft that will not need painting. If she paints the four walls and the ceiling, what is the area that she will paint?
_____

19. A slab for a house is shown. What is the area of the slab?

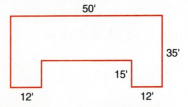

_____

20. Don must replace 20 sq ft of fencing. He can buy replacement wood strips that are 5 ft long and 4 in wide. How many of the replacement strips must he buy to do the job?

10.5 VOLUME

Volume is a measure of the amount of space a three-dimensional object occupies. When you worked with area, you were generally concerned with only two dimensions, usually length and width or base and height. A blackboard is a good example of a plane geometric figure with two dimensions. Consider a shoe box. The shoe box has three dimensions: length, width, and height. A shoe box can be filled with something. The amount that it takes to fill an object like a shoe box is the volume.

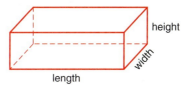

Volume is measured in cubic units. A cubic unit has the same measure on each edge. A cube of sugar is an example of a perfect cube. The following drawing illustrates the idea of a cube.

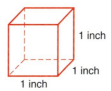

When calculating volume, you are trying to determine how many cubic units will fit into an object. There is a formula for finding the volume of each geometric solid.

A **rectangular solid** has length, width, and height and all faces are rectangles. The formula for finding the volume of a rectangular solid is

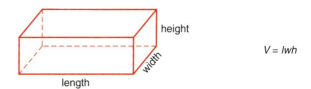

The *V* in volume formulas always represents volume.

EXAMPLE 1 What is the volume of a shoe box that is 6 inches wide, 14 inches long, and 5 inches high?

SOLUTION

$V = lwh$ *Write the formula for volume of a rectangle.*

$V = (14)(6)(5)$ *Substitute the given values.*

$V = 420$ cubic inches   *Solve the problem.*

The volume of the shoe box is 420 cubic inches (in^3). Cubic measure can be expressed as $unit^3$ (in^3 or cm^3 or m^3).

A **cube** is a rectangular solid with equal length, width, and height. The volume formula for a cube, then, is somewhat more simplified than the formula for a rectangular solid.

$V = e^3$

EXAMPLE 2 What is the volume of the cubic packing crate shown?

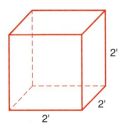

SOLUTION

$V = e^3$ *Write the formula for the volume of a cube.*

$V = 2^3$ *Substitute the given values.*

$V = (2)(2)(2)$ *Solve the problem.*

$V = 8 \text{ ft}^3$

The volume of the packing crate is 8 cubic feet (ft³).

A cylinder is a three-dimensional object having the shape of food cans or round glasses. Because it is round, the area of the base is used to help find the volume. The area of the base is multiplied by the height of the cylinder to find the volume.

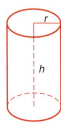

$V = \pi r^2 h$

EXAMPLE 3 Calculate the volume of a can if the radius of the base is 2.5 inches and the can is 6 inches tall.

SOLUTION

$V = \pi r^2 h$ *Write the formula for the volume of a cylinder since a can is a cylinder.*

$V = (3.14)(2.5)^2(6)$ *Substitute the given values.*

$V = (3.14)(2.5)(2.5)(6)$ *Solve the problem.*

$V = 117.75 \text{ in}^3$

The volume of the can is 117.75 cubic inches (in³).

A **cone** has a circular flat base and sloping sides that rise or descend to a single point. It is in the shape of an ice cream cone. The formula for its volume is similar to that for a cylinder, but because a cone is only part of a cylinder, you use a variation of the formula.

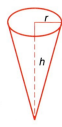

$V = \frac{1}{3}\pi r^2 h$

EXAMPLE 4 What is the volume of a cone with a radius of 3 cm and a height of 7 cm?

SOLUTION

$V = \frac{1}{3}\pi r^2 h$ *Write the formula for volume of a cone.*

$V = \frac{1}{3}(3.14)(3)^2(7)$ *Substitute the given values.*

$V = \frac{1}{3}(3.14)(3)(3)(7)$ *Solve the problem.*

$V = 65.94$ cm³

The volume of the cone is 65.94 cubic centimeters (cm³).

A **sphere** is any three-dimensional circular shape with the center an equal distance from all points on the surface. A ball is an example of a sphere. The formula for the volume of a sphere is based on the fact that the object is completely round.

$V = \frac{4}{3}\pi r^3$

EXAMPLE 5 What is the volume of a sphere that has a radius of 5 ft?

SOLUTION

$V = \frac{4}{3}\pi r^3$ *Write the formula for volume of a sphere.*

$V = \frac{4}{3}(3.14)(5)^3$ *Substitute the given values.*

$V = \frac{4}{3}(3.14)(5)(5)(5)$ *Solve the problem.*

$V = 523.3$ ft³

The volume of the sphere is 523.3 cubic feet (ft³).

> **a + b = x**
>
> # ALGEBRA CONNECTION
>
> This Algebra Connection is placed here to show you how far you have come in your study of mathematics. At the beginning of this book you learned how to write whole numbers and how to solve basic math problems such as
>
> $$26 + 84 = ?$$
>
> You learned early on that a variable could be used in place of the question mark.
>
> $$26 + 84 = x$$
>
> You know now that the variable represents a number. When you perform the math operations indicated in the problem, you find the number that the variable represents.
>
> When you learned how to operate with signed numbers and then with equations, you paved the path to solving even more difficult problems. In this chapter on geometry, you have worked with formulas with as many as four unknowns. Because of your previous work with order of operations, fractions, exponents, and solving equations, you are able to solve formulas like
>
> $$V = \frac{4}{3}\pi r^3$$
>
> You now have the skills to work increasingly difficult problems, whether in math or any other field of study.

Work the Comprehension Checkpoint problems to make sure you understand volume. If you have difficulty with these problems, review the appropriate material before attempting the Practice Sets.

COMPREHENSION CHECKPOINT

Determine the volume of each object described. Round all answers to the nearest tenth.

1. Cube: $e = 4$ cm _____

2. Rectangular solid: $l = 5$ m, $w = 3$ m, $h = 4$ m _____

3. Cylinder: $r = 6$ in, $h = 24$ in _____

4. Cone: $r = 8$ yd, $h = 12$ yd _____

5. Sphere: $r = 7$ ft _____

10.5A PRACTICE SET

Determine the volume of each figure. Round each answer to the nearest tenth.

1.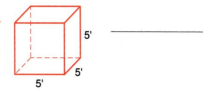

2.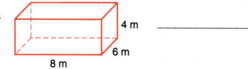

3.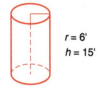

4.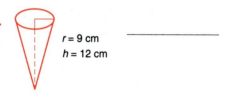

5.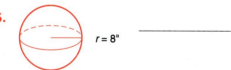

6.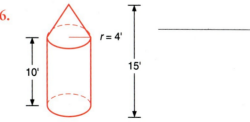

Applications

In each word problem, round the answer to the nearest tenth.

7. What is the volume of a cone if the radius is 8 cm and the height is 22 cm?

8. What is the volume of a cylinder if the radius is 16 ft and the height is 50 ft?

   _____

9. How many cubic feet is a cubic yard?

   _____

10. What is the volume of a rectangular solid if the length is 14 m, the width is 8 m, and the height is 5 m?

    _____

11. What is the volume of a cube if the length of each edge is 6 yd?

    _____

12. How many cubic centimeters are in a cubic meter?

    _____

13. What is the volume of a bowling ball if the radius is 4.5 in?

    _____

14. How much water would it take to fill a section of cylindrical pipe if the pipe is 6 ft long and has a diameter of 8 in?

    _____

15. A hotel in Dallas has a restaurant in a sphere at the top of a tower. What is the volume of the sphere if the radius is 28 ft?

    _____

16. What is the volume of a desk drawer that is 27 in long, 4 in deep, and 12 in wide?

    _____

17. How many cubic inches of ice would be in a snow cone if the cup is 6 in tall and the radius of the cup is 2.5 in?

    _____

18. What is the volume of a pickup truck bed if it is 2.75 m long, 2 m wide, and 0.8 m high?

    _____

19. What is the volume of a plastic pitcher if it is 10 in tall and has a circumference of 19 in?

    _____

20. A swimming pool is 10 yd long and 5 yd wide. It has a uniform depth of 4 ft for the first 6 yd and then drops to 8 ft deep for the final 4 yd. How many gallons of water would it take to fill the pool if one cubic foot holds 7.5 gallons of water?

    _____

10.5B PRACTICE SET

Determine the volume of each figure. Round each answer to the nearest tenth.

1.  _____

2. 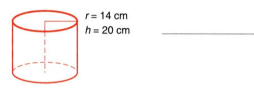 _____

3. 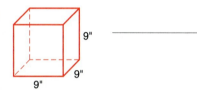 _____

4. 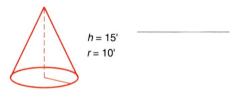 _____

5. 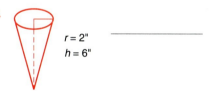 _____

6. 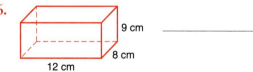 _____

Applications

In each word problem, round the answer to the nearest tenth.

7. What is the volume of a rectangular solid with length of 15 ft, width of 8 ft, and height of 7 ft?

_____

8. What is the volume of a sphere with radius of 10 m?

9. How many cubic inches will fit in a cubic foot?

10. What is the volume of a cylinder with radius of 3.5 ft and height of 10 ft?

11. What is the volume of a cone with radius of 6 cm and height of 30 cm?

12. A grain storage bin is 48 ft high and has a circumference of 37.68 ft. What is its volume?

13. A beach ball has a diameter of 16 in. What is the volume of the ball?

14. What is the volume of a cylindrical cup that is 60 cm tall and has a circumference of 38 cm?

15. A portable storage room is 12 ft long, 8 ft wide, and 6 ft tall. What is the maximum number of boxes that can be stored in the room if each box is a 2-ft cube?

16. Baseballs are shipped in boxes that are cubes. The ball touches each side of the box. If the radius of the ball is 1.25 in, how much of the volume of the box is not used by the ball?

17. Water weighs 62.4 pounds per cubic foot. What would be the weight of the water in a spherical storage tank if the radius of the tank is 8 ft?

18. What is the volume of a conical tent with a radius of 2 m and a height of 3 m?

19. One bag of topsoil contains 2 cubic feet. How many bags would be needed to cover an area 12 ft long and 6 ft wide with topsoil 2 in deep?

20. How many gallons of water will an aquarium hold if it is 48 in long, 18 in wide, and 24 in deep? (One cubic foot holds approximately 7.5 gallons of water.)

SUMMARY OF KEY CONCEPTS

KEY TERMS

geometry (10.1): The measurement of different shapes.

point (10.1): A position on a line.

line segment (10.1): The shortest distance between and connecting two points.

parallel lines (10.1): Lines that are the same distance apart throughout their length.

ray (10.1): A straight line extending indefinitely from a point.

angle (10.1): The result of two rays meeting at the same point.

vertex (10.1): The point where two rays meet to form an angle.

perpendicular lines (10.1): Lines that intersect at right angles.

right angle (10.1): An angle of 90°.

polygon (10.2): A closed geometric figure with three or more sides.

square (10.2): A polygon of four sides of equal length that intersect at right angles.

rectangle (10.2): A polygon of four sides in which opposite sides are equal and parallel and all angles are right angles.

parallelogram (10.2): A polygon of four sides with opposite sides equal and parallel. The angles are not right angles.

trapezoid (10.2): A polygon of four sides. Two sides are parallel but unequal in length.

triangle (10.2): A polygon of three sides.

right triangle (10.2): A triangle with two sides meeting to form one right angle.

equilateral triangle (10.2): A triangle with all sides equal in length.

scalene triangle (10.2): A triangle in which no two sides have the same length.

perimeter (10.2): The distance around the edge of a polygon.

subscript (10.2): A number (or letter) written below and to the right of another letter or number. Subscripts are used for identification purposes.

circle (10.3): A curved line connecting a series of points each of which is the same distance from a central point.

circumference (10.3): The distance around the circle (same as the perimeter of a polygon).

diameter (10.3): A line that connects two points on the circumference of a circle and passes through the center.

radius (10.3): A line from the center to a point on the circumference of a circle (one-half of a diameter).

area (10.4): The amount of space inside a polygon or circle.

volume (10.5): The amount of space that a three-dimensional object occupies.

rectangular solid (10.5): A three-dimensional object that has length, width, and height and all rectangular faces.

cube (10.5): A rectangular solid with equal length, width, and height.

cylinder (10.5): A three-dimensional object shaped like a can of food or a round glass.

cone (10.5): A three-dimensional object with a circular flat base and sloping sides that rise or descend to a single point.

sphere (10.5): A three-dimensional round object with the center an equal distance from all points on the surface.

KEY RULES

- **Formula for finding the perimeter of any polygon (10.2):**

$$p = s_1 + s_2 + s_3 + \cdots + s_n$$

where p = perimeter and the subscript n is the total number of sides in the polygon.

- **Formula for finding the radius of a circle (10.3):**

$$r = \frac{d}{2}$$

where r = radius and d = diameter.

- **Formula for finding the circumference of a circle (10.3):**

$$C = d\pi$$

where C = circumference and d = diameter.

- **Formula for finding diameter of a circle (10.3):**

$$\frac{C}{\pi} = d$$

where C = circumference and d = diameter.

- **Formulas for finding area (10.4):**

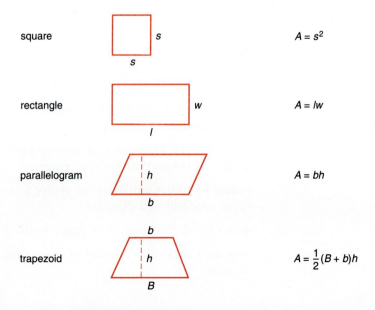

Summary of Key Concepts

triangle $A = \frac{1}{2}bh$

circle $A = \pi r^2$

▬ Formulas for finding volume (10.5):

rectangular solid $V = lwh$

cube $V = e^3$

cylinder $V = \pi r^2 h$

cone $V = \frac{1}{3}\pi r^2 h$

sphere $V = \frac{4}{3}\pi r^3$

CHAPTER 10

PRACTICE TEST 10A

1. Write the proper designation for the following angle.

     _____

2. Identify each set of lines as either parallel, perpendicular, or neither.

   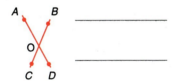

   a. _____

   b. _____

   c. _____

3. Identify the two sets of angles that are equal.

     _____

   _____

4. Give the name of each polygon.

   a.   a. _____

   b.   b. _____

   c.   c. _____

   d.   d. _____

   e.   e. _____

639

5. Draw and label one diameter and one radius in the circle.

6. Calculate the circumference of a circle if the diameter is 14 in. _____

7. Calculate the radius of a circle if the circumference is 20 ft. _____

8. Find the area of each polygon or circle.

    a. rectangle 12.6 cm by 9.5 cm   *a.* _____

    b. trapezoid, top 16", bottom 22", height 12"   *b.* _____

    c. parallelogram, base 30 m, height 8 m   *c.* _____

    d. circle, C = 42"   *d.* _____

    e. square, 25 m by 25 m   *e.* _____

    f. right triangle, legs 12' and 35'   *f.* _____

9. Calculate the volume of each geometric solid.

a. 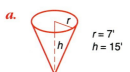   a. _____

b.    b. _____

c. 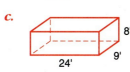   c. _____

d. 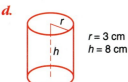   d. _____

e.   e. _____

10. What is the area of a rectangular field that is 45 yd long and 28 yd wide?

    _____

11. Calculate the volume of a cylindrical gasoline storage tank that is 30 ft high with a radius of 5 ft.

    _____

12. What is the perimeter of the section of land in the drawing?

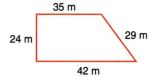

    _____

CHAPTER 10

PRACTICE TEST 10B

1. Write the proper designation for the following angle.

      _____

2. Identify each set of lines as either parallel, perpendicular, or neither.

   a.    a. _____

   b.    b. _____

   c.    c. _____

3. Identify the two sets of angles that are equal.

   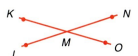   _____

   _____

4. Give the name of each polygon.

   a.   a. _____

   b.    b. _____

   c.   c. _____

   d.   d. _____

   e.   e. _____

5. Draw and label one diameter and one radius in the circle.

643

6. Calculate the circumference of a circle if the diameter is 24 in.

7. Calculate the radius of a circle if the circumference is 15 ft.

8. Find the area of each polygon or circle.

   a.  a. _____

   b. 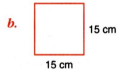 b. _____

   c.   c. _____

   d.  d. _____

   e. 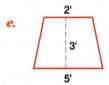 e. _____

   f. 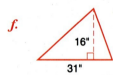 f. _____

9. Calculate the volume of each geometric solid.

   a.   a. _____

   b.  b. _____

   c.  c. _____

d.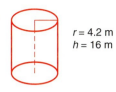

d. _____

e.

e. _____

10. What is the area of a circular mirror with a circumference of 36 in?

    _____

11. Calculate the perimeter of a blackboard that measures 8 ft by 5 ft.

    _____

12. What is the volume of a cylindrical pipe that is 18 in long and has a radius of 4 in?

    _____

SKILLS REVIEW

1. Write 746,283 in words.

   _____

2. Round 54,387 to the nearest hundred. _____

3. Round 973,896 to the nearest million. _____

4. Round 85.9549 to the nearest hundredth. _____

Perform the indicated operation. Express answers in lowest terms.

5. $\frac{5}{6} + \frac{7}{12} + \frac{3}{5} =$ _____

6. $\frac{5}{8} - \frac{5}{24} =$ _____

7. $\frac{6}{7} \times 2\frac{5}{6} =$ _____

8. $3\frac{2}{3} \div 1\frac{5}{12} =$ _____

9. $25.6 \times 18.74 =$ _____

10. $153.25 \div 0.25 =$ _____

11. Convert $\frac{2}{3}$ to a decimal. _____

12. Convert $\frac{3}{8}$ to a percent. _____

13. Convert 68% to a decimal. _____

14. Convert 40% to a fraction in lowest terms. _____

15. What is 18% of 400? _____

16. What is $8\frac{1}{3}$% of 96? _____

17. 30 is what percent of 240? _____

18. Write 0.5732 in words.

_____

19. Write 35.094 in words.

_____

20. Find the missing element in the proportion.

$\dfrac{5}{18} = \dfrac{x}{108}$ _____

21. Insert the appropriate symbol, < or >.

    a.   5 _____ −4   _____

    b.   −6.2 _____ −5.9   _____

    c.   0 _____ −3.8   _____

22. Indicate the absolute value of each term.

    a. $\left|4\dfrac{2}{3}\right|$ _____

    b. $|-4.5|$ _____

Perform the indicated operation.

23. $(8) + (-12) + (-9) + (16) =$ _____

24. $(-15) - (19) =$ _____

25. $(6)(-7)(-14)(-3) =$ _____

26. $(65) \div (-13) =$ _____

27. Solve: $3x + 8 = 41$ _____

28. Solve: $8(y - 2) + 5 = 29$ _____

29. Solve: $\dfrac{x}{5} - 8 = 24$ _____

30. Convert 8 pounds to ounces. _____

31. How many centimeters in 8 kilometers? _____

32. What is the area of a rectangle with length of 9 in and width of 5.75 in? _____

33. What is the volume of a sphere with radius of 5 ft? _____

34. What is the perimeter of a square with a side of 12 cm? _____

CHAPTER 10 SOLUTIONS

Skills Preview

1. ∠ACB or ∠C

2. a. perpendicular
   b. parallel
   c. neither

3. ∠PQT = ∠RQS
   ∠PQR = ∠TQS

4. a. rectangle
   b. square
   c. parallelogram
   d. trapezoid
   e. triangle

5. Answers will vary.

6. 43.96 in

7. 3.2 ft (rounded)

8. a. 196 sq in
   b. 54 sq yd
   c. 9 sq ft
   d. 360 sq m
   e. 20 sq cm
   f. 153.86 sq in

9. a. 540 cu in
   b. 64 cu cm
   c. 942 cu m
   d. 527.52 cu in
   e. 3,052.08 cu in

Section 10.1 Comprehension Checkpoint

1. ∠EFG (or ∠GFE)

2. ∠X

3. a. neither
   b. perpendicular
   c. parallel

4. ∠NOM = ∠POQ
   ∠PON = ∠QOM

10.1A Practice Set

1. ∠CBA; ∠B
2. ∠DEF; ∠E
3. ∠RQP; ∠Q
4. ∠RST; ∠S
5. ∠CAB; ∠A
6. perpendicular
7. neither
8. parallel
9. neither
10. parallel
11. ∠COD = ∠AOB
    ∠DOB = ∠COA
12. ∠POQ = ∠ROS
    ∠POR = ∠QOS
13. ∠XVW = ∠TVU
    ∠XVT = ∠WVU
14. ∠AOD = ∠BOC
    ∠AOB = ∠DOC
15. ∠OMN = ∠LMK
    ∠OML = ∠NMK

Section 10.2 Comprehension Checkpoint

1. a. triangle
   b. parallelogram
   c. rectangle
   d. square
   e. trapezoid
   f. right triangle

2. a. 45 cm
   b. 24 in
   c. 102 ft
   d. 84 m
   e. 120 yd

3. 84 m

10.2A Practice Set

1. trapezoid
2. rectangle
3. parallelogram
4. triangle

5. square

6. 8 cm + 8 cm + 8 cm + 8 cm = 32 cm

7. 42 m + 42 m + 29 m + 29 m = 142 m

8. 19 in + 24 in + 19 in + 12 in = 74 in

9. 30 yd + 30 yd + 18 yd + 18 yd = 96 yd

10. 50 ft + 42 ft + 24 ft = 116 ft

11. 12 in + 9 in + 12 in + 9 in = 42 in

12. 120 yd + 120 yd + 42 yd + 42 yd = 324 yd

13. 3.5 ft × 4 = 14 ft
    (14)($0.38) = $5.32 = total cost of edging

14. 3 ft + 3 ft + 6 ft 8 in + 6 ft 8 in = 18 ft 16 in = 19 ft 4 in

15. 25 ft + 25 ft + 18 ft + 18 ft = 86 ft total perimeter of the walls
    3 ft + 8 ft = 11 ft that will not need the molding
    86 ft − 11 ft = 75 ft of molding needed

16. 150 yd + 125 yd + 165 yd = 440 yd around the triangle
    1,760 yd per mile ÷ 440 yd = 4 trips around the field to go 1 mi
    3 miles × 4 trips per mile = 12 trips around the field to run 3 mi

17. An equilateral triangle has three equal sides. If the perimeter is 48 ft, divide 48 by 3 to find the length of each side.
    48 ft ÷ 3 = 16 ft = length of the side of the triangle

18. 40 yd + 40 yd + 30 yd + 30 yd = 140 yd = perimeter
    140 yd × 3 ft per yard = 420 ft = perimeter
    420 ft ÷ 4 ft = 105 fence posts

19. The dimensions of the pencil tray are 5 ft 6 in by 2 ft 8 in.
    ```
     5 ft  6 in
     5 ft  6 in
     2 ft  8 in
     2 ft  8 in
    ─────────────
    14 ft 28 in
    ```
    28 in = 2 ft 4 in
    14 ft + 2 ft 4 in = 16 ft 4 in = perimeter of tray

20. Perimeter of house = 160 ft
    Width of house = $\frac{160 - 48 - 48}{2} = \frac{64}{2} = 32$ ft
    32 ft + 8 ft + 8 ft = 48 ft = width of yard
    8 ft + 65 ft + 48 ft + 65 ft + 8 ft = 194 ft of fencing needed
    194 ÷ 3 = $64\frac{2}{3}$ yd, so he must buy 65 yards of fencing.
    65 × $8.90 = $578.50 = cost of the fencing

Section 10.3 Comprehension Checkpoint

1. diameter = 16 in
   circumference = 50.24 in

2. 6.4 in

3. 3.8 in

10.3A Practice Set

1. $C = d\pi$
   $C = (8)(3.14)$
   $C = 25.12$ cm

2. $C = d\pi$
   $C = (15)(3.14)$
   $C = 47.1$ in

3. $d = 2r$
   $d = 2(6)$
   $d = 12$ ft
   $C = d\pi$
   $C = (12)(3.14)$
   $C = 37.68$ ft

4. $d = 2r$
   $d = (2)(18)$
   $d = 36$ m
   $C = d\pi$
   $C = (36)(3.14)$
   $C = 113.04$ m

Solutions

5. $\dfrac{C}{\pi} = d$ $\dfrac{28}{3.14} = d$ 8.9 in = d
    $d \div 2 = r$
    $8.9 \div 2 = r$
    4.5 in = r

6. $\dfrac{C}{\pi} = d$ $\dfrac{42}{3.14} = d$ 13.4 ft = d
    $d \div 2 = r$
    $13.4 \div 2 = r$
    6.7 ft = r

7. $4\dfrac{7}{8} = 4.875$
    $\dfrac{C}{\pi} = d$ $\dfrac{4.875}{3.14} = d$ 1.6 in = d

8. $C = d\pi$
    $C = (12)(3.14)$
    $C = 37.7$ in

9. $d = \dfrac{C}{\pi}$ $d = \dfrac{6.25}{3.14}$ $d = 2$ in (rounded)

10. $d = 2r$
    $d = (2)(3)$
    $d = 6$
    $C = d\pi$
    $C = (6)(3.14)$
    $C = 18.8$ in

11. $d = \dfrac{C}{\pi}$ $d = \dfrac{9.5}{3.14}$ $d = 3$ in (rounded)
    $d \div 2 = r$
    $3 \div 2 = r$
    1.5 in = r

12. $d = \dfrac{C}{\pi}$ $d = \dfrac{25.12}{3.14}$ $d = 8$ ft

13. Radius of satellite orbit = 4,000 + 150 = 4,150 mi.
    $d = 2r$
    $d = (2)(4,150)$
    $d = 8,300$ mi
    $C = d\pi$
    $C = (8,300)(3.14)$
    $C = 26,062$ mi
    The satellite travels 26,062 mi in one revolution.

14. $d = \dfrac{C}{\pi}$ $d = \dfrac{78.5}{3.14}$
    $d = 25$ ft = diameter of the garden
    The diameter of the walking path is
    25 + 2 + 2 = 29 ft.
    The circumference of the walking path is:
    $C = d\pi$
    $C = (29)(3.14)$
    $C = 91.06 = 91.1$ ft (rounded)

15. 30 minutes is one-half hour.
    The minute hand will complete one-half of a circle in one-half hour.
    $d = 2r$ The length of the minute hand
    $d = (2)(24)$ is the radius of the circle.
    $d = 48$ in
    $C = d\pi$
    $C = (48)(3.14)$
    $C = 150.72$ in
    The minute hand would draw a circle 150.72 inches in circumference in one hour.
    $150.72 \div 2 = 75.4$ in = the length of the arc in one-half hour.

16. The distance from the airport is the radius of the circle the plane is making.
    $d = \dfrac{C}{\pi}$ $d = \dfrac{18}{3.14}$ $d = 5.73$ mi
    $d \div 2 = r$
    $5.73 \div 2 = r$
    2.9 mi = r

17. The diameter of the circle would be the minimum width of the board.
    20×8 in = 160 in = circumference of a circle
    $d = \dfrac{C}{\pi}$ $d = \dfrac{160}{3.14}$ $d = 50.955$ in
    The board should be 51 in wide.

18. One complete revolution of the tire means that the tire turns all the way around. The circumference of the tire is the distance the car will travel in one revolution.
    $d = 2r$
    $d = (2)(14)$
    $d = 28$ in
    $C = d\pi$
    $C = (28)(3.14)$
    $C = 87.9$ in = distance the car will travel in one revolution

19. 60 mph means that the car is going 1 mile per minute. 5,280 (number of feet in 1 mi) × 12 = 63,360 in traveled in one minute
63,360 inches per minute ÷ 87.9 inches per revolution = 720.8 revolutions per minute

20. Each curve is half of a circle. Since $\overline{AB} = \overline{BC}$ and both are 125 yd, the two curves together are a circle with a diameter of 125 yd. The circumference of the circle is
$C = d\pi$
$C = (125)(3.14)$
$C = 392.5$ yd
The distance traveled around one lap is 392.5 + 250 = 642.5 yd.
The distance traveled around 20 laps is (642.5)(20) = 12,850 yd = 7.3 mi.

Section 10.4 Comprehension Checkpoint

1. 36 sq in   **2.** 24 sq m   **3.** 140 sq yd   **4.** 21 sq ft   **5.** 112.5 sq ft   **6.** 113.04 sq ft

10.4A Practice Set

1. $A = bh$
$A = (17)(6)$
$A = 102$ sq m

2. $A = \frac{1}{2}(B + b)h$
$A = \frac{1}{2}(6 + 4)3$
$A = 15$ sq cm

3. $A = lw$
$A = (32)(16)$
$A = 512$ sq ft

4. $A = \pi r^2$
$A = (3.14)(18)^2$
$A = (3.14)(18)(18)$
$A = 1,017.36$ sq in

5. $A = s^2$
$A = 14^2$
$A = (14)(14)$
$A = 196$ sq m

6. $A = \frac{1}{2}bh$
$A = \frac{1}{2}(20)(15)$
$A = 150$ sq ft

7. $A = s^2$
$A = 12^2$
$A = (12)(12)$
$A = 144$ sq m

8. $A = lw$
$A = (8)(4.75)$
$A = 38$ sq in

9. $A = bh$
$A = (14)(7)$
$A = 98$ sq cm

10. $A = \frac{1}{2}bh$
$A = \frac{1}{2}(16)(17)$
$A = 136$ sq yd

11. $A = \frac{1}{2}(B + b)h$
$A = \frac{1}{2}(10 + 7)3$
$A = 25.5$ sq ft

12. $A = \pi r^2$
$A = (3.14)(15)^2$
$A = (3.14)(15)(15)$
$A = 706.5$ sq cm

Solutions

13. $A = lw$
$A = (94)(42)$
$A = 3{,}948$ sq ft

14. $A = \dfrac{1}{2}(B + b)h$
$A = \dfrac{1}{2}(80 + 70)5$
$A = 375$ sq ft

15. How much space will each tile cover? Each tile will cover 16 sq in ($A = s^2 = 4^2 = 16$ sq in). How many square inches are to be covered?
3 ft 4 in = 40 in
4 ft 8 in = 56 in
Area (of the wall) = lw
$A = (56)(40)$
$A = 2{,}240$ sq in
2,240 sq in ÷ 16 sq in = 140 tiles
It will take 140 tiles to cover the wall.

16. $A = \dfrac{1}{2}bh$
$A = \dfrac{1}{2}(20)(16)$
$A = 160$ sq ft

17. Area of the fountain = πr^2
$= (3.14)(10)(10)$
$= 314$ sq ft
The radius of the walkway is 10 ft + 5 ft = 15 ft.
$A = \pi r^2$
$A = \pi(15)(15)$
$A = 706.5$ sq ft = area of walkway and fountain
706.5 sq ft − 314 sq ft = 392.5 sq ft = area of walkway

18. Given the circumference, you can find the radius and calculate the area.
$d = \dfrac{C}{\pi} \quad d = \dfrac{500}{3.14} \quad d = 159.2$

$r = d \div 2$
$r = 159.2 \div 2$
$r = 79.6$
$A = \pi r^2$
$A = (3.14)(79.6)(79.6)$
$A = 19{,}895.5$ sq ft
The infield area is 19,895.5 sq ft.

19. $A = lw$
$A = (6)(3)$
$A = 18$ sq ft (each window)
$18 \times 9 = 162$
162 panes of glass would be needed.

20. The area of each circular piece he needs is:
$A = \pi r^2$
$A = (3.14)(1.5)(1.5)$
$A = 7.065$ sq in
The area of the piece of metal is:
$A = \dfrac{1}{2}(B + b)h$
$A = \dfrac{1}{2}(36 + 28)12$
$A = 384$ sq in
384 ÷ 7.065 = 54.4
The worker can cut a maximum of only 54 of the circular pieces from the scrap metal.

Section 10.5   Comprehension Checkpoint

1. 64 cu cm   **2.** 60 cu m   **3.** 2,713 cu in   **4.** 803.8 cu yd   **5.** 1,436 cu ft

10.5A   Practice Set

1. $V = e^3$
$V = 5^3$
$V = (5)(5)(5)$
$V = 125$ cu ft

2. $V = lwh$
$V = (8)(6)(4)$
$V = 192$ cu m

3. $V = \pi r^2 h$
   $V = (3.14)(6)^2(15)$
   $V = (3.14)(6)(6)(15)$
   $V = 1{,}695.6$ cu ft

4. $V = \frac{1}{3}\pi r^2 h$
   $V = \frac{1}{3}(3.14)(9)^2(12)$
   $V = \frac{1}{3}(3.14)(9)(9)(12)$
   $V = 1{,}017.4$ cu cm

5. $V = \frac{4}{3}\pi r^3$
   $V = \frac{4}{3}(3.14)(8)^3$
   $V = \frac{4}{3}(3.14)(8)(8)(8)$
   $V = 2{,}143.6$ cu in

6. Volume of cylinder:
   $V = \pi r^2 h$
   $V = (3.14)(4)^2(10)$
   $V = (3.14)(4)(4)(10)$
   $V = 502.4$ cu ft

   Volume of cone:
   $V = \frac{1}{3}\pi r^2 h$
   $V = \frac{1}{3}(3.14)(4)^2(5)$
   $V = \frac{1}{3}(3.14)(4)(4)(5)$
   $V = 83.7$ cu ft

   Combined volume:
   502.4 cu ft + 83.7 cu ft = 586.1 cu ft

7. $V = \frac{1}{3}\pi r^2 h$
   $V = \frac{1}{3}(3.14)(8)^2(22)$
   $V = \frac{1}{3}(3.14)(8)(8)(22)$
   $V = 1{,}473.7$ cu cm

8. $V = \pi r^2 h$
   $V = (3.14)(16)^2(50)$
   $V = (3.14)(16)(16)(50)$
   $V = 40{,}192$ cu ft

9. One cubic yard would be a cube that is 1 yd long on each edge. One yard is 3 ft. The edge of a cubic yard would be 3 ft. The number of cubic feet in a cubic yard would be $3 \times 3 \times 3 = 27$.

10. $V = lwh$
    $V = (14)(8)(5)$
    $V = 560$ cu m

11. $V = e^3$
    $V = (6)^3$
    $V = (6)(6)(6)$
    $V = 216$ cu yd

12. One meter = 100 centimeters, so a cubic meter would be 100 cm on each edge. The number of cubic centimeters in a cubic meter = $100 \times 100 \times 100 = 1{,}000{,}000$.

13. $V = \frac{4}{3}\pi r^3$
    $V = \frac{4}{3}(3.14)(4.5)^3$
    $V = \frac{4}{3}(3.14)(4.5)(4.5)(4.5)$
    $V = 381.5$ cu in

14. The radius of the pipe is 4 in ($8 \div 2 = 4$). The pipe is 72 inches long ($12 \times 6 = 72$).
    $V = \pi r^2 h$
    $V = (3.14)(4)^2(72)$
    $V = (3.14)(4)(4)(72)$
    $V = 3{,}617.3$ cu in

15. $V = \frac{4}{3}\pi r^3$
    $V = \frac{4}{3}(3.14)28^3$
    $V = \frac{4}{3}(3.14)(28)(28)(28)$
    $V = 91{,}905.7$ cu ft

16. $V = lwh$
    $V = (27)(4)(12)$
    $V = 1{,}296$ cu in

Solutions

17. $V = \frac{1}{3}\pi r^2 h$
    $V = \frac{1}{3}(3.14)(2.5)^2(6)$
    $V = \frac{1}{3}(3.14)(2.5)(2.5)(6)$
    $V = 39.3$ cu in

18. $V = lwh$
    $V = (2.75)(2)(0.8)$
    $V = 4.4$ cu m

19. First calculate the radius of the pitcher:
    $\frac{C}{\pi} = d$ $\frac{19}{3.14} = d$ $6 = d$ (rounded)
    $6 \div 2 = 3 = r$
    $V = \pi r^2 h$
    $V = (3.14)(3)^2(10)$
    $V = (3.14)(3)(3)(10)$
    $V = 282.6$ cu in

20. First convert all measurements to feet:
    10 yd = 30 ft (10 × 3 = 30)
    5 yd = 15 ft (5 × 3 = 15)
    6 yd = 18 ft (6 × 3 = 18)
    4 yd = 12 ft (4 × 3 = 12)

    The shallow end of the pool is a rectangular solid that is 18 ft by 15 ft by 4 ft. The volume of this end is:
    $V = lwh$
    $V = (18)(15)(4)$
    $V = 1,080$ cu ft

    The deep end is a rectangular solid that is 12 ft by 15 ft by 8 ft. The volume of the deep end is:
    $V = lwh$
    $V = (12)(15)(8)$
    $V = 1,440$ cu ft

    The volume of the pool is 1,080 cu ft + 1,440 cu ft = 2,520 cu ft.

    Each cubic foot holds 7.5 gallons of water. 2,520 × 7.5 = number of gallons of water needed. 18,900 = number of gallons of water needed to fill the pool.

Practice Test 10A

1. ∠ACB or ∠C

2. a. neither
   b. parallel
   c. perpendicular

3. ∠AOC = ∠BOD
   ∠AOB = ∠COD

4. a. trapezoid
   b. rectangle
   c. triangle
   d. square
   e. parallelogram

5. Answers will vary, but the diameter must go through the center and connect two points on the circumference. The radius must extend from the center to the circumference.

6. $C = 43.96$ in

7. 3.2 ft (rounded)

8. a. 119.7 sq cm
   b. 228 sq in
   c. 240 sq m
   d. 141 sq in (rounded)
   e. 625 sq m
   f. 210 sq ft

9. a. 769.3 cu ft
    b. 904.32 cu cm
    c. 1,728 cu ft
    d. 226.08 cu cm
    e. 8 cu in

10. 1,260 sq yd

11. 2,355 cu ft

12. 130 m

Skills Review   Chapters 1–10

1. seven hundred forty-six thousand, two hundred eighty-three
2. 54,400
3. 1,000,000
4. 85.95
5. $2\frac{1}{60}$
6. $\frac{5}{12}$
7. $2\frac{3}{7}$
8. $2\frac{10}{17}$
9. 479.744
10. 613
11. $0.\overline{6}$
12. 37.5%
13. 0.68
14. $\frac{2}{5}$
15. 72
16. 8
17. 12.5%
18. five thousand seven hundred thirty-two ten-thousandths
19. thirty-five and ninety-four thousandths
20. 30
21. a. >
    b. <
    c. >
22. a. $4\frac{2}{3}$
    b. 4.5
23. 3
24. −34
25. −1,764
26. −5
27. $x = 11$
28. 5
29. 160
30. 128 oz
31. 800,000
32. 51.75 sq in
33. 523.3 cu ft
34. 48 cm

APPENDIX A
ADDITION AND MULTIPLICATION TABLES

TABLE OF ADDITION FACTS

+	0	1	2	3	4	5	6	7	8	9
0	0	1	2	3	4	5	6	7	8	9
1	1	2	3	4	5	6	7	8	9	10
2	2	3	4	5	6	7	8	9	10	11
3	3	4	5	6	7	8	9	10	11	12
4	4	5	6	7	8	9	10	11	12	13
5	5	6	7	8	9	10	11	12	13	14
6	6	7	8	9	10	11	12	13	14	15
7	7	8	9	10	11	12	13	14	15	16
8	8	9	10	11	12	13	14	15	16	17
9	9	10	11	12	13	14	15	16	17	18

The addition facts table includes all sums for single-digit addition problems. Suppose you want to add 8 and 9. Find 8 in the left-hand column and go across to the right. Next, locate the 9 in the top row. Go down until the two lines cross. The number in the square at the intersection is the sum of 8 and 9, which is 17.

TABLE OF MULTIPLICATION FACTS

×	0	1	2	3	4	5	6	7	8	9
0	0	0	0	0	0	0	0	0	0	0
1	0	1	2	3	4	5	6	7	8	9
2	0	2	4	6	8	10	12	14	16	18
3	0	3	6	9	12	15	18	21	24	27
4	0	4	8	12	16	20	24	28	32	36
5	0	5	10	15	20	25	30	35	40	45
6	0	6	12	18	24	30	36	42	48	54
7	0	7	14	21	28	35	42	49	56	63
8	0	8	16	24	32	40	48	56	64	72
9	0	9	18	27	36	45	54	63	72	81

The multiplication facts table includes the products for all single-digit multiplication problems. Use this table just as you used the addition table. To multiply 4 and 7, find the 4 in the left-hand column and go across to the right. Find the 7 in the top row and go down. The square where the two lines meet contains the product of 4 and 7, which is 28.

APPENDIX B

TABLE OF PRIME FACTORS OF NUMBERS 1 THROUGH 100

APPENDIX B Table of Prime Factors of Numbers 1 through 100

	PRIME FACTORS		PRIME FACTORS		PRIME FACTORS		PRIME FACTORS
1	none	26	$2 \cdot 13$	51	$3 \cdot 17$	76	$2^2 \cdot 19$
2	2	27	3^3	52	$2^2 \cdot 13$	77	$7 \cdot 11$
3	3	28	$2^2 \cdot 7$	53	53	78	$2 \cdot 3 \cdot 13$
4	2^2	29	29	54	$2 \cdot 3^3$	79	79
5	5	30	$2 \cdot 3 \cdot 5$	55	$5 \cdot 11$	80	$2^4 \cdot 5$
6	$2 \cdot 3$	31	31	56	$2^3 \cdot 7$	81	3^4
7	7	32	2^5	57	$3 \cdot 19$	82	$2 \cdot 41$
8	2^3	33	$3 \cdot 11$	58	$2 \cdot 29$	83	83
9	3^2	34	$2 \cdot 17$	59	59	84	$2^2 \cdot 3 \cdot 7$
10	$2 \cdot 5$	35	$5 \cdot 7$	60	$2^2 \cdot 3 \cdot 5$	85	$5 \cdot 17$
11	11	36	$2^2 \cdot 3^2$	61	61	86	$2 \cdot 43$
12	$2^2 \cdot 3$	37	37	62	$2 \cdot 31$	87	$3 \cdot 29$
13	13	38	$2 \cdot 19$	63	$3^2 \cdot 7$	88	$2^3 \cdot 11$
14	$2 \cdot 7$	39	$3 \cdot 13$	64	2^6	89	89
15	$3 \cdot 5$	40	$2^3 \cdot 5$	65	$5 \cdot 13$	90	$2 \cdot 3^2 \cdot 5$
16	2^4	41	41	66	$2 \cdot 3 \cdot 11$	91	$7 \cdot 13$
17	17	42	$2 \cdot 3 \cdot 7$	67	67	92	$2^2 \cdot 23$
18	$2 \cdot 3^2$	43	43	68	$2^2 \cdot 17$	93	$3 \cdot 31$
19	19	44	$2^2 \cdot 11$	69	$3 \cdot 23$	94	$2 \cdot 47$
20	$2^2 \cdot 5$	45	$3^2 \cdot 5$	70	$2 \cdot 5 \cdot 7$	95	$5 \cdot 19$
21	$3 \cdot 7$	46	$2 \cdot 23$	71	71	96	$2^5 \cdot 3$
22	$2 \cdot 11$	47	47	72	$2^3 \cdot 3^2$	97	97
23	23	48	$2^4 \cdot 3$	73	73	98	$2 \cdot 7^2$
24	$2^3 \cdot 3$	49	7^2	74	$2 \cdot 37$	99	$3^2 \cdot 11$
25	5^2	50	$2 \cdot 5^2$	75	$3 \cdot 5^2$	100	$2^2 \cdot 5^2$

APPENDIX C

TABLE OF SQUARE ROOTS

APPENDIX C Table of Square Roots

Decimal approximations have been rounded to the nearest thousandth.

Number	Square Root	Number	Square Root	Number	Square Root	Number	Square Root
1	1	51	7.141	101	10.050	151	12.288
2	1.414	52	7.211	102	10.100	152	12.329
3	1.732	53	7.280	103	10.149	153	12.369
4	2	54	7.348	104	10.198	154	12.410
5	2.236	55	7.416	105	10.247	155	12.450
6	2.449	56	7.483	106	10.296	156	12.490
7	2.646	57	7.550	107	10.344	157	12.530
8	2.828	58	7.616	108	10.392	158	12.570
9	3	59	7.681	109	10.440	159	12.610
10	3.162	60	7.746	110	10.488	160	12.649
11	3.317	61	7.810	111	10.536	161	12.689
12	3.464	62	7.874	112	10.583	162	12.728
13	3.606	63	7.937	113	10.630	163	12.767
14	3.742	64	8	114	10.677	164	12.806
15	3.873	65	8.062	115	10.724	165	12.845
16	4	66	8.124	116	10.770	166	12.884
17	4.123	67	8.185	117	10.817	167	12.923
18	4.243	68	8.246	118	10.863	168	12.961
19	4.359	69	8.307	119	10.909	169	13
20	4.472	70	8.367	120	10.954	170	13.038
21	4.583	71	8.426	121	11	171	13.077
22	4.690	72	8.485	122	11.045	172	13.115
23	4.796	73	8.544	123	11.091	173	13.153
24	4.899	74	8.602	124	11.136	174	13.191
25	5	75	8.660	125	11.180	175	13.229
26	5.099	76	8.718	126	11.225	176	13.267
27	5.196	77	8.775	127	11.269	177	13.304
28	5.292	78	8.832	128	11.314	178	13.342
29	5.385	79	8.888	129	11.358	179	13.379
30	5.477	80	8.944	130	11.402	180	13.416
31	5.568	81	9	131	11.446	181	13.454
32	5.657	82	9.055	132	11.489	182	13.491
33	5.745	83	9.110	133	11.533	183	13.528
34	5.831	84	9.165	134	11.576	184	13.565
35	5.916	85	9.220	135	11.619	185	13.601
36	6	86	9.274	136	11.662	186	13.638
37	6.083	87	9.327	137	11.705	187	13.675
38	6.164	88	9.381	138	11.747	188	13.711
39	6.245	89	9.434	139	11.790	189	13.748
40	6.325	90	9.487	140	11.832	190	13.784
41	6.403	91	9.539	141	11.874	191	13.820
42	6.481	92	9.592	142	11.916	192	13.856
43	6.557	93	9.644	143	11.958	193	13.892
44	6.633	94	9.695	144	12	194	13.928
45	6.708	95	9.747	145	12.042	195	13.964
46	6.782	96	9.798	146	12.083	196	14
47	6.856	97	9.849	147	12.124	197	14.036
48	6.928	98	9.899	148	12.166	198	14.071
49	7	99	9.950	149	12.207	199	14.107
50	7.071	100	10	150	12.247	200	14.142

APPENDIX

D USING A CALCULATOR

APPENDIX D Using a Calculator

Modern technology has produced a wide assortment of handheld and larger electronic calculators. Whatever type of calculator you choose to use, you should consider it a supplemental tool to help you solve problems rather than a substitute for developing your basic math skills.

Any calculator capable of performing the four basic math operations—addition, subtraction, multiplication, and division—will be adequate for basic math problems. Some calculators will have a percent key. Some have a memory function that you can use while solving problems.

Different calculators will have different key arrangements. Be sure to check your user's manual to determine the proper way to use your calculator. Keys that all calculators have in common include the following:

Number keys. Number key pad for entering numbers. Notice that there is a key for a decimal point.

Function keys. There is a function key for each basic math operation.

Using the $=$ key instructs the calculator to perform the calculation you have entered.

There is also a display window that shows what data you have entered, or the answer to your computation.

Working a math problem using a calculator is a matter of pressing the numbers and the function keys in the proper order. The order is:

number, function key, number, $=$

Study Examples 1 through 7 to see how this works.

EXAMPLE 1 Add: 762 + 37

SOLUTION

Depress the keys in the following order:

7 6 2 $+$ 3 7 $=$

The answer, 799, appears in the display window immediately after the $=$ key is depressed.

EXAMPLE 2 Subtract: 82 − 67

SOLUTION

Depress the keys in the following order:

8 2 $-$ 6 7 $=$

The answer, 15, appears in the display window immediately after the $=$ key is depressed.

Appendix D Using a Calculator

EXAMPLE 3 Multiply: 653×47

SOLUTION

Depress the keys in the following order:

6 5 3 $\boxed{\times}$ 4 7 $\boxed{=}$

The answer, 30,691, appears in the display window immediately after the $\boxed{=}$ key is depressed.

EXAMPLE 4 Divide: $986 \div 58$

SOLUTION

Depress the keys in the following order:

9 8 6 $\boxed{\div}$ 5 8 $\boxed{=}$

The answer, 17, appears in the display window immediately after the $\boxed{=}$ key is depressed.

If the problem you need to solve includes a decimal point, you will have to enter the decimal point.

EXAMPLE 5 Multiply: 4.7×89.3

SOLUTION

Depress the keys in the following order:

4 $\boxed{\cdot}$ 7 $\boxed{\times}$ 8 9 $\boxed{\cdot}$ 3 $\boxed{=}$

The answer, 419.71, appears in the display window immediately after the $\boxed{=}$ key is depressed.

One advantage of using a calculator for this type of problem is that you do not need to point off the decimal point in the product. The calculator does it for you.

EXAMPLE 6 Divide: $65 \div 1.3$

SOLUTION

Depress the keys in the following order:

6 5 $\boxed{\div}$ 1 $\boxed{\cdot}$ 3 $\boxed{=}$

The answer, 50, appears in the display window immediately after the $\boxed{=}$ key is depressed.

Problems that require more than one operation require special attention on a handheld calculator if no memory is available.

EXAMPLE 7 Perform the calculation $1,840 - 16 \times 58$

SOLUTION

The order of operations tells us that the multiplication must be done before the subtraction. Multiply

1 6 $\boxed{\times}$ 5 8 $\boxed{=}$ 928

Write the product down on a piece of paper so that you will have it when you need it. Now you must begin again and work the second part of the problem.

1 8 4 0 $\boxed{-}$ 9 2 8 $\boxed{=}$ 912

This problem required two distinct operations to solve.

If you have keys like $\boxed{MT}$, $\boxed{MS}$, $\boxed{M+}$, and $\boxed{M-}$ on your calculator, then you have memory that you can use while solving problems. Again, different calculators will have different memory keys and capability. Check your manual to determine exactly how yours works. Consider Example 7 again.

EXAMPLE 7a Perform the calculation $1,840 - 16 \times 58$

SOLUTION

Perform the multiplication. After striking the $\boxed{=}$ key, depress the $\boxed{M+}$ key. This stores the result of the multiplication, 928, in memory. Now enter the following keys:

1 8 4 0 $\boxed{-}$ $\boxed{MT}$ $\boxed{=}$

This instructs the calculator to subtract the memory total from 1,840. When you depress the $\boxed{=}$ key, the answer, 912, is displayed.

Some calculators will also have a percent key. It looks like $\boxed{\%}$. This key allows you to perform percent problems without worrying about converting the percent to a decimal or a fraction.

EXAMPLE 8 What is 12% of 68?

SOLUTION

Remember that you are finding the amount here. That means that this is a multiplication problem. Depress the keys in this order:

6 8 $\boxed{\times}$ 1 2 $\boxed{\%}$

The display window will show 8.16. Notice that you did not have to strike the $\boxed{=}$ key. Also notice that the calculator did all the necessary conversions for you.

EXAMPLE 9 40 is 80% of what number?

SOLUTION

You are looking for the base. This is a division problem. Depress the keys in this order:

4 0 $\boxed{\div}$ 8 0 $\boxed{\%}$

The display window shows 50. This is your answer.

EXAMPLE 10   What percent is 28 of 112?

SOLUTION

You are looking for the rate. This is a division problem. Depress the keys in this order:

2 8 $\boxed{\div}$ 1 1 2 $\boxed{\%}$

The display is 25. Since you are looking for a rate, attach the % sign. 28 is 25% of 112.

Notice that the decimal point is in the correct position. There is no need for an adjustment as with manual calculations.

As you can see, a variety of problems can be worked with calculators. You may want to practice using your calculator by working some of the problems in the text. The rule of thumb, regardless of the type of calculator you have, is to know your calculator. Read your manual to determine the capabilities of the calculator you have chosen. Then use it wisely in solving problems.

INDEX

A

Absolute value, 387–88
  in adding signed numbers, 395–98
  in dividing signed numbers, 417–18
  in multiplying signed numbers, 409–10
Abstract number, 555
Addends, 17
Addition
  compared to multiplication, 32
  of decimal numbers, 199
  definition, 18
  of denominate numbers, 556
  of fractions with common denominators, 121–22
  of fractions with different denominators, 12–24
  of mixed numbers, 124–25
  in order of operations, 65–66
  perimeter of polygons, 596–97
  process, 17
  rules for, 74
  of signed numbers, 395–98
  solving equations by, 459–60
Addition property of equality, 459
  rules for, 497
Additive inverse concept, 398
Algebra, 630
  and arithmetic, 19
  and basic percent formula, 312
  and decimal numbers, 219
  equivalent values concept, 364
  and fractions, 115
  and liquid measurement, 540
  and proportions, 270
  rules for, 497
  terminology for, 451–54
  use of signed numbers, 398
Algebraic expression, 452–53
Aligning, 17–18
  decimal numbers, 199–200
Amount
  calculation, 299–300
  in percents, 292
Angle, 587
Apparent solution, 459–60
Area
  of circles, 619
  of geometric figures, 618–19
  of parallelograms, 616–17
  of polygons, 615–16
  of trapezoids, 618
  of triangles, 617

B

Base
  calculating, 311–12
  in percents, 292
Base number, 35
Basic percent formula, 292–93
  amount calculation, 299–300

Basic percent formula—*Cont.*
  applications, 327–28
  base calculation, 311–12
  rate calculation, 305–6
Borrowing, 25–26

C

Canceling, 139–40
  rules for, 162
Carried digits, 17
Celsius scale, 547–48
Centigram, 531–32
Centiliter, 540
Centimeter, 522
Circles, 607–9
  area of, 619
Circumference, 607–8
Coefficients
  distribution of, 481
  fractional, 473–74
Common denominators
  fractions with, 121–22
  in multiplication, 139–40
  in subtraction, 131
Common fractions, 93
  conversion of, 363–65
  conversion to decimal fractions, 349
  conversion to percents, 350–51
  converting decimal fractions to, 349–50
  converting percents to, 351–52
Composite numbers, 57–58
Cone, 629
Constant terms, 452
Conversion
  common fractions and decimals, 349
  percents and common fractions, 350–52
  rules for, 371
  in solving problems, 363–65
  terminating and repeating decimals, 357–58
Counting numbers; *see* Natural numbers
Cross multiplication, 263
Cube, 36, 627–28
Cubic units, 627–30
Cup, 539
Cylinder, 628

D

Day, 547
Decigram, 531–32
Deciliter, 540–41
Decimal digits, 185
Decimal fractions, 185
  conversion to common fractions, 349–50
  converting common fractions to, 349

Decimal-number division, 216–19
Decimal numbers, 185
  adding, 199
  conversion of, 363–65
  conversion to percents, 291
  dividing, 215–19
  multiplying, 207–8
  order of operations, 233–34
  place values, 185–86
  reading and writing, 186–87
  rounding, 193–94
  rules for, 239–40
  subtracting, 199–201
  terminating and repeating, 357–58
Decimal point, 185
  in scientific notation, 225–26
Decimeter, 522
Dekagram, 531–32
Dekaliter, 540–41
Denominate numbers, 522
  rules for, 566–67
  working with, 555–59
Denominator, 93
Determining digit, 11
Diameter, 607–8
Difference, 25
Distributive property
  of multiplication, 481–83
  in multistep equations, 490
  rules for, 497
Dividend, 41
  in decimal division, 215–18
  and prime factor, 58
Division
  of decimal numbers, 215–19
  of denominate numbers, 558–59
  for equivalent fractions, 112–15
  of fractions, 147
  of mixed numbers, 148–49
  in order of operations, 65–66
  in prime factorization, 57–58
  process, 41–45
  rules for, 74
  of signed numbers, 417–18
  solving equations by, 465–66
Division property of equality, 465
  rules for, 497
Divisors, 41
  in decimal division, 215–18
  decimal-number, 216–19
  multidigit, 43–45
  in prime factorization, 58
  single-digit, 41–43
  whole-number, 215–16

E

Ellipsis, 3, 357
English system of measurement, 520
Equality
  addition property of, 459
  division property of, 465
  multiplication property of, 473

I-1

Equations, 451
  equivalent, 459
  multistep, 489-91
  rules for, 497
  solutions by addition, 459-60
  solutions by division, 465-66
  solutions by multiplication, 473-75
Equilateral triangle, 595-96
Equivalent equations, 459
Equivalent fractions, 111-16
  rules for, 161
Equivalent values concept, 364
Expanded notation, 4, 481
Exponent, 35
Exponential notation, 35-36
  powers, 208
Exponents, in scientific notation, 225-26
Extremes, of a proportion, 263

F
Factors, 32-35
Fahrenheit scale, 547-48
Fluid ounce, 539
Foot, 520-21
Formula, 292, 451
Fractional coefficients, 473-74
Fraction bar, 93
Fractions; *see also* Common fractions *and* Decimal fractions
  adding, 121-25
  changed to higher terms, 111-12
  changed to lower terms, 112-13
  with common denominators, 121-22
  with different denominators, 122-24, 131-32
  dividing, 147
  equivalent, 111-16
  and greatest common factor, 114
  improper fractions and mixed numbers, 104-6
  multiplying, 139-40
  order of operations, 155-56
  reduced to lowest terms, 113-15
  rules governing, 161-63
  solutions to equations, 466
  subtracting, 131-34
  types of, 93-95
  use in ratios, 257-58
  use of conversion, 363-65

G
Gallon, 539
Geometry
  area of circles, 619
  area of geometric figures, 618-19
  area of parallelograms, 616-17
  area of polygons, 615-16
  area of trapezoids, 618
  area of triangles, 617
  of circles, 607-9
  of polygons, 595-98
  rules for, 636-37
  terminology for, 587-88
  of volume, 627-30
Gram, 531-32

Greatest common factor, 114
  rules for, 162

H
Hectogram, 531-32
Hectoliter, 540
Hectometer, 522
Higher terms, 111-12
  rules for, 162
Hour, 547

I
Improper fractions, 93-94
  changing whole numbers to, 105
  conversion to mixed numbers, 104
  converting mixed numbers to, 104-5
  in dividing mixed numbers, 148-49
  in multiplying mixed numbers, 140-41
  rules for, 161
Inch, 520-21
Increase or decrease problems, 319-21
Inequality symbols, 385-86
Inverting fractions, 147

K
Key rules
  for algebra, 497
  for converting fractions, 371
  for decimals, 239-40
  for denominate numbers, 566-67
  for fractions and mixed numbers, 161-63
  for geometry, 636-37
  for percent problems, 333
  for proportions, 277
  for signed numbers, 429
  units of measurement, 565-66
  for whole numbers, 74-75
Key terms, 73-74, 161, 239, 277, 333, 371, 429, 497, 565, 635-36
Kilogram, 531-32
Kiloliter, 540-41
Kilometer, 522

L
LCD; *see* Least common denominator
LCM; *see* Least common multiple
Least common denominator, 122-24
Least common multiple, 121-22
  rules for, 162
Length
  metric units, 522
  USCS units, 521
Like terms; *see* Similar terms
Line segment, 587
Liquid measurement, 539-42
Liter, 540
Lower terms, 112-13
  rules for, 162

M
Means, of proportions, 263
Measurement
  area of circles, 619

Measurement—*Cont.*
  area of parallelograms, 616-17
  area of polygons, 615-16
  area of trapezoids, 618
  area of triangles, 617
  of geometric areas, 618-19
  of lengths, 520-25
  of liquids, 539-42
  rules for, 565-66
  time and temperature, 547-48
  of volume, 627-30
  of weight, 531-33
Meter, 522
Metric system, 519
  Celsius scale, 547-48
  conversion to USCS, 523-25, 532-33, 541-42
  for lengths, 522-23
  liquid measurement, 540-41
  for weights, 531-32
Mile, 520-21
Milligram, 531-32
Milliliter, 540
Millimeter, 522
Minuend, 25-26
Minute, 547
Mixed decimal, 185
Mixed numbers, 93
  adding, 124-25
  conversion to improper fractions, 104-5
  converting improper fractions to, 104
  dividing, 148-49
  multiplying, 140-41
  rules for, 162-63
  subtracting, 132-35
Month, 547
Multidigit divisor, dividing by, 43-45
Multiplicand, 32
Multiplication
  compared to division, 41-42
  cross-multiplication, 263
  of decimal numbers, 207-8
  of denominate numbers, 557-58
  distributive property of, 481-83
  in division of fractions, 147
  in division of mixed numbers, 148-49
  for equivalent fractions, 111-12
  of fractions, 139-40
  of mixed numbers, 140-41
  in order of operations, 65-66
  process, 32-36
  rules for, 74
  rules for decimals, 239
  of signed numbers, 409-10
  solving equations by, 473-75
Multiplication facts, 32, 35
Multiplication property of equality, 473
  rules for, 497
Multiplier, 32
Multistep equations, 489-91

N
Natural-number factors, 57
Natural numbers, 3

Negative numbers
   addition of, 396
   compared to positive numbers, 385
   multiplication of, 409-10
Numbers; *see also* Denominate numbers
   additive inverse of, 388
   decimals, 185-87
   inequality symbols, 385-86
   less than zero, 383
   positive and negative, 385
   prime and composite, 57-59
   reciprocals of, 473-75
   in scientific notation, 225-26
   whole, 3
Numerator, 93
Numerical coefficient, 451

O

Operations; *see* Order of operations
Opposites, 388
Order of operations, 65-66
   for decimal numbers, 233-34
   rules for, 75
   rules for fractions, 155-56
   with signed numbers, 423-24
Ounce, 531

P

Parallel lines, 587
Parallelogram, 595-96
   area of, 616-17
Partial product, 34
Percents, 290
   applications, 327-28
   base calculation, 311-12
   basic formula, 292-93
   calculating the amount, 299-300
   conversion to decimals, 290-91
   conversion to common fractions, 351-52
   converting common fractions to, 350-51
   converting decimals to, 291
   increase or decrease, 319-21
   rate calculation, 305-6
   rules for, 333
Perimeter of polygons, 596-98
Perpendicular lines, 587
Pi (π)
   formula, 607-8
   numerical value, 357-58
Pint, 593
Place value, 3-4, 185-86
Point, 587
Polygons, 595-98
   area of, 615-16
Positive numbers, 383
   addition of, 395
   compared to negative numbers, 385
   multiplication of, 409-10
Pound, 531
Powers
   dividing by, 218
   of a number, 208
   in scientific notation, 225-26

Prime factorization, 57-59
   for least common multiple, 122
   rules for, 75
Prime factors, canceling, 114-15
Prime numbers, 57-59
Problems
   increase or decrease in percent, 319-21
   order of operations, 65-66
   using conversion, 363-65
Product, 32-35
Proper fractions, 93
Proportions, 263-64
   rules for, 277
   solving, 269-71

Q-R

Quart, 539
Quotient, 41
   in decimal division, 215-18
Radius, 607-8
   and area of circles, 619
Rate
   calculating, 305-6
   in percents, 292
Ratios, 257-58
   and proportions, 263-64
   rules for, 277
Ray, 587
Reciprocal of a number, 147, 473-75
Rectangle, 595-96
Rectangular solid, 627
Remainder, 41
Repeated addition; *see* Multiplication
Repeating decimals, 357-58
Right angle, 587
Right triangle, 595-96
Root; *see* Solution
Rounding
   of decimal numbers, 193-94
   process, 11
   repeating decimals, 357-58
   rules for, 74
   rules for decimals, 239
Rounding place, 11
Rule of proportions, 263, 277

S

Scalene triangle, 595-96
Scientific notation
   numbers without exponents, 226
   rules for, 239-40
   writing numbers in, 225-26
Second, 547
Sets of numbers, 3
Signed numbers
   addition of, 395-98
   and algebra, 398
   concept, 385
   division of, 417-18
   multiplication of, 409-10
   order of operations, 423-24
   rules for, 429
   subtraction of, 403-4
Similar terms, 452
Simplifying
   denominate numbers, 555
   an expression, 452-53

Single-digit number, dividing by, 41-43
Solution (or Root), 451
   apparent, 459-60
   to equations by addition, 459-60
Sphere, 629
Square, 36, 595-96
Subscript, 596
Subtraction
   of decimal numbers, 199-201
   of denominate numbers, 556-57
   of fractions with common denominators, 131
   of fractions with different denominators, 131-32
   in order of operations, 65-66
   process, 25-26
   rules for, 74
   of signed numbers, 403-4
Subtrahend, 25-26
Sum, 17

T

Tablespoon, 539
Teaspoon, 539
Temperature measurement, 547-48
Term, 452
Terminating decimals, 357-58
Terms of a proportion, 263
Three-dimensional objects, 627-30
Time measurement, 547
Ton, 531
Total; *see* Sum
Trapezoids, 595-96
   area of, 618
Triangles, 595-96
   area of, 617

U-V

United States Customary System, 519
   conversion to metrics, 523-25, 532-33, 541-42
   length units, 520-22
   liquid measures, 539-40
   weight measures, 531
Unlike terms, 452
Variable, 451
Vertex, 587
Volume, 627-30

W-Z

Week, 547
Weight measurement, 531-33
Whole-number division, 215-16
Whole numbers, 3
   changed to improper fractions, 105
Word problems
   basic percent formula, 327-28
   rules for, 74-75
   strategies for solving, 51-52
Written form of a number, 5
Yard, 520-21
Year, 547
Zero
   numbers less than, 383
   as placeholder, 208